Field, Thermionic, and Secondary Electron
Emission
Spectroscopy

Field, Thermionic, and Secondary Electron Emission Spectroscopy

A. Modinos
University of Salford
Salford, England

Plenum Press • New York and London

69893469

Library of Congress Cataloging in Publication Data

Modinos, A., 1938–
    Field, thermionic, and secondary electron emission spectroscopy.

    Bibliography: p.
    Includes index.
    1. Field emission. 2. Thermionic emission. 3. Secondary electron emission. 4. Electron spectroscopy. 5. Surfaces (Physics) I. Title.
QC700.M63   1983   530.4'1   83-21311
ISBN 0-306-41321-3

PHYS

© 1984 Plenum Press, New York
A Division of Plenum Publishing Corporation
233 Spring Street, New York, N.Y. 10013

Printed in the United States of America

To Penny and Evris

PREFACE

Apart from their well-known technological applications, field, thermionic, and secondary electron emission can be and are often used as spectroscopies, i.e., as a means of investigating the electronic properties of the emitting surface. It is the aim of this book to present the theory of field, thermionic, and secondary electron emission spectroscopies from a unifying point of view and to demonstrate the applicability of the theory to the analysis of electron emission data from individual surfaces. The book is reasonably self-contained, although certain sections require some knowledge of the quantum theory of scattering, and may serve as an introduction to the subject. At the same time it presents a report on recent advances which will be of interest to the specialist.

The traditional theory of field and thermionic emission, based on the free-electron model for metals, its applications, and its limitations, are discussed in Chapters 1 and 2. Those aspects of the theory of solids and their surfaces which are necessary for a quantitative theory of field, thermionic, and secondary electron emission spectroscopies are discussed in Chapters 3 and 4. The theory of these spectroscopies and discussion of relevant experimental data takes up the remaining six chapters of the book.

It is assumed, unless otherwise explicitly stated, that the emitting surface is a single-crystal plane. In the case of field emission from adsorbate-covered metal surfaces (Chapter 6) or field emission through singly adsorbed atoms (Chapter 7) it is assumed that the substrate surface is a single-crystal plane of the metal under consideration. It is also assumed that the emitted current density is sufficiently small, so that the space charge in front of the emitting surface is negligible.

A research grant from the Leverhulme Trust covering expenses relating to the preparation of this book is gratefully acknowledged.

A. Modinos
University of Salford

CONTENTS

Chapter 1.   ELECTRON EMISSION FROM FREE-ELECTRON METALS

1.1.   The Free-Electron Theory of Metals   1
1.2.   Scattering by One-Dimensional Barriers   3
1.3.   Electron Emission from a Metal Surface   7
1.4.   Field Emission   13
1.5.   Thermionic Emission   18
1.6.   Thermal-Field Emission   23
1.7.   Energy Distribution of the Emitted Electrons   24
　　　 1.7.1.   Energy Distribution of Field-Emitted Electrons   24
　　　 1.7.2.   Energy Distribution of the Emitted Electrons in the Thermal-Field
　　　　　　　 Region   30
　　　 1.7.3.   Energy Distribution of Thermally Emitted Electrons   31

Chapter 2.   WORK FUNCTION AND OTHER EMISSION MEASUREMENTS

2.1.   Introduction   35
2.2.   Thermionic Data Based on Schottky and Richardson Plots   35
2.3.   Field Emission Microscopy   42
2.4.   The Local Field at the Surface of a Field Emitter   48
2.5.   Field Emission Work Functions   50
2.6.   Work Function Anisotropy   54
2.7.   Energy Analyzers   55

**Chapter 3.   THE CRYSTALLINE SOLID AS A STACK OF ATOMIC LAYERS
　　　　　　　　WITH TWO-DIMENSIONAL PERIODICITY**

3.1.   Introduction   59
3.2.   The Independent-Electron Theory of Solids   59
3.3.   Two- and Three-Dimensional Lattices in Real and Reciprocal Space   67
3.4.   Complex Band Structure   71
3.5.   Scattering by a Layer of Muffin-Tin Atoms   76
　　　 3.5.1.   Scattering by an Individual Muffin-Tin Atom   77
　　　 3.5.2.   Multiple Scattering within the Layer   80
　　　 3.5.3.   Properties of the $\mathbf{X}$ Matrix   87

Chapter 4.   ELECTRON STATES IN A SEMI-INFINITE METAL

4.1.   Introduction   91
4.2.   Bound One-Electron States   92

4.2.1. Bulk States   94
4.2.2. Surface States and Resonances   99
4.3. The Local Density of States   110
4.4. Scattering States   113
4.4.1. Scattering States with Source at $z = -\infty$ 113
4.4.2. Scattering States with Source at $z = +\infty$ (LEED States)   114
4.4.3. Surface Resonances ($E > 0$)   115
4.5. Lifetime Effects   117

Chapter 5. *FIELD EMISSION SPECTROSCOPY OF METALS*

5.1. Introduction   121
5.2. Energy Distribution of the Emitted Electrons. Basic Formulas   121
5.2.1. A Green's Function Method for Calculating the Total Energy
        Distribution of the Emitted Electrons   129
5.2.2. Calculation of $\langle \chi_{q\parallel} | G(E) | \chi_{q\parallel} \rangle$ 132
5.3. Calculation of the Reflection Matrix R^c   136
5.4. Measured and Calculated Energy Distributions of Field-Emitted Electrons   141
5.4.1. The (100) Plane of Tungsten   141
5.4.2. The (110) Plane of Tungsten   148
5.4.3. The (100) Plane of Molybdenum   152
5.4.4. The (111) Plane of Copper   154
5.4.5. Field Emission Energy Distribution from the Platinum Group
        Metals   155
5.5. Field Emission from Ferromagnetic Metals   155
5.6. Spin-Polarized Field Emission from the (100) and (111) Planes of Nickel .   158
5.7. Many-Body Effects   164
5.8. Photofield Emission   167

Chapter 6. *FIELD EMISSION FROM ADSORBATE-COVERED SURFACES*

6.1. Introduction   169
6.2. Field Emission from Adsorbate-Covered Surfaces. Some Experimental
      Results   170
6.2.1. Inert Atoms on Tungsten   170
6.2.2. Metallic Adsorbates on Tungsten: $\Delta\phi(\theta)$ and $B(\theta)$ 182
6.2.3. Metallic Adsorbates on Tungsten: Energy Distributions   185
6.2.4. Hydrogen on Tungsten   188
6.3. Calculation of the Energy Distribution of the Emitted Electrons from
      Adsorbate-Covered Metal Surfaces   192
6.4. Inelastic Electron Tunneling   199
6.5. Measurements of Desorption Energies and Diffusion Coefficients   203

**Chapter 7. *FIELD EMISSION SPECTROSCOPY OF SINGLY ADSORBED
           ATOMS***

7.1. Introduction   205

7.2. Adatom Resonances 207
7.3. Energy Distribution of Electrons Field-Emitted from the Neighborhood of a Single Adatom 218

Chapter 8.   FIELD EMISSION FROM SEMICONDUCTOR SURFACES

8.1. Introduction 237
8.2. Band Bending in the Zero-Emitted-Current Approximation 237
8.3. Electronic Structure of Semiconductor Surfaces 244
8.4. Field Emission in the Zero-Internal-Current Approximation 252
8.5. Field Emission from Germanium (100) 266
8.6. Replenishment of the Surface States 272
8.7. Internal Voltage Drop and Saturation Effects 282

Chapter 9.   THERMIONIC EMISSION SPECTROSCOPY OF METALS

9.1. Basic Formulas 295
9.2. The Herring–Nichols Formula for the Elastic Reflection Coefficient 301
9.3. Calculation of μ_0 304
9.4. Effect of Thermal Vibration of the Atoms on the Elastic Reflection Coefficient 306
9.5. Calculation of the Surface Barrier Scattering Matrix Elements 309
9.6. Deviations from the Schottky Line 313
9.7. Thermionic Emission from Cu(100): Theoretical Results 319

Chapter 10.   SECONDARY ELECTRON EMISSION SPECTROSCOPY

10.1. Introduction 327
10.2. Bulk Density of States Effects in Secondary Electron Emission Spectra .. 332
10.3. Quantitative Analysis of the Fine Structure in Secondary Electron Emission Spectra 341

APPENDIX A: DETAILED BALANCE

A.1. Introduction 347
A.2. Elastic Reflection Coefficient 348
A.3. Inelastic Reflection Coefficient 350

APPENDIX B: PHYSICAL CONSTANTS, UNITS, AND CONVERSION FACTORS 353

REFERENCES 355

INDEX 371

ELECTRON EMISSION FROM FREE-ELECTRON METALS

1.1. THE FREE-ELECTRON THEORY OF METALS

The free-electron theory of metals proposed by Sommerfeld (1928) has provided the basis for practically all quantum-mechanical theories of electron emission from metals until relatively recently, and some important formulas derived on the basis of this theory are still extensively used in the analysis of relevant experimental data. It is, therefore, appropriate to begin our present study with a brief exposition of the free-electron theory of these phenomena. At the same time we shall point out the limitations of this theory which have become evident by the accumulation of experimental data which it cannot explain.

The free-electron theory of metals is based on the assumption that the conduction band electrons behave, effectively and for many practical purposes, as if they were free particles. In this model the electron states (in the conduction band) are described by plane waves

$$\psi_{\mathbf{k}}(\mathbf{r}) = \frac{1}{L^{3/2}} \exp\left(i\mathbf{k} \cdot \mathbf{r}\right) \tag{1.1}$$

where $L^3 \equiv V$ is the volume of the metal, which is assumed to be very large, and $\mathbf{k}$ (the wave vector) is given by

$$\mathbf{k} = \frac{2\pi}{L}\left(n_x, n_y, n_z\right) \tag{1.2}$$

where n_x, n_y, n_z are positive or negative integers. The energy of the electron in the state $\psi_{\mathbf{k}}$ is given by

$$E_{\mathbf{k}} = \hbar^2 k^2 / 2m \tag{1.3}$$

where the zero of energy coincides with the bottom of the conduction band. For a complete description of an electron state we must also specify the spin orientation of the electron. For given **k** we have two states (spin up, spin down) with the same spatial wave function [given by Eq. (1.1)]. Having this in mind and letting L go to infinity we find that the number of electron states per unit volume with energy between E and $E + dE$ is given by

$$\rho(E)\ dE = \frac{2}{(2\pi)^3} \int\!\!\int\!\!\int_E^{E+dE} d^3k \tag{1.4a}$$

$$= \frac{1}{2\pi^2}\left(\frac{2m}{\hbar^2}\right)^{3/2} E^{1/2}\ dE \tag{1.4b}$$

The probability of an electron state with energy E being occupied (at equilibrium) is given by the Fermi–Dirac distribution function

$$f(E) = \frac{1}{1 + \exp\left[(E - E_F)/k_B T\right]} \tag{1.5}$$

where k_B is Boltzmann's constant, T is the absolute temperature, and E_F is the Fermi level. The latter is determined by the requirement

$$\int_0^\infty \rho(E)f(E)\ dE = N_c \tag{1.6}$$

where N_c is the number of conduction band electrons per unit volume and it is in effect an empirical parameter. For a metal, E_F is practically independent of temperature (see, e.g., Peierls, 1955) and can, in most cases, be replaced by its value at $T = 0$. The latter is given by

$$E_F = \frac{\hbar^2}{2m}(3\pi^2 N_c)^{2/3} \tag{1.7}$$

Apart from the above general formulas we shall, for the subsequent development of the theory of electron emission, require an expression for the number of electrons which, under equilibrium conditions, cross a unit area (we shall assume without any loss of generality that this is parallel to the xy plane) from left to right, per unit time, with total energy between E and $E + dE$ and normal energy, defined by

$$W = \frac{\hbar^2 k_z^2}{2m} \tag{1.8}$$

between W and $W + dW$. We denote this quantity by $N(E,W) \; dE \; dW$. We have

$$N(E,W) \; dE \; dW = \frac{2f(E)}{(2\pi)^3} \iiint_{\{E,W\}} v_z d^3 k \qquad (1.9)$$

where

$$v_z \equiv \frac{1}{\hbar} \frac{\partial E_{\mathbf{k}}}{\partial k_z} = \frac{\hbar k_z}{m} \qquad (1.10)$$

is the velocity of the electron normal to the unit area under consideration, and $\{E,W\}$ indicates that only states with total energy between E and $E + dE$ and normal energy between $W + dW$ corresponding to $v_z > 0$ are to be included in the integration.

One can easily show that

$$N(E,W) \; dE \; dW = \frac{m}{2\pi^2\hbar^3} f(E) \; dE \; dW \qquad (1.11)$$

1.2. SCATTERING BY ONE-DIMENSIONAL BARRIERS

Consider a potential barrier $V(z)$ such as the one shown in Fig. 1.1. Let an electron with total energy E and wave vector $\mathbf{k}_{\parallel} = (k_x, k_y)$ be incident on this

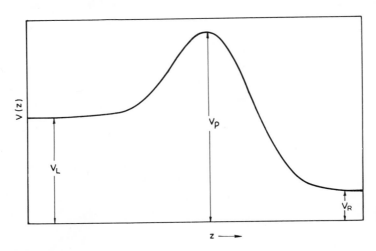

FIGURE 1.1

One-dimensional potential barrier.

barrier from the left. The electron will be partly reflected by the barrier and partly transmitted through it. We can calculate the reflection and transmission coefficients, as defined by Eqs. (1.17) and (1.18), as follows. Since the potential depends only on the z coordinate, the electronic wave function has the form

$$\psi = e^{i\mathbf{k}_\| \cdot \mathbf{r}_\|}u(z) \tag{1.12}$$

where $\mathbf{r}_\| \equiv (x,y)$ and $u(z)$ satisfies the following equation:

$$-\frac{\hbar^2}{2m}\frac{d^2u}{dz^2} + V(z)u = Wu(z) \tag{1.13}$$

where the normal energy $W = E - \hbar^2 k_\|^2/2m > V_L$ and may lie above or below the top of the barrier, denoted by V_p in Fig. 1.1. In the region of constant potential ($V = V_R$) to the right of the barrier the wave function has the following form:

$$u(z) = C \exp{(iq_R z)} \tag{1.14}$$

with

$$q_R = \left[\frac{2m}{\hbar^2}(W - V_R)\right]^{1/2}$$

which represents a transmitted wave propagating in the positive z direction. In the region of constant potential ($V = V_L$) to the left of the barrier the wave function has the form

$$u(z) = A \exp{(iq_L z)} + B \exp{(-iq_L z)} \tag{1.15}$$

with

$$q_L = \left[\frac{2m(W - V_L)}{\hbar^2}\right]^{1/2}$$

The first term in the above equation represents the incident wave and the second term represents the reflected wave. The incident, reflected, and transmitted current densities are calculated in the usual manner from the formula (see, e.g., Landau and Lifshitz, 1958)

$$j_{z,\alpha} = \frac{\hbar}{2mi}\left(f_\alpha^* \frac{\partial}{\partial z}f_\alpha - f_\alpha \frac{\partial}{\partial z}f_\alpha^*\right) \tag{1.16}$$

where f_α stands for the incident, reflected, and transmitted wave, respectively. The transmission coefficient is defined by

$$D(W) \equiv \frac{j_{z,\text{tra}}}{j_{z,\text{inc}}} = \frac{q_R}{q_L} \left| \frac{C}{A} \right|^2 \tag{1.17}$$

and the reflection coefficient by

$$R(W) \equiv \frac{j_{z,\text{ref}}}{j_{z,\text{inc}}} = \left| \frac{B}{A} \right|^2 \tag{1.18}$$

Explicit analytic expressions for D and R can be obtained only for a few specific potential barriers. On the other hand, one can obtain $D(W)$ and $R(W)$ numerically, and without much difficulty, for *any* barrier which depends only on the z coordinate. This can be done as follows. One starts from the known wave function [Eq. (1.14)] at some point to the right of the barrier where the potential is constant. The numerical value of C is unimportant and can be put equal to unity. One then proceeds to integrate numerically the Schrödinger equation, Eq. (1.13), moving from right to left in a step-by-step procedure using one of a number of methods which are available for this purpose (see, e.g., Buckingham, 1962). By carrying the integration into the region of constant potential to the left of the barrier, we obtain the wave function there and hence the values of A and B in Eq. (1.15), and hence $D(W)$ and $R(W)$. We note that when the potential barrier decreases slowly and monotonically to the right of the barrier top, one does not need to start the integration from the region of constant potential, which, thus, becomes irrelevant to the transmission problem. Under the above conditions there always exists a region, $z \geq z_r$, to the right of the barrier peak where the ordinary WKB approximation is valid (see, e.g., Landau and Lifshitz, 1958), and there the wave function appropriate to a transmitted wave has the form

$$u(z) = \frac{C}{[\lambda(z)]^{1/2}} \exp\left[i \int_{z_r}^{z} \lambda(z) \, dz \right] \tag{1.19}$$

where

$$\lambda(z) \equiv \left\{ \frac{2m}{\hbar^2} [W - V(z)] \right\}^{1/2} \tag{1.20}$$

Obviously $W > V(z)$ for $z > z_r$. In this case the transmission coefficient is given by

$$D(W) = \frac{1}{q_L} \left| \frac{C}{A} \right|^2 \tag{1.21}$$

If we take advantage of Eq. (1.19) we can start the numerical integration at z_r. Again, the numerical value of C is unimportant and can be chosen arbitrarily.

Although the above procedure for calculating the transmission coefficient is easy to carry out and should be used when the situation demands an accurate analysis of specific experimental data, it is not appropriate for a general analysis of electron emission phenomena. For that purpose a less accurate method leading to an analytic expression for the transmission coefficient (at least for certain values of the physical parameters) is much preferable. Such a method, and one which has been extensively used in the study of electron emission, has been developed by Miller and Good (1953a). Their method is a generalization of the ordinary WKB method and it leads to the WKB results under certain limiting conditions. Their theory applies to the case of a slowly varying potential barrier, with only two classical turning points. This means that there are only two real roots of the equation

$$W - V(z) = 0 \qquad (1.22)$$

when W lies below the top of the barrier. We denote them by z_1 and z_2 and we assume that $z_1 < z_2$. Also, it is assumed that when W goes above the top of the barrier, the two real roots of Eq. (1.22) go unambiguously into two complex roots, the one being the complex conjugate of the other since $V(z)$ is real. We denote by z_1 the one with the positive imaginary part, and by z_2 the one with the negative imaginary part. The method of Miller and Good is not applicable if $[W - V(z)]$ has additional zeros or singularities in the vicinity of z_1 or z_2. When the conditions for the applicability of their method are satisfied as, for example, in the case of the barrier shown in Fig. 1.1, the transmission coefficient is given to a good approximation by the following formula:

$$D(W) = \{1 + \exp[-A(W)]\}^{-1} \qquad (1.23)$$

where

$$A(W) = 2i \int_{z_1}^{z_2} \lambda(\xi) \, d\xi \qquad (1.24)$$

where $\lambda(\xi)$ is given by Eq. (1.20). The normal energy W may lie above or below the top of the barrier. In performing the ξ integration in Eq. (1.24) one must remember that z_1 and z_2 and hence ξ are complex when W lies above the top of the barrier. In performing the ξ integration that branch of $\lambda(\xi)$ is taken for which $-\pi < \arg[\lambda^2(\xi)] \leq \pi$. Accordingly, when ξ is real we get $\arg[\lambda(\xi)] = 0$ for $W > V(\xi)$ and $\arg[\lambda(\xi)] = \pi/2$ for $W < V(\xi)$. One can easily show that $A(W)$ is always real and that it is positive for energies above the top of the barrier and negative for energies below the top of the barrier. It is also obvious that when the energy W lies well below the barrier top, we have $\exp[-A(W)] \gg 1$ and Eq. (1.23) reduces to the well-known formula

$$D(W) \simeq \exp\left[-2\int_{z_1}^{z_2} |\lambda(z)| \, dz\right] \qquad (1.25)$$

which is obtained by the ordinary WKB method (see, e.g., Landau and Lifshitz, 1958). We note that when the incident energy is equal to the barrier height we obtain $A = 0$ and the WKB formula, Eq. (1.25), gives $D = 1$, whereas the Miller and Good formula, Eq. (1.23), gives $D = 0.5$. A comparison with exact results made by Miller and Good (1953a) for a barrier similar to the one shown in Fig. 1.1 suggests that Eq. (1.23) underestimates the transmission coefficient by 10% or so in the energy region around the top of the barrier. The approximation improves considerably as we move away from the barrier top and is very accurate for energies sufficiently away from it. The accuracy at a given energy will, of course, depend on the shape of the barrier.

1.3. ELECTRON EMISSION FROM A METAL SURFACE

Consider a semi-infinite metal occupying the half-space from $z = -\infty$ to $z \simeq 0$. According to the free-electron theory of metals an electron inside the metal sees a constant (zero) potential. On the other hand, we know from experience that at zero temperature a certain minimum energy, equal to the work function ϕ, must be supplied to the metal before an electron can escape from it. We also know from classical electrostatics that an electron situated at a finite distance from the plane surface of a perfect conductor is attracted to it, by the well-known "image" force: $-e^2/4z^2$. (The minus sign arises because we have assumed that the z axis points outwards from the metal.) Hence, the potential energy of the electron on the vacuum side of the metal–vacuum interface is asymptotically given by

$$V(z) \simeq E_F + \phi - e^2/4z \qquad (1.26)$$

In recent years, a lot of effort has been put into microscopic studies of the potential barrier at a metal–vacuum interface (see, e.g., Lang, 1973; Inkson, 1971; Appelbaum and Hamann, 1972a; Harris and Jones, 1973; 1974). All investigators agree that Eq. (1.26) is valid for $z > $ (3 Å or so). The detailed shape of the barrier for $z < 3$ Å depends on the metal surface under consideration. In the traditional (free-electron) theories of electron emission it is assumed that Eq. (1.26) remains valid right up to a point z_c such that $V(z_c) = 0$. When an external electric field F is applied to the surface, a term $-eFz$ is added on the right of Eq. (1.26). In electron emission experiments $F > 0$. In summary, the potential energy field seen by an electron in an electron emission experiment is approximated by

$$V(z) = E_F + \phi - \frac{e^2}{4z} - eFz \qquad \text{for } z > z_c$$
$$= 0 \qquad \text{for } z < z_c \tag{1.27}$$

where z_c is determined by $V(z_c) = 0$. In Fig. 1.2 we show this potential barrier for a typical value of $E_F + \phi$ and for a value of the applied field appropriate to a field emission experiment. We note that the peak of the barrier described by Eq. (1.27) occurs at

$$z_m = (e/4F)^{1/2} \tag{1.28}$$
$$= 1.9/F^{1/2} \text{ Å} \qquad \text{(with } F \text{ in V/Å)}$$

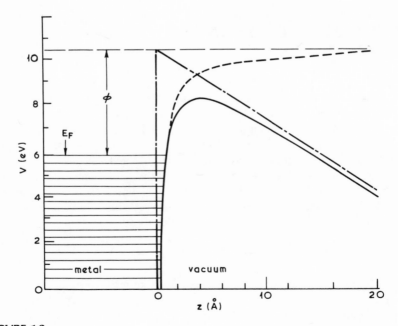

FIGURE 1.2

The surface potential barrier seen by an electron in a field emission experiment (solid line). The contributions of the image potential and of the applied field ($F = 0.3$ V/Å) are shown by the broken and the broken-solid line, respectively.

and has the value

$$V_{max} = E_F + \phi - (e^3F)^{1/2} \qquad (1.29)$$
$$= E_F + \phi - 3.79F^{1/2} \text{ eV} \qquad \text{(with } F \text{ in V/Å)}$$

In order to construct a theory of electron emission from metal surfaces we proceed as follows. We assume (this is justified because the electric field is negligible inside the metal) that the number of electrons with normal energy between W and $W + dW$ impinging on the surface barrier from within the metal when a current is emitted is practically identical with its equilibrium value given by

$$N(W,T) \, dW = \frac{mdW}{2\pi^2\hbar^3} \int_W^\infty f(E) \, dE \qquad (1.30)$$
$$= \frac{mk_BT}{2\pi^2\hbar^3} \ln\left[1 + \exp\left(-\frac{W - E_F}{k_BT}\right)\right] dW$$

Each of these electrons has a probability $D(W)$ to be transmitted through the surface potential barrier. Hence, the emitted current density, i.e., the number of electrons emitted per unit area per unit time multiplied by the magnitude of the electronic charge, is given by

$$J(F,T) = e \int_0^\infty N(W,T)D(W) \, dW \qquad (1.31)$$

where T and F denote as usual the temperature and the applied field, respectively. Murphy and Good (1956) have shown that if one assumes the validity of the surface barrier given by Eq. (1.27) and if one uses Eq. (1.23) for the calculation of the transmission coefficient for this barrier, one obtains, starting from Eq. (1.31), an expression for $J(F,T)$ valid for any values of F and T, which in turn, leads to explicit and relatively simple formulas for $J(F,T)$ for certain regions of F and T. As we shall see, the Murphy and Good theory leads in a systematic way to the Fowler–Nordheim formula for field emission, i.e., emission at very low temperatures with a strong applied field, and to the Richardson and Schottky formulas for thermionic emission, i.e., emission at high temperatures with a weak or zero applied field.

The transmission coefficient for the potential barrier of Eq. (1.27) is, according to Eqs. (1.23) and (1.24), given by

$$D(W,F) = \{1 + \exp[Q(W)]\}^{-1} \qquad (1.32)$$

where

$$Q(W) \equiv -2i \int_{z_1}^{z_2} \lambda(\xi) \, d\xi \tag{1.33}$$

$$\lambda(\xi) = \left[\frac{2m}{\hbar^2} \left(W - E_F - \phi + \frac{e^2}{4\xi} + eF\xi \right) \right]^{1/2} \tag{1.34}$$

and z_1, z_2 are the roots of the equation

$$\lambda^2(\xi) = 0 \tag{1.35}$$

For $W < V_{max}$, with V_{max} given by Eq. (1.29), z_1 and z_2 are both real and $z_1 < z_2$. For $W > V_{max}$, z_1 and z_2 are complex, the one being the complex conjugate of the other with $\text{Im}(z_1) > 0$ and $\text{Im}(z_2) < 0$. We note that $\lambda(\xi)$, as defined by Eq. (1.34), has a singularity at $\xi = 0$. This does not affect the applicability of Eq. (1.32) when $W < V_{max}$, but it does so for energies sufficiently above V_{max} when $\text{Re}(z_1) = \text{Re}(z_2)$ approaches zero. When

$$W > W_l \equiv V_{max} + \left(1 - \frac{1}{\sqrt{2}} \right) (e^3 F)^{1/2} \tag{1.36}$$

Eq. (1.32) may not be applicable. Fortunately, for energies above W_l the transmission coefficient is very close to unity in practice, and one can put

$$D(W,F) \simeq 1 \qquad \text{for } W > W_l \tag{1.37}$$

when evaluating the current density according to Eq. (1.31). The calculation of $D(W,F)$ for $W < W_l$, reduces to the calculation of the integral defined by Eq. (1.33). It turns out that this integral can be put into the following form (Murphy and Good, 1956):

$$Q(W) = \frac{4}{3} \sqrt{2} \left(\frac{F\hbar^4}{m^2 e^5} \right)^{-1/4} y^{-3/2} v(y) \tag{1.38}$$

$$y \equiv (e^3 F)^{1/2} / |\phi + E_F - W| \tag{1.39}$$

$$v(y) \equiv -(y/2)^{1/2} \{ -2E[(y-1)^{1/2}/(2y)^{1/2}]$$
$$+ (y+1)K[(y-1)^{1/2}/(2y)^{1/2}] \} \qquad \text{for } y > 1 \tag{1.40}$$

$$v(y) = (1+y)^{1/2} \{ E[(1-y)^{1/2}/(1+y)^{1/2}]$$
$$- yK[(1-y)^{1/2}/(1+y)^{1/2}] \} \qquad \text{for } y < 1 \tag{1.41}$$

$E[k]$ and $K[k]$ are the well-known elliptic integrals

$$K[k] = \int_0^{\pi/2} \frac{d\theta}{(1 - k^2 \sin^2 \theta)^{1/2}} \tag{1.42}$$

$$E[k] = \int_0^{\pi/2} (1 - k^2 \sin^2 \theta)^{1/2} \, d\theta \tag{1.43}$$

We note that $y > 1$ corresponds to energies above the barrier peak ($W > V_{max}$) while $y < 1$ corresponds to energies below the barrier peak ($W < V_{max}$). In Table 1.1 are listed the values of $v(y)$ for $y < 1$, as calculated by Burgess, Kroemer, and Houston (1953). Using Eqs. (1.30) to (1.38) we obtain the following general expression for the emitted current density:

$$
\begin{aligned}
J(F,T) &= e \int_0^\infty N(W,T) D(W,F) \, dW \\
&= \frac{emk_BT}{2\hbar^3\pi^2} \left[\int_0^{W_l} \frac{\ln\{1 + \exp[-(W - E_F)/k_BT]\} \, dW}{1 + \exp[Q(W)]} \right. \\
&\quad \left. + \int_{W_l}^\infty \ln\left[1 + \exp\left(-\frac{W - E_F}{k_BT}\right)\right] dW \right\}
\end{aligned}
\tag{1.44}
$$

We shall see that under certain conditions the integrals in the above equation can be performed analytically, but before we do that, we would like to point out the following. Evaluating the transmission coefficient on the basis of Eq. (1.32) implies that an electron incident on the surface barrier from within the metal can be reflected *only* by the potential field within the region $z_1 \leq z \leq z_2$ when $W < V_{max}$, and from the immediate vicinity of the barrier peak when $W > V_{max}$. That would be the case, if the potential barrier varied *slowly* to the left of the barrier peak (as it does on the right of it), and if it joined smoothly to the constant potential (zero) inside the metal. This is clearly not the case for the potential barrier of Eq. (1.27), which rises steeply at the interface ($z \simeq z_c$) and whose derivative dV/dz has an obvious discontinuity at $z = z_c$. At first sight one may argue that the steep rise and the discontinuity of the derivative at $z = z_c$ are artificial and that by neglecting them one is nearer to the actual situation. It turns out, however, that even in a free-electron-like metal there are oscillations in the potential near the surface. These oscillations are shown in Fig. 1.3 for a specific case studied by Lang and Kohn (1970). Because of these oscillations a weak reflecting region does exist at the metal–vacuum interface even for a free-electron-like metal. When looked upon from this point of view, the discontinuity in the potential of Eq. (1.27) at $z = z_c$ acquires a real physical significance in so far as it simulates the effect of the above-mentioned oscillations. Reflection at the metal–vacuum

TABLE 1.1

Values of $v(y)$, $s(y)$, and $t(y)$

y	$v(y)$	$t(y)$	$s(y)$
0.00	1.0000	1.0000	1.0000
0.05	0.9948	1.0011	0.9995
0.10	0.9817	1.0036	0.9981
0.15	0.9622	1.0070	0.9958
0.20	0.9370	1.0111	0.9926
0.25	0.9068	1.0157	0.9885
0.30	0.8718	1.0207	0.9835
0.35	0.8323	1.0262	0.9777
0.40	0.7888	1.0319	0.9711
0.45	0.7413	1.0378	0.9637
0.50	0.6900	1.0439	0.9554
0.55	0.6351	1.0502	0.9464
0.60	0.5768	1.0565	0.9366
0.65	0.5152	1.0631	0.9261
0.70	0.4504	1.0697	0.9149
0.75	0.3825	1.0765	0.9030
0.80	0.3117	1.0832	0.8903
0.85	0.2379	1.0900	0.8770
0.90	0.1613	1.0969	0.8630
0.95	0.0820	1.1037	0.8483
1.00	0.0000	1.1107	0.8330

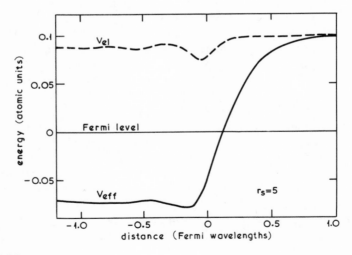

FIGURE 1.3

Effective one-electron potential V_{eff} near the metal surface; V_{el} is the electrostatic contribution to this potential. Jellium model, $r_s = 5$. The parameter r_s is defined by $r_s = [4\pi N_c/3]^{-1/3}$, where N_c is the electron density in the bulk of the metal; the Fermi wavelength equals $2\pi/k_F = 3.274r_s$. (From Lang and Kohn, 1970.)

interface is responsible, for example, for the well-known phenomenon of the periodic deviations from the Schottky line in thermionic emission (see Section 1.5). We must emphasize, however, that electron reflection at the metal–vacuum interface (at $z \simeq 0$) and phenomena which depend upon it cannot be properly analyzed within the free-electron approximation since, in most instances, the reflection of the electron at the interface is intimately connected with the crystalline character of the metal and with the band-structure of its energy levels (see, e.g., Chapter 9). On the other hand, those quantities which depend, essentially, only on the height of the surface barrier and its asymptotic behavior for $z > (3$ Å or so) can be evaluated on the basis of Eq. (1.44). We shall now show that for certain regions of the temperature and the applied field the integrals in Eq. (1.44) can be performed analytically.

1.4. FIELD EMISSION

It is obvious from Eq. (1.30) that for sufficiently low temperatures the function $N(W,T)$ diminishes rapidly for $W > E_F$. On the other hand, for a typical surface barrier ($\phi \simeq 4$ eV, $F \simeq 0.4$ V/Å), $D(W,F)$ calculated according to Eqs. (1.32) and (1.38) diminishes rapidy for $W < E_F$. Under these conditions the second integral in Eq. (1.44) contributes a negligible amount to $J(F,T)$ and the integrand in the first integral vanishes except in the immediate neighborhood of the Fermi level. Therefore, we can drop the second integral in Eq. (1.44) and we can replace the limits in the first integral by $-\infty$ and $+\infty$, respectively. Also, for typical values of the parameters ϕ and F ($\phi > 3$ eV, $F < 0.6$ V/Å)

$$\exp\left[Q(W)\right] \gg 1 \tag{1.45}$$

and, therefore, the unity in the denominator of the first integral in Eq. (1.44) can be dropped. We obtain

$$J(F,T) = \frac{emk_BT}{2\pi^2\hbar^3} \int_{-\infty}^{+\infty} \exp\left[-Q(W)\right] \ln\left[1 + \exp\left(-\frac{W - E_F}{k_BT}\right)\right] dW \tag{1.46}$$

Since most of the contribution to the above integral comes from the immediate neighborhood of the Fermi level, we may replace the exponent in the transmission coefficient by the first two terms of the following Taylor expansion around E_F:

$$-Q(W) = -b_0 + c_0(W - E_F) - f_0(W - E_F)^2 + \cdots \tag{1.47}$$

where

$$b_0 \equiv \frac{4}{3}\left(\frac{2m}{\hbar^2}\right)^{1/2}\frac{\phi^{3/2}}{eF}\,v(y_0) \tag{1.48a}$$

$$= 0.683v\left(3.79\,\frac{F^{1/2}}{\phi}\right)\frac{\phi^{3/2}}{F} \tag{1.48b}$$

In Eq. (1.48b) ϕ is given in eV and F in V/Å; the same applies to Eqs. (1.49b), (1.50b), and (1.52b):

$$c_0 \equiv 2\left(\frac{2m}{\hbar^2}\right)^{1/2}\frac{\phi^{1/2}}{eF}\,t(y_0) \tag{1.49a}$$

$$= 1.025(\phi^{1/2}/F)t(3.79\,F^{1/2}/\phi)\;(\mathrm{eV})^{-1} \tag{1.49b}$$

$$f_0 \equiv \frac{1}{2}\left(\frac{2m}{\hbar^2}\right)^{1/2}\frac{v(y_0)}{eF\phi^{1/2}}\left(1-\frac{e^3F}{\phi^2}\right)^{-1} \tag{1.50a}$$

$$= 0.256(F\phi^{1/2})^{-1}v(3.79\,F^{1/2}/\phi)(1-14.36F/\phi^2)^{-1}\;(\mathrm{eV})^{-2} \tag{1.50b}$$

where

$$t(y) \equiv v(y) - \frac{2}{3}\,y\,\frac{dv}{dy} \tag{1.51}$$

and

$$y_0 \equiv (e^3F)^{1/2}/\phi \tag{1.52a}$$

$$= 3.79F^{1/2}/\phi \tag{1.52b}$$

Numerical values of $t(y)$, as calculated by Burgess, Kroemer, and Houston (1953), are listed in Table 1.1. We note that under field-emission conditions $y < 1$. Keeping only the first two terms in Eq. (1.47) and substituting into Eq. (1.46) we obtain

$$J(F,T) = \frac{emk_BT}{2\pi^2\hbar^3}\,e^{-b_0}\int_{-\infty}^{+\infty}e^{c_0(W-E_F)}\ln\left[1+\exp\left(-\frac{W-E_F}{k_BT}\right)\right]dW \tag{1.53}$$

We note that when $c_0k_BT < 1$ the integrand in the above equation decreases exponentially for $W < E_F$ and for $W > E_F$ as does the original expression under the integral sign of Eq. (1.46), and that the two expressions coincide in the neighborhood of the Fermi level. The integral in Eq. (1.53) can be evaluated analytically and the result is

$$J(F,T) = \frac{e^3 F^2}{16\pi^2 \hbar\, \phi t^2(y_0)} \frac{\pi c_0 k_B T}{\sin{(\pi c_0 k_B T)}} \exp\left[-\frac{4}{3e}\left(\frac{2m}{\hbar^2}\right)^{1/2} v(y_0)\frac{\phi^{3/2}}{F} \right]$$

(1.54)

It can be shown (Murphy and Good, 1956) that Eq. (1.54) is a valid approximation when the following two conditions are satisfied:

$$\phi - (e^3 F)^{1/2} > \pi^{-1}\left[\frac{\hbar^4}{m^2 e^2}\right]^{1/4} (eF)^{3/4} + k_B T(1 - c_0 k_B T)^{-1} \quad (1.55)$$

$$1 - c_0 k_B T > (2f_0)^{1/2} k_B T \quad (1.56)$$

The field emission region of F and T as determined by Eqs. (1.55) and (1.56) is shown in Fig. 1.4 for $\phi = 4.5$ eV. For points on the boundary of the field emission region the error in the current density as given by Eq. (1.54) when compared with an exact evaluation of the same quantity according to Eq. (1.44) varies between 15% and 40%.

At very low temperatures, i.e., when $\pi c_0 k_B T \ll 1$, Eq. (1.54) reduces to the

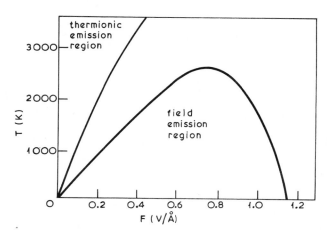

FIGURE 1.4

Thermionic emission and field emission regions of temperature and applied field for a 4.5 eV work function. (From Murphy and Good, 1956.)

well-known Fowler–Nordheim equation (Fowler and Nordheim, 1928; Nordheim, 1928) given by

$$J(F) = A'F^2 \exp\left[-B'\phi^{3/2}/F\right] \tag{1.57}$$

where

$$A' \equiv e^3 \bigg/ \left[16\pi^2\hbar\,\phi t^2\left(\frac{(e^3F)^{1/2}}{\phi}\right)\right] \tag{1.58}$$

and

$$B' \equiv \frac{4}{3e}\left(\frac{2m}{\hbar^2}\right)^{1/2} v\left(\frac{(e^3F)^{1/2}}{\phi}\right) \tag{1.59}$$

Substituting the appropriate numerical values for the various constants we find

$$J(F) = \frac{1.537 \times 10^{10}F^2}{\phi t^2(3.79F^{1/2}/\phi)} \exp\left[-\frac{0.683\phi^{3/2}}{F}\,v\left(\frac{3.79F^{1/2}}{\phi}\right)\right]\frac{A}{\text{cm}^2} \tag{1.60}$$

where ϕ is given in eV and F in V/Å. This is one of the most important equations in electron emission theory and has found many and varied applications. These have been described in considerable detail by a number of authors (Good and Müller, 1956; Dyke and Dolan, 1956; Gomer, 1961; Swanson and Bell, 1973). We shall discuss some of these applications in Chapter 2.

At this stage we would like to point out that, if one plots $\ln(J/F^2)$ versus $1/F$, the corresponding curve, known as the Fowler–Nordheim (FN) plot, is practically a straight line in the narrow region of field strength of a typical field-emission experiment (0.3 V/Å $\leq F \leq 0.5$ V/Å). This is a fact which has been verified experimentally by many authors (see, e.g., Fig. 1.5). The slope of an FN plot is, according to Eq. (1.60), given by

$$S_{\text{FN}} = \frac{d\ln(J/F^2)}{d(1/F)} = -0.683s\left(\frac{3.79F^{1/2}}{\phi}\right)\phi^{3/2} \tag{1.61}$$

where

$$s(y) = v(y) - \frac{y}{2}\frac{dv}{dy} \tag{1.62}$$

Numerical values of $s(y)$ are listed in Table 1.1. One can see from this table that $s(y)$ is practically constant within a narrow region of the applied field. Hence, in

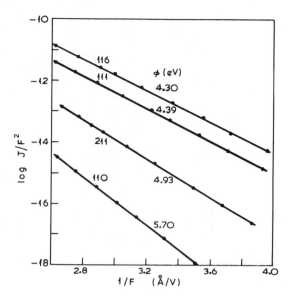

FIGURE 1.5

Fowler–Nordheim plots of field emission from the (116), (111), (211), and (110) planes of tungsten. (From Müller, 1955.)

principle, one can determine the work function from the slope of the FN plot. In practice, an absolute determination of the work function is hindered by the fact that the absolute value of the local field at the emitting plane cannot be obtained with sufficient accuracy (see Sections 2.4 and 2.5). We note also that the proportionality constant in front of $\phi^{3/2}$ in Eq. (1.61) depends to some degree on our choice of surface potential barrier [i.e., Eq. (1.27)]. We mention, for example, that for the triangular barrier

$$V(z) = E_F + \phi - eFz \qquad \text{for } z > 0 \qquad (1.63)$$
$$= 0 \qquad \text{for } z > 0$$

one obtains (see, e.g., Gadzuk and Plummer, 1973)

$$S_{\text{FN}} \text{ (triangular barrier)} = -0.683\phi^{3/2} \qquad (1.64)$$

For typical values of ϕ and F, $s(3.79F^{1/2}/\phi) = 0.95$ according to Table 1.1, hence the proportionality constant between the measured S_{FN} and $\phi^{3/2}$ changes by about 5% when the potential barrier of Eq. (1.27) is replaced by the barrier of Eq. (1.63). Any realistic surface potential barrier must lie somewhere between these two barriers and asymptotically (i.e., for $z \geq 3$ Å or so) it must coincide with the image barrier. It is clear that for such a barrier Eq. (1.61) is valid to a very good degree of approximation. We must emphasize, however, that the magnitude of the emitted current density (i.e., the vertical position of the FN plot) is much more sensitive to the choice of barrier. For example, replacing the image barrier of Eq. (1.27) by the triangular barrier of Eq. (1.63) may lead to a reduction in the emitted current density of ten times or more. At the present time it is very difficult to measure with sufficient accuracy the magnitude of the emitted current density (see Section 2.3).

1.5. THERMIONIC EMISSION

At high temperatures and weak applied fields most of the contribution to the integral in Eq. (1.44) comes from a very narrow region of energy at the top of the surface potential barrier. Hence, one is permitted to substitute in Eq. (1.44) for $N(W,T)$ and $D(W,F)$ series expansions of these quantities valid for $W \simeq V_{max}$. Since $V_{max} - E_F \gg k_B T$ even at the highest possible temperatures,

$$\ln \left[1 + \exp \left(-\frac{W - E_F}{k_B T} \right) \right] \simeq \exp \left(-\frac{W - E_F}{k_B T} \right) \tag{1.65}$$

Using standard properties of the elliptic functions, one can expand $y^{-3/2} v(y)$ around the value $y = 1$ which corresponds to $W = V_{max}$ according to Eqs. (1.29) and (1.39). Keeping only the first term in this expansion one obtains

$$\tfrac{4}{3}\sqrt{2} y^{-3/2} v(y) \simeq -\pi \epsilon' \tag{1.66}$$

where

$$\epsilon' \equiv \frac{y - 1}{y} = 1 - \frac{\phi + E_F - W}{(e^3 F)^{1/2}} \tag{1.67}$$

Hence, for $W < W_l$, with W_l defined by Eq. (1.36), we have

$$1 + \exp \left[Q(W) \right] \simeq 1 + \exp \left[-\left(\frac{F \hbar^4}{m^2 e^5} \right)^{-1/4} \pi \epsilon' \right] \tag{1.68}$$

For $W > W_l$ the expression on the right-hand side of Eq. (1.68) is approximately equal to unity, so that no significant error is introduced if one divides the integrand in the second integral of Eq. (1.44) by this quantity. Consequently, the two integrals in Eq. (1.44) can be replaced by the following single integral:

$$J(F,T) = \frac{emk_B T}{2\pi^2 \hbar^3} e^{-\phi/k_B T}$$

$$\times \int_0^\infty \frac{\exp\left(-\dfrac{W - E_F - \phi}{k_B T}\right) dW}{1 + \exp\left\{-\left(\dfrac{F\hbar^4}{m^2 e^5}\right)^{-1/4} \pi \left[1 + \dfrac{W - E_F - \phi}{(e^3 F)^{1/2}}\right]\right\}} \quad (1.69)$$

When the conditions, Eqs. (1.72) and (1.73), appropriate to thermionic emission are satisfied, the integrand in the above formula is effectively zero except in the neighborhood of the barrier top and, therefore, the lower limit of the integral in Eq. (1.69) can be replaced by $-\infty$. When this is done, the integral can be performed analytically and the result is

$$J(F,T) = \frac{em(k_B T)^2}{2\pi^2 \hbar^3} \left(\frac{\pi h_0}{\sin \pi h_0}\right) \exp\left[-\frac{\phi - (e^3 F)^{1/2}}{k_B T}\right] \quad (1.70)$$

where

$$h_0 \equiv \left(\frac{F\hbar^4}{m^2 e^5}\right)^{1/4} \frac{(e^3 F)^{1/2}}{\pi k_B T} \quad (1.71)$$

It can be shown (Murphy and Good, 1956) that Eq. (1.70) is a valid approximation, when the following two conditions are satisfied:

$$\ln\left[(1 - h_0)/h_0\right] - \frac{1}{h_0(1 - h_0)} > -\pi\left(\frac{me}{\hbar^2}\right)^{1/2} (eF)^{-3/4}[\phi - (e^3 F)^{1/2}] \quad (1.72)$$

$$\ln\left[(1 - h_0)/h_0\right] - \frac{1}{(1 - h_0)} > -\pi\left(\frac{me^3}{\hbar^2}\right)^{1/4} (eF)^{-1/8} \quad (1.73)$$

The thermionic emission region of F and T as determined by Eqs. (1.72) and (1.73) is shown in Fig. 1.4 for $\phi = 4.5$ eV. On the boundary of this region the error in the current density as given by Eq. (1.70) when compared with an exact evaluation of the same quantity on the basis of Eq. (1.44) varies between 15% and 40%.

We note in passing that formulas similar to those of Murphy and Good for the thermionic current [Eq. (1.70)], and also for the field-emission current [Eq. (1.54)], had been derived by different methods by Sommerfeld and Bethe (1933) and by Guth and Mullin (1941; 1942). A unified theory of electron emission similar to that of Murphy and Good has also been given by Christov (1978).

For very weak applied fields, i.e., when $\pi h_0 \ll 1$, Eq. (1.70) reduces to the well-known Schottky formula (Schottky, 1914; 1919; 1923)

$$ J(F,T) = A_R T^2 \exp\left[- \frac{\phi - (e^3 F)^{1/2}}{k_B T} \right] \qquad (1.74) $$

where

$$ A_R \equiv \frac{emk_B^2}{2\pi^2 \hbar^3} = 120 \text{ A cm}^{-2} \text{ deg}^{-2} \qquad (1.75) $$

We note that

$$ \Delta\phi \equiv (e^3 F)^{1/2} = 3.79 F^{1/2} \text{ eV} \qquad (1.76) $$

with F expressed in V/Å, represents the lowering of the barrier height brought about by the application of the field [see Eq. (1.29)]. In a typical thermionic-emission experiment $\Delta\phi \leq 0.1$ eV. We note that the position z_m of the peak of the barrier, calculated from Eq. (1.28), lies sufficiently away from the surface ($z_m \gg 4$ Å) so there is no ambiguity about the validity of the image law there.

According to Eq. (1.74), if we plot $\ln (J)$ against $F^{1/2}$, we should obtain a straight line, whose slope is given by $m_S = e^{3/2}/k_B T$. Such plots are known as Schottky plots or Schottky lines. All thermionic emission measurements from single crystal planes that have been performed so far, under conditions such that space charge effects are negligible, have confirmed that the above prediction of the Schottky theory is valid to a very good degree of approximation (see Chapter 2). However, when very accurate measurements of the variation of the emitted current with the applied field are made, one finds small oscillations about the Schottky line which increase in amplitude and period as F increases. Although such periodic deviations from the Schottky line have been observed in emission from single-crystal planes (see Fig. 1.6), most measurements of these deviations have, so far, been taken from polycrystalline emitters (Seifert and Phipps, 1939; Nottingham, 1956; Haas and Thomas, 1972). The quantity shown in Fig. 1.6 is

$$ \Delta J(F) = \left\{ \ln\left[\frac{J(F)}{J_0} \right] - SF^{1/2} \right\} \Big/ \ln (10) \qquad (1.77) $$

where J_0 is the zero-field current density, obtained by extrapolation to zero field of the experimental Schottky line whose slope S is, within the limits of the exper-

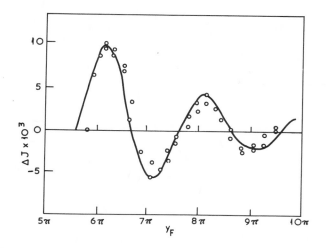

FIGURE 1.6

Periodic deviations from the Schottky line obtained from the (111) plane of tungsten. The experimental points are denoted by circles. $y_F \equiv 357.1 F^{-1/4}$ for field F in V/cm. (From Stafford and Weber, 1963.)

imental error, equal to the theoretical slope $m_S = e^{3/2}/k_B T$. The periodic deviations from the Schottky line cannot be accounted for by the Schottky formula (Eq. 1.74) or for that matter by Eq. (1.70). The factor

$$\frac{\pi h_0}{\sin \pi h_0} \simeq 1 + \frac{1}{6} (\pi h_0)^2 \tag{1.78}$$

in the latter equation suggests a small nonperiodic deviation from the Schottky line proportional to $F^{3/2}$.

　　The failure of Eq. (1.70) to predict any periodic deviations from the Schottky line is due to our approximate treatment of the transmission probability [Eq. (1.32)], which, as we have already pointed out, fails to take into account the partial reflection of the electron in the *immediate* vicinity of the metal–vacuum interface. It does not necessarily imply a failure of the free-electron theory. The first theories (Guth and Mullin, 1941; Juenker, Colladay, and Coomes, 1953; Miller and Good, 1953b) of the periodic deviations from the Schottky line assumed the validity of the surface potential barrier described by Eq. (1.27) and consisted of a more or less accurate evaluation of the transmission coefficient for this barrier. The results of these theories are qualitatively correct in spite of the inadequacy of the assumed model potential. The essential physics of the phenomenon is well understood (see, e.g., Herring and Nichols, 1949) and can be summarized as follows.

The quantum-mechanical wave representing an electron incident on the surface barrier from within the metal, with normal energy in the vicinity of the top of the barrier, is to some extent scattered backward and forward between the two reflecting regions, corresponding roughly to the interface at $z = z_c$ and the top of the barrier at $z = z_m$, before leaving the metal. This leads to interference effects between the partially scattered waves, which are very sensitive to the position z_m of the barrier peak which is inversely proportional to $F^{1/2}$ [Eq. (1.28)]. For some values of the applied field $(z_m - z_c)$ is approximately equal to an integral multiple of the average electron wavelength and consequently the scattered waves between the two reflecting regions are in phase, leading to an increase in the amplitude of the wave function there and hence to an increase in the transmitted current. For other values of the applied field the opposite is true, leading to a decrease in the transmitted current. Obviously, a quantitative study of the periodic deviations from the Schottky line must be based on accurate measurements from single-crystal planes, and in the analysis of such data it will be necessary to take into account the detailed electronic properties (band structure of the energy levels, etc.) of the metal surface under consideration. As we shall see in Chapter 9, a more realistic theory of thermionic emission from metal surfaces leads to the following equation:

$$J(F,T) = \bar{t}(T,F)A_R T^2 \exp\left[-\frac{\phi - (e^3 F)^{1/2}}{k_B T} \right] \qquad (1.79)$$

where $\bar{t}$, a dimensionless quantity (an average transmission coefficient), depends on the particular properties of the emitting surface under consideration. Within the range of applied field and temperature normally used in a thermionic emission experiment, $\bar{t}$ depends only slightly on the field and the temperature, at least in most cases, as evidenced by the smallness of the periodic deviations from the Schottky line and by the fact that their amplitude is practically independent of the temperature (Nottingham, 1956).

If we put $F = 0$ in Eq. (1.79) we obtain the well-known equation

$$J_0(T) = A_R \bar{t}_0 T^2 \exp\left(-\phi/k_B T\right) \qquad (1.80)$$

where $\bar{t}_0 \equiv \bar{t}(T,0)$. The above equation is known in the literature as the Richardson–Laue–Dushman equation, although sometimes it is referred to simply as the Richardson equation. An equation of this form was originally suggested by Richardson (1902; 1912) and by Von Laue (1918a,b) prior to the discovery of quantum mechanics. Dushman (1923) gave the first quantum mechanical theory of thermionic emission, but, since at his time the spin of the electron had not been discovered, his formula differed from Eq. (1.74) by a factor of (1/2). The theoretical derivation of this equation was put on a firm foundation after the work of Fowler and Nordheim (1928) and Nordheim (1928).

1.6. THERMAL-FIELD EMISSION

Electron emission at intermediate temperatures and applied fields which do not belong either to the field emission region or the thermionic emission region is sometimes referred to as thermal-field emission (TF emission). Christov (1966; 1978) was able, starting from Eq. (1.44), to derive analytic expressions which give the emitted current density with sufficient accuracy for any values of T and F. These expressions reduce to Eq. (1.54) and Eq. (1.70) in the field-emission and thermionic-emission regions, respectively. In the TF region Christov's expressions contain factors with a complicated dependence on F and T, and a computer program is usually required for a calculation of the current density using his formulas. In Fig. 1.7 we show the results of such a calculation by Bermond, Lenoir, Prulhiere, and Drechsler (1974). The same authors measured the emitted current from various planes of tungsten for different values of the applied field and for temperatures between 300 and 1500 K. Their experimental results are in good qualitative agreement with the predictions of the theory.

A set of curves and numerical tables giving the emitted current density for various values of field and temperature in the TF region can be found in the

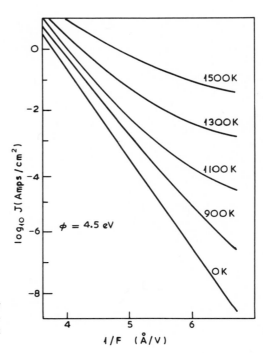

FIGURE 1.7

Thermal-field emission. Current density as a function of the applied field at various temperatures for a 4.5 eV work function. (From Bermond *et al.*, 1974.)

review article of Dyke and Dolan (1956). In the calculations of these authors, which are based on Eq. (1.31), the transmission coefficient for the image barrier is calculated by the ordinary WKB method. Although the latter method overestimates the transmission probability in the vicinity of the barrier top (see Section 1.2), the current density versus field characteristic obtained in the above manner for different temperatures is not qualitatively different from that obtained on the basis of Eq. (1.44). It is worth remembering, however, that the magnitude of the emitted current density obtained on the basis of the free-electron theory may be very different from the actual current density even when the shape of the current versus field curve is given correctly by this theory, and, therefore, the formulas given in the present chapter cannot be used for the determination of one or the other experimental parameter from the measured magnitude of the current density.

1.7. ENERGY DISTRIBUTION OF THE EMITTED ELECTRONS

The number of electrons emitted per unit area per unit time with total energy between E and $E + dE$ will be denoted by $j(E)\, dE$. The quantity $j(E)$ is known as the total energy distribution (TED) of the emitted electrons. This quantity, when calculated according to the free-electron theory of metals, will be denoted by $j_0(E)$. We have

$$j_0(E) = \int_0^E N(E,W)D(W,F)\, dW \qquad (1.81)$$

where $N(E,W)$ is the supply function given by Eq. (1.11) and $D(W,F)$ is the transmission coefficient for the potential barrier at the metal–vacuum interface. According to our assumptions (see Section 1.3), the potential barrier is given by Eq. (1.27) and the corresponding transmission coefficient $D(W,F)$ by Eqs. (1.32)–(1.38).

1.7.1. ENERGY DISTRIBUTION OF FIELD-EMITTED ELECTRONS

In the field emission region, as defined by Eqs. (1.55) and (1.56), we can drop the unity in the brackets of Eq. (1.32), as we have done in our discussion of the emitted current density (Section 1.4). We obtain

$$j_0(E) = \frac{m}{2\pi^2\hbar^3} f(E) \int_0^E e^{-Q(W)}\, dW \qquad (1.82)$$

where $Q(W)$ is given by Eq. (1.38). In the present case practically all the contribution to the integral in Eq. (1.82) comes from the region $W \simeq E$. There we may replace $Q(W)$ by the first two terms of its Taylor series expansion around E. We have

$$-Q(W) \simeq -b_E + \frac{1}{d_E} (W - E) \tag{1.83}$$

where

$$b_E \equiv \frac{4}{3} \left(\frac{2m}{\hbar^2} \right)^{1/2} \frac{(\phi + E_F - E)^{3/2}}{eF} v(y_E) \tag{1.84}$$

$$\frac{1}{d_E} \equiv 2 \left(\frac{2m}{\hbar^2} \right)^{1/2} \frac{(\phi + E_F - E)^{1/2}}{eF} t(y_E) \tag{1.85}$$

$$y_E \equiv (e^3 F)^{1/2} / (\phi + E_F - E) \tag{1.86}$$

and $v(y_E)$ and $t(y_E)$ are the functions tabulated in Table 1.1. We note that the above expansion differs from the one given by Eq. (1.47) in that the present expansion is made around E and not around E_F as in Eq. (1.47). If we substitute Eq. (1.83) into Eq. (1.82) the resulting integrand, like the exact integrand in Eq. (1.81), is effectively zero except in the neighborhood of $W = E$, hence, we can replace the lower limit of the integral by $-\infty$. Then the integration can be performed analytically and we find

$$j_0(E) = \frac{m d_E f(E)}{2\pi^2 \hbar^3} \exp \left[-\frac{4}{3} \left(\frac{2m}{\hbar^2} \right)^{1/2} \frac{(\phi + E_F - E)^{3/2}}{eF} v(y_E) \right] \tag{1.87}$$

If, instead of the expansion around E [Eq. (1.83)], we substitute in Eq. (1.82) the expansion around E_F given by Eq. (1.47), we find the following formula for the energy distribution of the emitted electrons:

$$j_0'(\epsilon) = \frac{J_0}{e d_0} \frac{\exp (\epsilon / d_0)}{1 + \exp (\epsilon / p\, d_0)} \tag{1.88}$$

where

$$\epsilon \equiv E - E_F \tag{1.89}$$
$$p \equiv k_B T / d_0 \tag{1.90}$$
$$1/d_0 \equiv c_0 \quad \text{[as given by Eq. (1.49)]} \tag{1.91}$$

and J_0 equals the emitted current density at zero temperature as given by Eq. (1.57). The above expression for the energy distribution was originally derived by

Young (1959). It is obvious that Young's formula can only be used for energies close to the Fermi level (approximately within 0.5 eV from it). This is the region of energy where practically all of the current comes from. In this respect, it is worth observing that the emitted current density $J(F,T)$, as given by Eq. (1.54), is equal to

$$J(F,T) = e \int_{-\infty}^{+\infty} j_0'(\epsilon) \, d\epsilon \qquad (1.92)$$

We note, however, that modern energy analyzers are capable of measuring the energy distribution down to 2 eV below the Fermi level and the possibility exists that the accessible energy range will be expanded further (Gomer, 1978). For a calculation, within the free electron approximation, of the energy distribution in this extended energy region one must use Eq. (1.87). It is also important to remember that Eq. (1.87), similarly Eq. (1.88), is valid only in the field emission region as defined by Eqs. (1.55) and (1.56). In practice, for a typical emitter ($\phi \simeq 4$ eV), this implies that $p < 0.7$, where p is the parameter defined by Eq. (1.90). In Fig. 1.8a we show a set of theoretical TED curves calculated by Swanson and Crouser (1967a) on the basis of Eq. (1.88) for various values of the parameter p. The same authors measured the energy distribution of the emitted electrons from various crystallographic planes of tungsten for the same range of values of p. As an example, we show in Fig. (1.8b) their normalized experimental TED curves for the (112) plane. We note that the energy range covered by the experiments of Swanson and Crouser is limited to $|\epsilon| \simeq 0.5$ eV, and it is, therefore, legitimate to analyze these results on the basis of Young's formula. It is also clear that over this limited region of energy the experimental results, shown in Fig. 1.8b, are in reasonably good agreement with the predictions of the free-electron theory. Swanson and Crouser obtained TED curves similar to those in Fig. 1.8b from the (111), (116), and (310) planes of tungsten but not from the (100) plane. The TED curve for this plane, shown in Fig. 1.8c, was found to be significantly different from the one predicted by Eq. (1.88). This finding of Swanson and Crouser presented the first serious experimental challenge of the adequacy of the Fowler–Nordheim theory of field emission. With the subsequent development of more efficient energy analyzers it became apparent that, taken over a wider energy region, the TED curves of most tungsten planes reveal considerable fine structure. This is demonstrated in Fig. 1.9, which shows the results of TED measurements taken by Plummer and Bell (1972) and by Varburger, Penn, and Plummer (1975) from several planes of tungsten. The quantity shown in this figure is known as the enhancement factor and is defined by

$$R(E) = \frac{j(E)}{j_0(E)} \qquad (1.93)$$

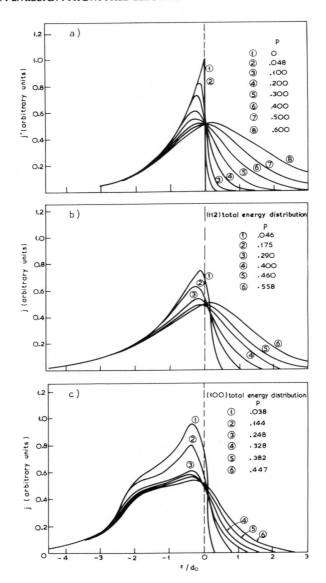

FIGURE 1.8

(a) Theoretical total energy distributions [Eq. (1.88)] at various values of p. (b) Experimental total energy distributions from the (112) plane of tungsten at various values of p, where $d_0 = 0.146$ eV and $F = 3.48 \times 10^7$ V/cm. (c) Experimental total energy distribution from the (100) plane of tungsten at various values of p, where $d_0 = 0.174$ eV and $F = 4.08 \times 10^7$ V/cm; $\epsilon \equiv E - E_F$. (From Swanson and Crouser, 1967a.)

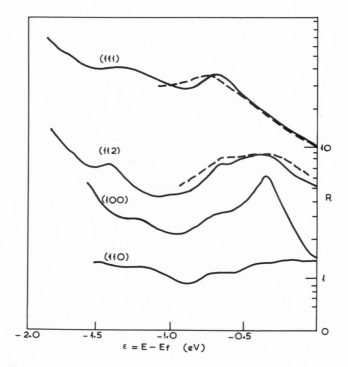

FIGURE 1.9

The enhancement factor R for four planes of tungsten at $T = 78$ K. The free electron distribution, Eq. (1.87), was calculated using the following work functions: $\phi(110) = 5.25$ eV, $\phi(100) = 4.64$ eV, $\phi(112) = 4.90$ eV, and $\phi(111) = 4.45$ eV. The broken curves show the dependence of the enhancement factor for the (111) and (112) planes on the electric field; for the other two planes R is practically independent of the field. The approximate electric fields were as follows: 0.27 V/Å for the broken (111) curve, 0.33 V/Å for the solid (111) curve; 0.29 V/Å for the broken (112) curve, 0.38 V/Å for the solid (112) curve; 0.37 V/Å for the (100) curve; 0.33 V/Å for the (110) curve. (From Vorburger *et al.*, 1975.)

where $j(E)$ is the measured TED of the field-emitted electrons and $j_0(E)$ is the corresponding free-electron distribution calculated according to Eq. (1.87). If $j(E)$ were free-electron-like, then $R(E)$ would be a constant. We see that this is not the case and that R for the (100), (111), and (112) planes of tungsten shows significant structure in the energy region covered by the Plummer and Bell experiments. We note that the R factors could be determined experimentally only within a constant (for a given plane) scaling factor because of difficulties associated with an accurate measurement of the emitting area and collection efficiency of the analyzer. The curves in Fig. 1.9 have been arbitrarily shifted vertically in order to be presented in the same figure. We emphasize, however, that the variation of R with E is unambiguous and in this respect the experimental curves shown in Fig. 1.9 are reproducible (Gadzuk and Plummer, 1973). We must also emphasize that R factors with significant structure are not a peculiarity of tungsten and that such structure has been observed in a number of other cases (see, e.g., Dionne and Rhodin, 1976). These observations show quite clearly that the fine structure in the energy distribution of the emitted electrons depends on the electronic properties of the emitting surface and that, when properly analyzed, it can provide valuable information about these properties. It is obvious that for a proper analysis of these data one has to go beyond the free-electron theory, described in the present chapter.

In principle, if the measured TED deviates considerably from the one predicted by the free-electron theory [Eq. (1.87)], the total emitted current density may not be accurately represented by the Fowler–Nordheim theory. However, the region of energy that contributes most to the emitted current is not different from that predicted by the Fowler–Nordheim theory, i.e., $|\epsilon| < 0.5$ eV, and if R does not have any pronounced structure within this region, then although the emitted current density will, in general, be different from that calculated on the basis of Eq. (1.57) by a factor of $R(E_F)$, a plot of $\ln (J/F^2)$ versus $1/F$ will still give a straight line with a slope given by Eq. (1.61). On the other hand, if the R factor has a pronounced structure, e.g., a peak, in the vicinity of the Fermi level, the preexponential term in Eq. (1.57) may have to be replaced by an expression with a stronger dependence on the applied field. Whether this will lead to any noticeable deviation from linearity in a FN plot of $\ln (J/F^2)$ versus $1/F$ over the limited range of applied field 0.3 V/Å $< F < 0.5$ V/Å used in most field emission experiments is another matter. It is worth noting that in the case of the (100) plane of tungsten, where a peak exists at 0.37 eV below the Fermi level, no deviation from linearity has been observed in the corresponding FN plot and no deviation in the slope of this plot from the expected theoretical value, given by Eq. (1.61), which can be attributed to the existence of the above peak in the TED curve can be conclusively established on the basis of existing data.

At this point we should also emphasize that when the emitted current density is very high ($J > 10^6$ A/cm^2), which for a typical emitter ($\phi \simeq 4.5$ eV) implies an applied field $F > 0.5$ V/Å, the effects of space-charge accumulation in front of the emitter, which we have assumed to be negligible, become significant and they must be taken into account in the calculation of the emitted current density for a given cathode (emitter)–anode voltage (Barbour, Dolan, Trolan, Martin, and Dyke, 1953) and in the calculation of the total energy distribution. The latter appears to be much wider than the Fowler–Nordheim TED curve when space charge becomes significant (Bell and Swanson, 1979). In the present volume we shall assume, unless otherwise stated, that the emitted current density is sufficiently low, so that space-charge effects are negligible.

1.7.2. ENERGY DISTRIBUTION OF THE EMITTED ELECTRONS IN THE THERMAL-FIELD REGION

In the present case, the transmission coefficient must be calculated using Eqs. (1.32)–(1.39) without any further approximation and the integral in Eq. (1.81) must be calculated numerically. Such calculations have been done by a number of authors for various values of field and temperature (El-Kareh, Wolfe, and Wolfe, 1977; Gadzuk and Plummer, 1971a; Bell and Swanson, 1979). Unfortunately, at the present time, we do not have the necessary TED measurements from single-crystal planes of metals at sufficiently high temperatures ($T > 1000$ K) and applied fields ($F \gtrsim 0.1$ V/Å) to check the predictions of the free-electron theory in the TF region, and to find out, in particular, whether fine structure, such as the one shown in Fig. 1.9, exists in TED curves at high temperatures.

In Fig. 1.10a we show the results of TED measurements in the TF region taken by Gadzuk and Plummer (1971a) from a tungsten emitter (not necessarily from a single-crystal plane) heated to 1570 K and for emitter-to-anode voltages between 400 and 1600 V corresponding to applied fields in the region, approximately, of 0.1 to 0.4 V/Å. The experimental curves are in very good qualitative

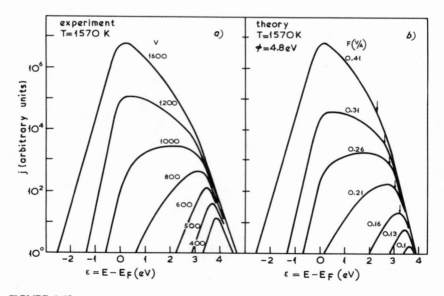

FIGURE 1.10

(a) Experimental total energy distribution of emitted electrons as a function of applied emitter-to-anode voltage for a tungsten emitter at $T = 1570$ K. (b) Theoretical total energy distribution for a 4.8-eV work function and $T = 1570$ K. (From Gadzuk and Plummer, 1971a.)

agreement with theoretical curves, shown in Fig. 1.10b, calculated by the same authors on the basis of Eqs. (1.81) and (1.32)–(1.39), assuming an effective work function $\phi = 4.8$ eV. At high energies, above the top of the barrier the shape of the TED curve is essentially determined by the Fermi–Dirac distribution function which for $\epsilon \gg k_B T$ is proportional to exp $(-\epsilon/k_B T)$ and hence the slope of $\log_{10} j_0(\epsilon)$ is negative and equals $-(\log_{10} e)/k_B T$. A change in the slope occurs at an energy near the top of the barrier, indicated by an arrow in Fig. 1.10b, where the emission changes from transmission over the barrier to tunneling. For low applied fields the transition from a negative slope for energies above the barrier top to a positive slope for energies below the barrier top is quite sharp. The transition is much "softer" for higher applied fields as expected from Eq. (1.32), for the transmission coefficient. It is worth noting how the relative contribution of each energy region changes with the applied field at a given temperature. At low fields it comes mostly from a narrow region at the top of the barrier. At intermediate fields comparable emission comes from the region of the Fermi level and the top of the surface barrier. At high fields emission comes mostly from the region of the Fermi level. It is obvious that the FWHM (full width at half-maximum) of the TED curve will have a maximum at intermediate fields. This is demonstrated quite clearly in Fig. 1.11. The curves shown in this figure were calculated by Bell and Swanson (1979) on the basis of Eq. (1.81) in the manner described at the beginning of this section.

Another interesting aspect of thermal-field TED distributions is the following. As pointed out by Gadzuk and Plummer (1971a) the shape of such a distribution near the top of the barrier is bound to be sensitive to the shape of this barrier. For the higher fields shown in Fig. 1.10 the position z_m of the barrier peak given by Eq. (1.28) is quite small, $z_m \simeq 3$–4 Å. Accordingly, any discrepancy between theory and experiment that is due to the inadequacy of the image law potential at small distances ought, in principle, to be reflected in a deviation from the theoretical TED curve that manifests itself most strongly at the higher fields. Of course, a quantitative analysis along the above lines presupposes an accurate measurement of the field F at the emitter surface and an independent knowledge of the work function. At present, on the evidence of Fig. 1.10 we can only say that the experimental results are consistent with our assumption that the image potential barrier remains valid for distances approaching 3–4 Å from the metal–vacuum interface.

1.7.3. ENERGY DISTRIBUTION OF THERMALLY EMITTED ELECTRONS

At the high temperatures ($T > 1500$ K) and low applied fields ($F < 0.001$ V/Å) normally used in thermionic emission experiments [this is the region where Eq. (1.79) is valid], only electrons with energy above the top of the barrier con-

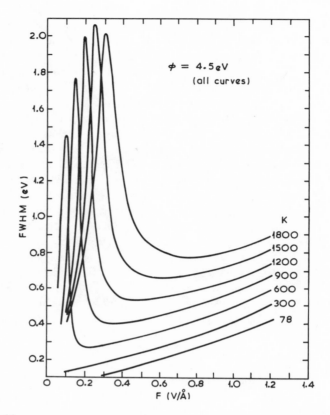

FIGURE 1.11

Theoretical values of the FWHM of the total energy distribution of the emitted electrons as a function of applied field at various temperatures for a 4.5-eV work function. (From Bell and Swanson, 1979.)

tribute significantly to the emitted current. Let us assume that to a first approximation $D(W,F)$ is given by

$$D(W,F) = 0 \qquad \text{for } W < V_{max} \qquad (1.94)$$
$$= 1 \qquad W > V_{max}$$

For energies above the top of the barrier, the unity in the denomination of Eq. (1.5) can be dropped and the Fermi–Dirac distribution reduces to a Maxwellian one, i.e.,

$$f(\mathbf{E}) \propto e^{-\mathbf{E}/k_B T} \qquad (1.95)$$

where E is the energy measured from the top of the barrier. Using the above equations and Eq. (1.81), we find the following formula for the energy distribution of the thermally emitted electrons from a free-electron metal:

$$j_0(E) \propto E e^{-E/k_B T} \tag{1.96}$$

Reflection of electrons by the barrier top (neglected in the above approximation) may result in a small deficiency of electrons at very low energies ($E \simeq 0$) which, if need be, can be calculated with sufficient accuracy. A more significant deviation either in the form of a large deficiency that cannot be accounted for in the above manner, or a "structure" in the TED curve must be due to the non-free-electron character of the emitter and as such it will depend on the specific properties of the emitting surface at high temperatures (see Chapter 9). No measurements exist of the energy distribution, as defined at the beginning of Section 1.7, for thermally emitted electrons. The only relevant experimental data are those of Hutson (1955), who measured the very similar distribution—we shall denote it by $j(E_t)$, where $E_t \equiv E - E_x$ and E_x denotes the kinetic energy of the emitted electron in the x direction which lies in the plane of the emitting surface. The corresponding

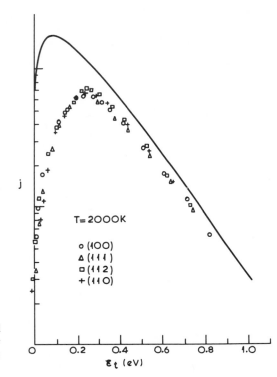

FIGURE 1.12

Experimental energy (E_t, as defined in the text) distributions of thermally emitted electrons from four tungsten planes. (From Hutson, 1955.)

quantity, calculated for a free-electron metal and with the transmission coefficient approximated by Eq. (1.94), will be denoted by $j_0(E_t)$. We find

$$j_0(\mathbf{E}_t) \propto \mathbf{E}_t^{1/2} e^{-E_t/k_B T} \qquad (1.97)$$

which is very similar to Eq. (1.96). In Fig. 1.12 we show the experimental results for $j(E_t)$, taken by Hutson from different planes of tungsten at $T = 2000$ K and a given anode potential. In this figure the distributions are plotted on a logarithmic scale, so that the curves corresponding to different planes can be superposed by vertical and horizontal displacement without the use of scale factors. We note that in the experiment the energy is measured relative to a fixed point, so the horizontal shifts necessary to bring the various curves into alignment give the differences between the work functions of these planes at the given temperature. (For details of this aspect of Hutson's measurements the reader is referred to the original paper.) The solid curve in Fig. 1.12 is a theoretical energy distribution calculated according to Eq. (1.97). We see that the experimental curves do not depend on the crystallographic orientation in agreement with Eq. (1.97) and approach the theoretical curve for sufficiently large values of E_t. On the other hand, they show a marked deficiency at lower energies relative to the theoretical curve which cannot be understood within the free-electron model of the emitter. It has been suggested by Smith (1955) that the true distribution might be in agreement with Eq. (1.97) and that the "deficiency" in the distribution observed by Hutson might be due to a number of possible factors affecting the resolution of the energy analyzer. From the theoretical point of view, it seems rather unlikely that a deviation from "free-electron" behavior, which by implication depends on the specific properties of the emitting surface, will at the same time be independent of crystallographic direction as appears to be the case for the results shown in Fig. 1.12. Unfortunately, no other measurements of energy distributions of thermally emitted electrons have been reported during the last 25 years.

WORK FUNCTION AND OTHER EMISSION MEASUREMENTS

2.1. INTRODUCTION

A variety of methods have been employed over many years for the experimental determination of the work function of metal and semiconductor surfaces. There exist, at the present time, a number of review articles relating to work function measurements on single-crystal planes (Riviere, 1969; Fomenko, 1966; Haas and Thomas, 1972; Swanson and Bell, 1973; Hölzl and Shulte, 1979). Most of these review articles contain extensive lists of the available experimental data along with a detailed description of various methods of measurement. Some of these methods are based on the formulas for field emission and thermionic emission given in Chapter 1. We shall describe, very briefly, some of these methods. We are concerned only with measurements taken from single-crystal planes of metals and under conditions such that space-charge effects are negligible. Some aspects of electron emission from adsorbate-covered metal surfaces will be discussed in Chapter 6. We note that although quite often the same methods are used for measuring the work function and other "emission parameters" of adsorbate covered surfaces as for clean uniform surfaces, a quantitative analysis of the data in the former case involves additional assumptions (if the same formulas are to be used), which may or may not be justified. We emphasize that the formulas in Chapter 1 apply only to emission from plane and uniform metal surfaces.

2.2. THERMIONIC DATA BASED ON SCHOTTKY AND RICHARDSON PLOTS

Nichols (1940) obtained the first measurements of a thermionic current emitted from a single-crystal plane. The apparatus used by him and, in modified form, by subsequent workers (Smith, 1954; Hughes, Levinstein, and Kaplan, 1959) is shown schematically in Fig. 2.1. The emitter, wire A, consists of a single long crystal, which can be rotated in front of the slit S cut in the cylindrical anode B.

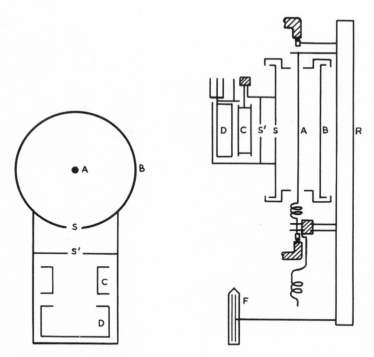

FIGURE 2.1

Cross section and top view of tube used by Nichols (1940) and, in modified form, by Smith (1954) and others to measure the thermionic constants for various crystallographic directions. The single-crystal wire A is mounted in the rotor R so that it can be rotated in front of the slit system SS'.

Provided a sufficiently high voltage is applied between the filament A and the anode B the thermal electrons proceed radially and in straight lines from the emitter to the anode. Only electrons emitted from a small area, corresponding to a selected crystallographic direction normal to the axis of the wire, and whose trajectories are in line with the slit system SS' pass through to the collector and are measured. By rotating the filament, the thermionic current from different crystallographic planes can be obtained. An alternative method of measurement is based on the use of a large monocrystal as the thermionic emitter in an essentially planar diode arrangement (Azizov and Shuppe, 1966; Protopopov, Mikheeva, Sheinberg, and Shuppe, 1966; Protopopov and Strigushchenko, 1968). The apparatus used by these authors is shown schematically in Fig. 2.2. In the present case a single crystal is cut into a cylinder of approximately 10 mm diameter with one end closed by a flat surface parallel to the crystallographic plane under consideration. The electrons emitted from this flat surface are focused onto the collector while those emitted from the sides of the cylinder are suppressed. For a detailed description of the apparatus and of the means of minimizing the experimental error associated with each of the above techniques, the reader is referred to the original papers. In Fig. 2.3 we show an illustrative set of Schottky plots taken by Smith (1954), using a cylindrical diode, from various planes of tungsten. Note that in this figure, it is the current and not the current density that is plotted versus $F^{1/2}$. Similar plots have been obtained by Azizov and Shuppe (1966) using a planar diode arrangement and by many other workers. Smith notes in his paper that for applied fields larger than 2600 V/cm no deviation from the Schottky line, aside from the expected small periodic deviation, has been observed. The latter were too small to be measured with any accuracy in Smith's experiments. For lower applied fields (not shown in Fig. 2.3) the experimental curves bend downward. This may be due to large-scale patchiness, if such exists, and to space charge limitation of the emitted current. The slopes of the straight lines shown in Fig. 2.3 agree with the theoretical slope $m_S = e^{3/2}/k_B T$, predicted by the Schottky formula, Eq. (1.74), within the limits of the experimental error (8%). The "zero-field" value of the emitted current is obtained by extrapolating the Schottky line to zero field. This extrapolation implies that $\bar{t}$ in Eq. (1.79) is practically independent of the applied field and is, therefore, justified only when the "periodic" deviation from the

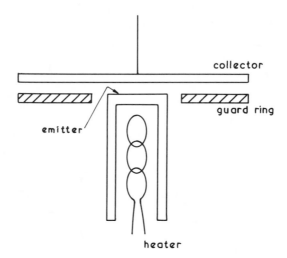

FIGURE 2.2

Planar diode arrangement (schematic) used in the thermionic emission experiments of Azizov and Shuppe (1966).

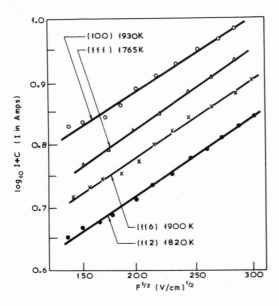

FIGURE 2.3

Schottky plots for four faces of tungsten. $C = 8, 9, 7.9, 8.0$ for the (100), (111), (116), and (112) planes, respectively. (From Smith, 1954).

Schottky line (its amplitude diminishes as the field goes to zero) is small and symmetric about this line. The zero-field value of the emitted current, obtained in the above manner, corresponds to the quantity defined by Richardson's formula, Eq. (1.80), and will, therefore, be denoted by $J_0(T)$. Using the values of $J_0(T)$ obtained in this manner over a range of temperatures, a plot of $\ln(J_0/T^2)$ versus $1/T$ may be obtained. A plot of this kind, known as a Richardson plot or a Richardson line, is shown in Fig. 2.4. It is a straight line in the temperature range of a typical experiment. The Richardson work function ϕ^* is equal, by definition, to the slope of the Richardson line times $(-k_B)$. We have

$$\phi^* = - k_B \frac{d}{d(1/T)} \ln (J_0/T^2) \tag{2.1}$$

Substituting Eq. (1.80) into Eq. (2.1) we obtain

$$\phi^* = \phi + (k_B T^2) \frac{1}{\bar{t}_0} \frac{d\bar{t}_0}{dT} - T \frac{d\phi}{dT} \tag{2.2}$$

where ϕ is the actual work function at temperature T.

A dependence of $\bar{t}_0$ on the temperature may arise because of the shift, a small fraction of an eV in a typical experiment, in the average energy of the emitted electrons with the temperature and also because the reflection of the electrons at the metal–vacuum interface depends on the amplitude of the vibrating atoms of the metal, which in turn depends on the temperature (see Chapter 9). The consequent variation in $\bar{t}_0$ is very small and, therefore, when thermionic data are analyzed on the basis of Eq. (2.2), the second term in this equation is usually neglected.

The temperature coefficient of the work function $d\phi/dT$ can be measured experimentally by a number of methods (Hutson, 1955; Shelton, 1957; Van Oostrom, 1963; Swanson and Crouser, 1967a; Haas and Thomas, 1977). In order to give an idea of the order of magnitude of $d\phi/dT$ we give in Table 2.1 two sets of experimental data taken by different workers from different crystals using different experimental techniques. The results for tungsten were obtained by Swanson and Crouser (1967a) from field-emission data (see Section 2.5) while those of copper were obtained by Haas and Thomas (1977) using incident electron beam techniques. We note that the above results for tungsten are not very different from the earlier results of Hutson (1955) based on measurements of the energy distribution of thermal electrons at different temperatures in the region 1700–2000 K. The results of Hutson, as corrected by Smith (1955), are shown in the second

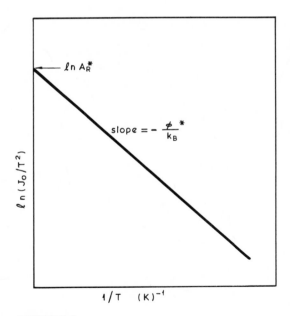

FIGURE 2.4

A Richardson plot.

TABLE 2.1

The Temperature Coefficient of the Work Function for Some Faces of Tungsten and Copper

Tungsten[a]	$d\phi/dT$ (eV/K) 100 K $< T <$ 900 K	Copper[b]	$d\phi/dT$ (eV/K) 100 K $< T <$ 900 K
112	-1.43×10^{-4}	110	-1.7×10^{-4}
100	-1.09×10^{-4}	100	-2.4×10^{-4}
111	3.5×10^{-5}	111	-8.0×10^{-5}
116	5.0×10^{-5}	112	-1.6×10^{-4}

[a]Swanson and Crouser (1967a)
[b]Haas and Thomas (1977)

column of Table 2.2. We see that only in the case of the (100) plane are the results very different. This may be partly due to the non-free-electron character in the field emission energy distribution from this plane (see Fig. 1.8c), which may introduce a small, temperature-dependent, deviation in the slope of the experimental plot of $\log_{10} (J/F^2)$ versus $1/F$, not accounted for by the Fowler–Nordheim formula on the basis of which the values in the second column of Table 2.1 have been calculated. An interesting feature of the tungsten results is the fact that for some planes the temperature coefficient of the work function is positive and for others negative, in contrast to the copper results, where this coefficient is negative for every plane. This is true not only for the planes shown in Table 2.1 but for every other Cu plane for which $d\phi/dT$ has been measured (Haas and Thomas, 1977). At the moment we do not have a proper theory for the variation of the work function with temperature. A qualitative analysis of the various contribu-

TABLE 2.2

Thermionic Constants for Some Faces of Tungsten[a]

Face	$\phi = \phi^* + T\, d\phi/dT$ (eV)	$A_R \bar{t}_0$ (A cm^{-2} deg^{-2})
111	$4.38 + 4.9 \times 10^{-5}T$	92
112	$4.68 - 6.7 \times 10^{-5}T$	55
100	$4.52 - 0.1 \times 10^{-5}T$	104
116[b]	$4.30 + 4.9 \times 10^{-5}T$	94

[a]Smith (1954; 1955); Hutson (1955).
[b]These data were actually taken in a direction between the (114) and (116).

tions to the temperature coefficient of the work function has been given by Herring and Nichols (1949).

It is obvious from the above discussion that the last term in Eq. (2.2) cannot be neglected. Hence, the actual work function ϕ at temperature T is in general different from the Richardson work function ϕ^*. We have

$$\phi \simeq \phi^* + T\frac{d\phi}{dT} \tag{2.3}$$

We note that in many reports of thermionic measurements of the work function it is ϕ^* and not ϕ that is recorded by the authors. The intercept of the experimental Richardson line with the ordinate axis (see Fig. 2.4) is usually denoted by $\ln A_R^*$. It is easy to show that

$$\ln A_R^* = \ln (A_R \bar{t}_0) + \frac{\phi^* - \phi}{k_B T} \tag{2.4}$$

where $(\phi^* - \phi)$ is given by Eq. (2.2). If we drop the second term in Eq. (2.2), we find

$$A_R \bar{t}_0 = A_R^* \exp\left(\frac{1}{k_B}\frac{d\phi}{dT}\right) \tag{2.5}$$

Hence, if we know the temperature coefficient of the work function, we can determine from the Richardson plot both the work function ϕ and the average transmission coefficient $\bar{t}_0$. We recall that A_R is the universal constant given by Eq. (1.75). In the third column of Table 2.2 we show the values of $A_R \bar{t}_0$ obtained in the above manner by Smith (1955) for various planes of tungsten. These are based on measurements of ϕ^* and A_R^* by Smith (1954) and independent measurements of $d\phi/dT$ by Hutson (1955). The corresponding values of $\phi = \phi^* + T(d\phi/dT)$ for each plane are shown in the second column of Table 2.2.

The Richardson work function ϕ^* of the low index planes of tungsten has been measured by a number of different workers. A critical evaluation of these measurements up to the year 1969 has been given by Riviere (1969). His summary of the experimental results and his remarks including a best estimate for the work function of some planes is given in Table 2.3, which has been taken from his article. We note, also, that work function values obtained by other methods—e.g., field emission, photoemission, and other methods—are in good agreement with the thermionic values. In general the agreement between the values obtained by essentially different methods is more or less the same as that obtained by different workers using the same or variations of the same method. The truth of the

TABLE 2.3
Measured Values of ϕ^* (eV) for Some Faces of Tungsten*

Ref.	110	111	112	116	100	Remarks
Brown et al. (1950)					4.59 ± 0.02	ϕ^*(110) too low
Smith (1954)	4.58	4.39	4.65	4.29	4.52	
Hughes et al. (1959)	5.27 ± 0.05		5.24 ± 0.05			
Sytaya et al. (1962)	5.30 ± 0.06				4.66 ± 0.06	
Sultanov (1964)	5.33 ± 0.03	4.40 ± 0.03		4.30 ± 0.03		
Azizov et al. (1966)	5.40 ± 0.05	4.42 ± 0.03	4.80 ± 0.05		4.55 ± 0.05	
Fine et al. (1965)	5.18 ± 0.08					
Best estimate	5.30 ± 0.12	4.40 ± 0.03		4.30 ± 0.03	4.58 ± 0.08	

[a]Riviere (1969).

above statement can be checked by inspection of the available lists of work function data (Riviere, 1969; Haas and Thomas, 1972; Hölzl and Schulte, 1979). Unfortunately, the existing thermionic data on A_R^* and $d\phi/dT$ are not sufficient for an accurate estimate of $A_R \bar{t}_0$. Azizov and Shuppe (1966), using the same values for $d\phi/dT$ as Smith (shown in the second column of Table 2.2), found values for $A_R \bar{t}_0$ for various tungsten planes which are significantly different from the corresponding values found by Smith. In a subsequent paper by Shuppe and his co-workers (Protopopov et al., 1966), the opinion is expressed that the values for $A_R \bar{t}_0$ obtained by these authors, from Richardson's plots for tungsten, molybdenum, and tantalum, were not sufficiently accurate—due to the presence of residual impurities—for a reliable determination of $\bar{t}_0$.

2.3. FIELD EMISSION MICROSCOPY

In order to obtain the very high fields necessary for electron emission at low temperatures, the emitting surface, though "microscopically" flat, must lie in a region which, when viewed on a larger scale, has a high curvature resulting in a very high applied field at the emitting surface (cathode), when a reasonably high voltage is applied between the cathode and the anode. For this purpose field emit-

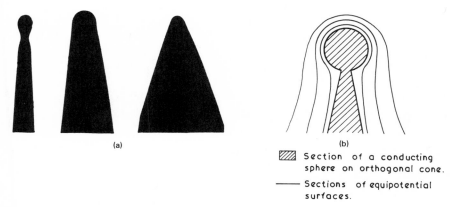

(a)   (b)

▨ Section of a conducting
sphere on orthogonal cone.

── Sections of equipotential
surfaces.

FIGURE 2.5

(a) Typical field emitter geometries from electron micrographs. (b) Conducting sphere on orthogonal cone and the corresponding equipotential surfaces (solid lines) when the sphere is charged. (From Dyke and Dolan, 1956).

ters are made from fine wires etched to a sharp tip as shown in Fig. 2.5a. The thermodynamically stable state of the tip surface consists of flat areas, corresponding to well-defined crystallographic planes, which blend together to form a curved surface (Herring and Nichols, 1949). The latter can be approximated by a hemispherical surface of radius r_t (see, e.g., Fig. 2.5b). The radius of a field-emitter tip may vary from 100 to 2000 Å. For a typical emitter $r_t \simeq 1000$ Å.

In Fig. 2.6 we show a schematic diagram of a field emission diode. The anode is nearly spherical (the exact shape of the anode is not important) with the emitter tip at its center of curvature. The electrons emitted from the tip have very little initial kinetic energy and their trajectories are determined by the spatial distribution of the accelerating applied field, which is approximately spherical in the immediate vicinity of the hemispherical surface of the tip. This is evidently so for a metallic emitter, for in this case the surface of the tip is an equipotential surface and the electric field is normal to it. The magnitude of the field diminishes approximately in proportion to the square of the radial distance from the tip, so that the electrons acquire their final velocity within a few tip radii from the emitting surface, following from then onward radial trajectories to the anode (Gomer, 1961; Gadzuk and Plummer, 1973). The latter is covered with a phosphor coating, a fluorescent screen, which activated by the impinging electrons generates a magnified image of the tip surface. The magnification factor is proportional to the ratio of the distance between the tip and the screen over the tip radius and in practice is of the order of 10^5. This is the basis of the field emission microscope

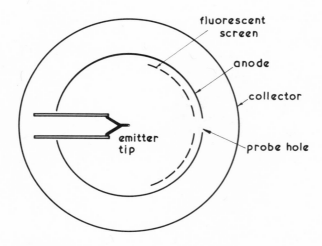

FIGURE 2.6

Schematic diagram of a field emission diode.

invented by Müller (1937). Since the principles of field emission microscopy as well as many examples of its utility can be found in a number of excellent review articles and books (Good and Müller, 1956; Dyke and Dolan, 1956; Gomer, 1961; Swanson and Bell, 1973; Gadzuk and Plummer, 1973), we need not reproduce the details of the arguments in the present volume. We shall simply summarize the essential results of field emission microscopy, referring the reader to one or the other of the above sources for the details.

The trajectory (schematic) of a field-emitted electron is shown by the solid line in Fig. 2.7. We note that this figure is not drawn to scale and that in reality the tip to screen distance, a few centimeters, is approximately 10^5 times larger than the tip radius. The initial radial and transverse components of the electron velocity are denoted by u_r and u_t, respectively. Because u_t is different from zero the electron arrives at the screen at a point displaced by an amount $D/2$ from the forward direction (as determined by the radial component of the initial velocity). A detailed calculation (Gomer, 1961) shows that

$$D/2 \simeq 2\langle u_t \rangle t \tag{2.6}$$

where $\langle u_t \rangle$ denotes the average (initial) transverse velocity of the emitted electrons, and t is given by

$$t \simeq x(2 \text{ eV/m})^{-1/2} \tag{2.7}$$

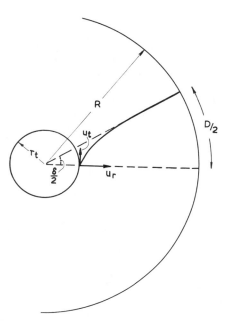

FIGURE 2.7

The trajectory (schematic) of a field-emitted
electron.

where x denotes the distance from the tip to the screen and V is the applied voltage
between the cathode–emitter and the anode. In order to estimate $\langle u_t \rangle$ one may
write

$$\langle u_t \rangle = (2\langle E_t \rangle/m)^{1/2} \qquad (2.8)$$

where $\langle E_t \rangle$ is the average transverse energy at the emitting surface of the field-
emitted electrons. A straightforward calculation of this quantity, based on the
Fowler–Nordheim theory of field emission, gives

$$\langle E_t \rangle = \frac{0.433F}{v(3.79F^{1/2}/\phi)\phi^{1/2}} \quad \text{eV} \qquad (2.9)$$

where v is the elliptic function tabulated in Table 1.1, ϕ is an effective work function, and F is the field at the tip surface in V/Å. Empirically it is found that

$$F = \beta V \equiv V/kr_t \qquad (2.10)$$

where $k \simeq 5$ at the apex of the emitter and increases with polar angle (see, e.g.,
Fig. 2.9).

It is clear from Fig. 2.7 that the half-width of a given region on the tip will appear to an observer looking at the screen extended by an amount $\delta/2$ given by

$$\frac{\delta}{2} = \frac{D}{2M} \tag{2.11}$$

where M is the magnification factor of the field emission microscope. If the tip was perfectly spherical M would be equal to x/r_t. In reality the presence of the emitter shank compresses the field lines towards the axis of the tip, thus reducing the magnification. It is found that for practical emitters

$$M = \frac{1}{\gamma}\frac{x}{r_t} \tag{2.12}$$

where $\gamma \simeq 1.5$. Finally, using the above equations, we find that the resolution of a field-emission microscope is given by

$$\delta \simeq 2.62\gamma(r_t/kv\phi^{1/2})^{1/2} \text{ Å} \tag{2.13}$$

where r_t is expressed in Å. For a typical emitter, this gives a resolution of 25 Å or so (Gomer, 1961; 1978).

Apart from the statistical limitation given by Eq. (2.13), there exists another limitation to the resolution of a field emission microscope due to Heisenberg's uncertainty principle which states that, if an electron originates from a region of width δ_0, it must by necessity have a minimum transverse velocity $u_t \simeq \hbar/(2m\delta_0)$, which thus puts an intrinsic limit to the resolution of any electron microscope. However, when this factor is taken into account, the resulting estimate for the resolution is not significantly different from that given by Eq. (2.13), (Good and Müller, 1956; Gomer, 1961).

Returning to Fig. 2.6, we see that if a "probe hole" is opened in the screen (anode), in the manner originally suggested by Müller (1955), then the electrons which pass through this hole, to be collected by an appropriate collector, originate from a region on the tip which on the one hand is sufficiently small (with a diameter of 100 Å or so) to be part of a single-crystal plane and on the other hand it is sufficiently large compared to the resolution of the microscope. That this is, in fact, the case is demonstrated quite clearly in Fig. 2.8, which shows a micrograph of a tungsten emitter, taken by Müller (1955), with the image of the probe hole appearing as a dark hole in the center of the (012) plane. The dark regions in the micrograph correspond to planes of high work function (low emission) and the light regions to planes of low work function (high emission). The indexing of the planes can be established from the symmetry of the pattern which depends of course on the orientation of the tip (see, e.g., Dyke and Dolan, 1956; Gomer,

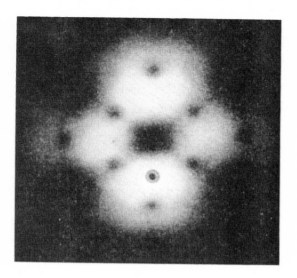

FIGURE 2.8

Field emission micrograph of a tungsten emitter with the (110) plane in the center of the pattern, and the probe hole adjusted to the (012) plane. (From Muller, 1955).

1961). In practice, the crystallographic plane to be measured is adjusted onto the probe hole by shifting either the cathode or the anode depending on the apparatus being used. For accurate measurements the probe hole must be adjusted to the center of the plane and cover only a fraction of this plane as shown in Fig. 2.8. We note, in passing, that similar micrographs can be obtained in thermionic emission experiments using the cylindrical diode described in Section 2.2 (see, e.g., Nichols, 1940; Haas and Thomas, 1972).

The absolute value of the density of the field-emitted current can, in principle, be determined by dividing the collected current by the size of the hole projected back onto the tip surface. For this purpose, however, an accurate knowledge (measurement) of the magnification factor M is required, which is difficult to obtain in practice. In certain cases it may be possible to measure the emitting area directly. Ehrlich and Plummer (1978) managed to do so in the case of field emission from tungsten (110). They determined the total area of the crystal plane under consideration from field ion micrographs which made possible a direct counting of the atoms at the periphery of the plane. The fraction of the plane, covered by the probe hole, which contributed the measured current, was then determined to a good approximation by field emission micrographs like the one shown in Fig. 2.8. We shall discuss Ehrlich and Plummer's results in more detail in Section 5.4.2.

2.4. THE LOCAL FIELD AT THE SURFACE OF A FIELD EMITTER

A central difficulty, which always arises in a quantitative analysis of field emission measurements, derives from the awkward shape of the emitter, which does not permit a straightforward determination of the local field at the emitting surface for a given applied voltage between the cathode–emitter and the anode. An analytic determination of the field distribution in the space between the cathode and the anode is possible only when the cathode and anode surfces are equipotential surfaces of relatively simple geometrical shape. In the case of metal field emitters we can safely assume that the emitter surface is an equipotential surface and so, of course, is the anode, in which case the problem reduces to that of solving Laplace's equation for a given applied voltage between cathode and anode. Dyke, Trolan, Dolan, and Barnes (1953) observed that the equipotential surfaces surrounding a sphere on an orthogonal cone (Fig. 2.5b) fit reasonably well the geometry of a practical emitter (Fig. 2.5a). They calculated the electrostatic field for different parameters of the sphere–cone cathode, with the anode surface approximated by a paraboloid. The variation of the local field, at the emitter surface, with the angular distance θ from the emitter apex, calculated according to this model, is shown in Fig. 2.9. The quantity β is defined by Eq. (2.10) and $\beta_0 = \beta(\theta = 0)$. The two dashed curves correspond to two emitter profiles with a slight and a pronounced constriction, respectively. The solid line corresponds to an average emitter. The circles represent experimental data for a tungsten emitter obtained by Swanson and Crouser (1967a) by measuring the variation in the slope of the FN plot, of $\log_{10}(I/V^2)$ versus $1/V$ (I denotes the current) for various (310) planes along the [100] zone of a (310) oriented emitter. According to Eq. (1.61) and Eq. (2.10), we have

$$m_f = \frac{d}{d(1/V)}\log_{10}(I/V^2) = -0.296\phi^{3/2}s\left(\frac{3.79F^{1/2}}{\phi}\right)\bigg/\beta \quad (2.14)$$

where β is measured in Å^{-1}, F in V/Å, and ϕ in eV. Since the work function for each of the (310) planes along the [100] zone, corresponding to different values of θ, has the same value, and because s is practically constant (see Table 1.1), the variation $\beta(\theta)/\beta_0$ may be obtained from the corresponding variation in the measured m_f. We note that the curve for $\beta(\theta)/\beta_0$ obtained in the above manner may not be always reliably used for an estimation of the local field at another plane [in the present example a plane other than (310)], if the extension of the latter is significantly different from that of the planes used in determining $\beta(\theta)/\beta_0$. The local field at the center of a flat plane with a size larger than average will be smaller than the one estimated in the above manner, by some amount which may

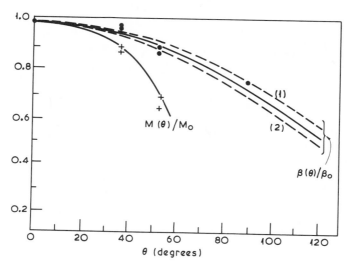

FIGURE 2.9

Variation of the local field factor $\beta(\theta)$ (circles) and of the magnification factor $M(\theta)$ (crosses) with angular distance θ from the emitter apex. $\beta_0 = \beta(\theta = 0)$; $M_0 = M(\theta = 0)$. The broken lines (1) and (2) have been calculated for two sphere-on-orthogonal-cone models imitating two emitter profiles with a pronounced and slight constriction, respectively. The solid line corresponds to an average emitter. (From Swanson and Crouser, 1967a.)

not be easy to estimate with sufficient accuracy (Müller, 1955). Ideally, one would like the flat segments on the field emitter, corresponding to the various crystallographic planes, to be of the same size with a diameter an order of magnitude smaller than the emitter radius r_t, so that the emitter surface is macroscopically smooth. At the same time the extension of the flat planes must be sufficiently large so that an electron emitted from the center of a given plane does not feel the effects of the edges of this plane. This is very important since a theoretical analysis of the data is possible only on the assumption that the emitting plane extends to infinity (see Chapters 1 and 5). In some instances, exemplified by the (110) plane of tungsten, the size of the emitting plane may be of critical importance. The low work function of W(110), in relation to its surrounding planes on the emitter tip, demands a relatively large size, if distortion of the electrostatic potential at the central region of this plane is to be avoided. On the other hand, a large size leads to a relatively low applied field at the center of this plane and makes its evaluation, in the manner we describe here, very difficult (Müller, 1955; Young and Müller, 1962; Young and Clark, 1966a, 1966b; Lea and Gomer, 1971; Swanson and Crouser, 1967a; Todd and Rhodin, 1973; Polizzotti and Erhlich, 1980).

We note that experimental curves, of $\beta(\theta)/\beta_0$, like that of Fig. 2.9, determine only the variation of the applied field on the emitter surface and not its absolute magnitude. Obviously, if ϕ for a given plane is known (measured independently by another method), then one can determine β and hence F at this plane from the measured value of m_f using Eq. (2.14). It is worth noting, in this respect, that the elliptic function s in Eq. (2.14) depends on $F = \beta V$. Fortunately, this dependence is weak and an accurate solution for β may be obtained efficiently by an iterative process. Once the magnitude of β for a given plane has been fixed in the above manner the values of β for other planes can be obtained from the experimental $\beta(\theta)/\beta_0$ curve. More direct methods for measuring the local field at the emitter surface, under conditions appropriate to a field emission experiment, are not presently available. A method valid in the thermal-field region has been suggested by Bermond (1975). Another method, valid under field-ion operating conditions, has been proposed by Sakurai and Muller (1973). The first method claims an accuracy of 5% and the second 3%. For an order of magnitude calculation of the field at the surface of a field-emission tip one may use empirical formulas which have been developed for that purpose (see, e.g., Gomer, 1961; Dyke and Dolan, 1956). Equation (2.10) is one example of such a formula.

2.5. FIELD EMISSION WORK FUNCTIONS

It is obvious, from the discussion in Sections 2.3 and 2.4, that, at present, one may be able to obtain from the experimental FN plots, and on the basis of Eq. (2.14), the relative magnitude of the work function of different planes but not, directly, the absolute magnitude of the work function of an individual plane. These relative values can, of course, be translated into absolute values if the work function of one of the planes is known independently. The FN work function values shown in Fig. 2.10 like those in Fig. 1.5, were obtained in this manner. In general, the work function values obtained in this way are in reasonably good agreement with the values obtained by other methods (Riviere, 1969; Hölzl and Shulte, 1979). This is demonstrated for the case of tungsten in Fig. 2.10. The circles denote FN values and the squares denote experimental values by other methods (thermionic emission, photoemission, contact potential difference, etc.), obtained from a variety of sources (Todd and Rhodin, 1973; Haas and Thomas 1972; Riviere, 1969; Hölzl and Schulte, 1979). In the case of the closed packed planes a critical analysis of the data (Todd and Rhodin, 1973) suggests that the work function for these planes must lie somewhere in the region indicated by the corresponding brackets in Fig. 2.10. Although there exists, at least for some planes, a significant spread in the experimental values, the data obtained by the FN method are consistent within the limits of experimental error with the data

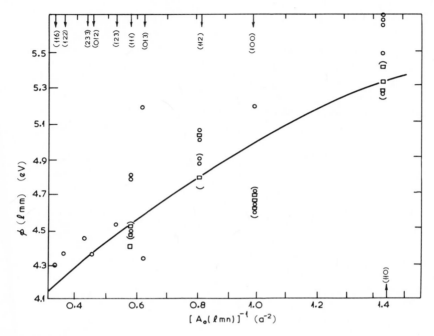

FIGURE 2.10

Work function of different faces of tungsten. The circles denote values obtained from field emission measurements. The squares are experimental values obtained by other methods (thermionic emission, contact potential difference, photoelectric threshold, surface ionization). $A_0(lmn)$ is the area of the surface unit cell ($a = 3.16$ Å). The solid line is the result of a semiempirical theory (Modinos, 1978b).

obtained by the other methods, with the exception of the (110) plane where field-emission data analyzed on the basis of Eq. (2.14) give, on average, a work function value which exceeds that obtained by other (non-field-emission) methods by 0.5 eV or so. The discrepancy between the FN value of the work function for W(110) and the value for the same quantity obtained by other methods, is due to the difficulty associated with the accurate determination of the local field factor β for this plane for the reasons we have already mentioned in the previous section. The studies of Todd and Rhodin (1973) and of Polizzoti and Erhlich (1980) support this point of view. An alternative method of analyzing field emission data (to be described below) also gives values for $\phi(110)$ in reasonable agreement with data obtained by the other methods (Plummer and Bell, 1972). Smith (1970) suggested that the redistribution of the electronic charge at the surface of the emitting plane, induced by the high applied field, might lead to an increase in the apparent work function. However, detailed calculations (Theophilou and Modinos, 1972; Theophilou, 1972; Lang and Kohn, 1973) have shown that the increase in the apparent work function due to the applied field is relatively small (<0.1 eV) for the fields ordinarily used in a field emission experiment.

We note that Eq. (2.14) is valid at very low temperatures. At higher temperatures the slope of the FN plot must be determined according to Eq. (1.54), which is valid in the extended field emission region as defined by Eqs. (1.55) and (1.56). We find, instead of Eq. (2.14), the following equation:

$$m_f(T) = \frac{d}{d(1/V)} \log_{10} (I/V^2) = -0.296\phi^{3/2}s\left(\frac{3.79F^{1/2}}{\phi}\right) \bigg/ \beta(T) + \delta$$

$$(2.15)$$

where

$$\delta \equiv V(1 - \pi p \cot \pi p)/2.3 \qquad (2.16)$$

where p is the parameter defined by Eq. (1.90) and $\beta(T)$ is the local field factor defined by Eq. (2.10) at temperature T. We have

$$\beta(T)/\beta(T = 0) \simeq 1 - \Delta r_t/r_t \qquad (2.17)$$

where Δr_t is the increase in the tip radius due to thermal expansion.

One can find the variation of the work function with temperature from the variation of m_f with the temperature, provided the thermal coefficient of expansion is known. We have

$$\frac{\phi(T)}{\phi(T = 0)} = \left[\frac{m_f(T) - \delta}{m_f(T = 0)}\right]^{2/3} \left[\frac{\beta(T)}{\beta(T = 0)}\right]^{2/3} \qquad (2.18)$$

where in evaluating δ one may replace V in Eq. (2.16) by its midrange value. Swanson and Crouser (1967a) employed the above procedure to determine the temperature coefficient of the work function of various tungsten planes. Their results are shown in the second column of Table 2.1.

In order to overcome the difficulties associated with the determination of the local field factor and for the purpose of determining absolute rather than relative values of the work function, Young and Miller (1962) suggested an alternative method for the determination of the work function which avoids the need for an accurate knowledge of the applied field. Their method, which was subsequently elaborated upon by Young and Clark (1966a; 1966b), is based on the assumption that the energy distribution of the field-emitted electrons is adequately represented by Eq. (1.88). For energies below the Fermi level ($E_F - E > k_B T$), $f(E) \simeq 1$, in which case we obtain

$$S_{ED} \equiv \frac{\partial}{\partial \epsilon} \log_{10} j_0'(\epsilon) = \frac{0.434}{d_0} \qquad (2.19)$$

where, according to Eqs. (1.91) and Eq. (2.10),

$$\frac{1}{d_0} = (1.025) \frac{\phi^{1/2}}{\beta V} t(3.79 F^{1/2}/\phi) \qquad (2.20)$$

where t is given by the third column of Table 1.1. Combining Eq. (2.19) with Eq. (2.14) we obtain

$$\phi = -1.5 \left[\frac{t(y_0)}{s(y_0)} \right] \frac{m_f}{S_{ED} V} \qquad (2.21)$$

which shows that we can determine ϕ by measuring m_f, S_{ED} and the applied voltage V. The value of $y_0 = 3.79 F^{1/2}/\phi$ can be determined by an iterative process in view of the slow variation of both t and s with y_0.

It is obvious that the above method will give reliable values of the work function *only* when the energy distribution in the region where S_{ED} is measured (usually within 0.5 eV or so from the Fermi level) is free-electron-like, i.e., when $R(\epsilon)$ as defined by Eq. (1.93) is constant. For example, it is obvious from Fig. 1.9 that it ought to give reliable results for the W(110) plane, and it does so (Plummer and Bell, 1972) in spite of earlier evidence to the contrary (Young and Clark, 1966b). It is equally obvious that the method cannot be used for a determination of the work function of W(100). We note that in applying Eq. (2.21) it would be wrong to evaluate S_{ED} at a single energy even when the TED curve appears to be free-electron-like. Optimum results are obtained if one identifies S_{ED} with the slope of a straight line which best fits the experimental plot of $\log_{10} j(\epsilon)$ versus ϵ in the energy region extending from the Fermi level to 0.5 eV or so below it (Swanson and Crouser, 1967a; Lea and Gomer, 1971).

Vorburger, Penn, and Plummer (1975) proposed a modified version of the above method which takes into account the deviation of the energy distribution from free-electron behavior as described by the enhancement factor $R(\epsilon)$. When this is done, one finds, instead of Eq. (2.21), the following formula:

$$-\frac{V S_{ED}}{m_f} = 1.5 \left[\frac{t(y_E)}{s(y_0)} \right] \frac{(\phi - \epsilon)^{1/2}}{\phi^{3/2}} - \left(\frac{\partial \ln R}{\partial \epsilon} \right) \frac{V}{m_f} \qquad (2.22)$$

where y_0 is given by Eq. (1.52) and y_E by Eq. (1.86). Note that the quantities on the left-hand side (LHS) of this equation V, m_f, and S_{ED} (evaluated at a given ϵ $= E - E_F$) can all be measured. If $R(\epsilon)$ is in practice independent of the applied field, then if we plot the LHS of Eq. (2.32) versus V, the resulting plot will be a straight line whose slope is proportional to $\partial \ln R / \partial \epsilon$ and its zero field intercept with the ordinate axis will be equal to $t(y_E)(\phi - \epsilon)^{1/2}/[s(y_0)\phi^{3/2}]$. From the latter we can extract the value of ϕ (the numerical values for y_E and y_0 can be determined by an iterative process in the usual manner). If $R(\epsilon)$, as we have assumed, is independent of the applied field, then the work function evaluated in the above manner must be independent of the energy ϵ. The results reported by Vorburger *et al.* (1975) show that this is indeed the case for the W(100) and W(013) planes. The corresponding value of the work function for the W(100) plane agrees, within the limits of experimental error, with independent estimates of this quantity. This is not the case, however, for the (013) plane for reasons which are not well understood. The results obtained by the same authors for the W(111) and W(112) are more difficult to analyze since $R(\epsilon)$ for these two planes depend to some degree on the applied field (see Fig. 1.9). When this is the case the extrapolation to zero field [see Eq. (2.22)] cannot be done in a systematic manner.

2.6. WORK FUNCTION ANISOTROPY

The experimental facts relating to work function anisotropy (dependence of the work function on the crystallographic direction) are demonstrated quite clearly by the tungsten results shown in Fig. 2.10. We note that this anisotropy comes from the (electrostatic) dipole layer contribution to the work function which constitutes a relatively small fraction of the surface potential barrier at the metal-vacuum interface (see, e.g., Fig. 1.3). The number on the abscissa axis of this figure gives the inverse of the area $A_0(lmn)$ of the unit cell of the two-dimensional lattice associated with the (lmn) plane of tungsten in units of a^{-2} (a is the lattice constant; tungsten has a bcc structure with atoms at the center and corners of a cube of side $a \simeq 3.16$ Å). We see that there exists a definite correlation between $A_0^{-1}(lmn)$ and the work function $\phi(lmn)$, which implies an almost linear dependence of the surface dipole layer potential on the number of atoms per unit area of the surface plane. The work function of the W(100) plane is a notable exception to this rule. At present, we do not have a satisfactory theory for a quantitative analysis of the work function anisotropy based on first principles, although a promising beginning was made by Smoluchowski (1941). Most of the calculations (self-consistent or otherwise) of the surface potential and work function have, so far, been based on the jellium model of the metal (the positive ions are replaced by a uniform distribution of positive charge terminated abruptly at the surface) which cannot give rise to any anisotropy in the work function. A perturbation treatment of the effect of the ion lattice on the surface potential of simple (non-transition) metals (a self-consistent jellium potential represents the unperturbed state) shows that the work function increases in most cases, with the density of surface atoms (Lang, 1973). The jellium model is, of course, not applicable to transition metals where the scattering of the electrons by the ionic cores is too strong to be treated by perturbation. Realistic self-consistent calculations of the surface potential of transition metals are possible but lengthy (see, e.g., Kerker, Ho, and Cohen, 1978; Posternak, Krakauer, Freeman, and Koelling, 1980). It is obvious that, in order to establish any systematic trends in the variation of the work function with the density of surface atoms and other parameters, one requires such calculations for many different crystallographic planes (a very difficult task indeed). Modinos (1978b) suggested a semiempirical theory, appropriate for transition metals, based on simple considerations of geometry and on bulk properties of the metal, in an attempt to account for the observed anisotropy in the work function of tungsten. His theory predicts an almost linear dependence on the density of surface atoms (solid line in Fig. 2.10), in good agreement with the experimental data on tungsten with the exception of the (100) plane which seems to be a rather special case. For some reason or other [this may be related to the existence of a band of surface states on W(100) near the Fermi level], the electronic charge distribution on this plane is sensitive to the details of the surface potential and must, therefore, be calculated self-consistently (Posternak et al., 1980). Theoretical work on the work function and other surface properties has been reviewed by a number of authors (Lang, 1973; Hölzl and Shulte, 1979; Inglesfield and Holland, 1981).

2.7. ENERGY ANALYZERS

The trajectory of a field-emitted electron from the emitter (the spherical surface of a field emission tip) to the anode of a typical field emission diode is shown schematically in Fig. 2.7. It indicates that by the time the electron has reached the anode its transverse energy (the energy associated with the transverse component of the initial velocity at the emitting plane) has been transformed into radial energy. In fact, detailed calculations show (Gomer, 1961; Gadzuk and Plummer, 1973) that practically all (99%) of the transverse energy is transformed into radial energy within a radial distance from the tip surface of ten tip radii or less. We recall that the tip-to-anode distance is of the order of 10^5 tip radii. It is obvious that only the *total* energy of the electron is conserved during its flight from the emitting surface to the anode and so, only the total energy of the electron can be measured in an experiment using a conventional field emission diode characterized by the spherical geometry of Fig. 2.6. The energy distribution of the emitted electrons can be measured by the positioning of an energy analyzer in between the probe hole of the anode and the collector. Present day analyzers belong to two general categories. The first category consists of those analyzers which use a variable retarding potential barrier to collect from the emitted current that fraction, denoted by i_c, carried by electrons with total energy exceeding a given value $E = E_F + \phi_c - eV_r$ (see Fig. 2.11). One measures i_c as a function of the retarding potential V_r and the energy distribution, $j(E) = di_c/dV_r$, is obtained by differentiation which can be done electronically. Since the number of low-energy electrons is exponentially less than the number of those having the highest energy near the Fermi level (99.9% of the total current comes from within a region of about 0.5 eV from the Fermi level), the range of energies accessible by this method is limited by shot noise to a region of 0.5 eV or so from the Fermi level (Gomer, 1978).

The second category of field-emission energy analyzers make use of an electrostatic deflection analyzer (Kuyatt and Plummer, 1972). This operates on the following principle. Consider the hemispherical element of Fig. 2.12 and denote the applied voltage between the two hemispheres by V_H. One can show (see, e.g., Kuyatt and Simpson, 1967) that, for a given value of V_H only electrons incident normally (or almost normally) on the entrance aperture A, within a very narrow region of energy, will go through the element and come out from exit B after being

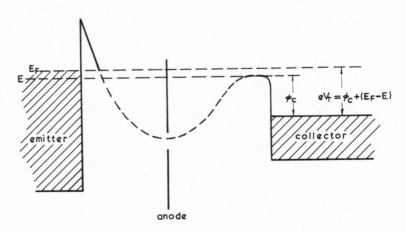

FIGURE 2.11
Schematic description of energy analysis by retardation.

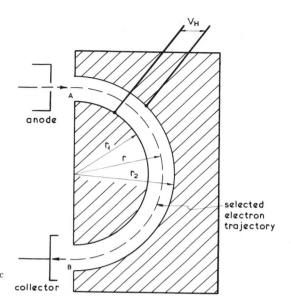

FIGURE 2.12

Schematic diagram of an electrostatic
deflection analyzer.

deflected by 180°. For this to be the case, the electronic energy E must lie within
the region: $E_a - \Delta E/2 \lesssim E \lesssim E_a + \Delta E/2$, where

$$E_a \simeq eV_H[(r_2/r_1) - (r_1/r_2)]^{-1} \qquad (2.23)$$

where r_1 is the radius of the inner hemisphere and r_2 of the outer sphere. The
resolution ΔE is of the order of

$$\Delta E \simeq E_a w/r \qquad (2.24)$$

where w is the width of the input and exit apertures in the diametrical plane of
the hemispheres and r is the mean radius of the hemispheres. It is seen that the
energy resolution is proportional to the energy of the electron. Therefore, the electrons whose kinetic energy at the anode is of the order of 10^3 eV are in practice
retarded to 1–2 eV before entering the analyzer. In this manner, and with the
presently available electrostatic analyzers, one can measure the energy distribution
down to 2 eV below the Fermi level, with an energy resolution of $\Delta E \simeq 0.02$–
0.03 eV (Gadzuk and Plummer, 1973). The lower limit of the accessible energy
range is to a large degree determined by the electron-optics system which slows
down the electrons prior to their entry into the analyzer, and partly by aperture

scattering. In principle, it is possible to avoid the need for retardation and to minimize electron scattering at the apertures by the use of magnetic analyzers. Gomer (1978) suggests that the use of magnetic analyzers may, eventually, extend the energy range accessible to field emission studies to 5–6 eV below the Fermi energy. That will be a remarkable achievement.

We note that, although we talked specifically about field emission energy analyzers in the present section, essentially the same methods can be employed in the energy analysis of the emitted electrons in thermionic, secondary electron emission, etc.

THE CRYSTALLINE SOLID AS A STACK OF ATOMIC LAYERS WITH TWO-DIMENSIONAL PERIODICITY

3.1. INTRODUCTION

We have already established that for a complete analysis of the experimental data relating to electron emission one must go beyond the free-electron model of Sommerfeld. Our knowledge of the electronic properties of metals and other crystalline solids has advanced considerably during the last 20 years and a lot is presently known about the electronic energy levels and the nature of the electronic wave functions in an extended crystalline solid. During the last ten years considerable progress has also been made in our understanding of the electronic properties of solid surfaces, and, in many cases, quantitative agreement between theory and various experimental measurements has been achieved (see, e.g., Woodruff, 1981; Inglesfield and Holland, 1981). In this and the following chapter we introduce the basic concepts and some important results from the theory of the electronic properties of crystalline solids and their surfaces which are necessary for a systematic analysis of electron emission phenomena.

3.2. THE INDEPENDENT-ELECTRON THEORY OF SOLIDS

This theory is based on the assumption, justified by empirical evidence, that the actual interactions between the electrons and nuclei in a solid (crystalline or otherwise) may be replaced, to a good degree of approximation, by a one-electron potential field which describes the interaction of any one electron with an average self-consistent field produced by the other electrons and the nuclei in the solid.

One commonly used method for constructing a (local) potential of this kind is known as the $X\alpha$ method (Slater, 1974). According to this method, the one-electron potential energy field is given by

$$V_{\uparrow(\downarrow)}(\mathbf{r}) = V_c(\mathbf{r}) + V_{X\alpha\uparrow(\downarrow)}(\mathbf{r}) \tag{3.1}$$

where $\uparrow(\downarrow)$ refers to an electron of spin up (down). The first term in Eq. (3.1) denotes the Coulomb energy of the electron in the electrostatic field produced by all the nuclei and electrons in the solid. We have

$$V_c(\mathbf{r}) = -e^2 \sum_j \frac{Q_j}{|\mathbf{r} - \mathbf{R}_j|} + e^2 \int \frac{n(\mathbf{r}') \, d^3r'}{|\mathbf{r} - \mathbf{r}'|} \tag{3.2}$$

$$n(\mathbf{r}) = n_\uparrow(\mathbf{r}) + n_\downarrow(\mathbf{r}) \tag{3.3}$$

where $n_\uparrow$ and $n_\downarrow$ denote the local density of electrons of spin up and spin down, respectively. The first term in Eq. (3.2) represents the interaction of the electron with the positively charged nuclei which are assumed stationary. $\mathbf{R}_j$ denotes the position (known from experiment) and Q_j the charge (in units of e) of the jth nucleus in the solid. The second term in Eq. (3.2) represents the Coulomb energy of the electron in the electrostatic field produced by *all* electrons in the solid. It includes a self-energy contribution, which ought not to be included, by allowing the electron to interact with its own contribution to the electrostatic field. This self-energy contribution to the potential energy field is cancelled by the second term in Eq. (3.1), which is given by

$$V_{X\alpha\uparrow(\downarrow)}(\mathbf{r}) = -3\alpha e^2 \left[(3/4\pi) n_{\uparrow(\downarrow)}(\mathbf{r}) \right]^{1/3} \tag{3.4}$$

where α is an adjustable parameter.

The above contribution to the potential energy derives from the exclusion principle according to which an electron with a given spin $\uparrow(\downarrow)$ at a point $\mathbf{r}$ keeps other electrons with the same spin out of its immediate neighborhood. A region is thus created around the electron (at $\mathbf{r}$) known as the Fermi hole, because of its intimate relation with Fermi statistics, where the equivalent of one electronic charge is missing. If we assume that the electronic density $n_{\uparrow(\downarrow)}(\mathbf{r})$ varies slowly with the position we may approximate the Fermi hole by a sphere of radius R given by

$$R = \left[\frac{4\pi}{3} n_{\uparrow(\downarrow)}(\mathbf{r}) \right]^{-1/3} \tag{3.5}$$

The missing charge of one electron within the above sphere is equivalent to a positive charge e uniformly distributed within this sphere. It follows that the electrostatic (exchange) interaction of an electron with its Fermi hole is approximately equal to the potential energy of an electron at the center of a uniformly charged sphere of radius R, given by Eq. (3.5), and total charge e. The latter energy equals $-3e^2/(2R)$ and is therefore described by Eq. (3.4) with $\alpha \simeq (4\pi/3)^{2/3}/2$. A more appropriate value of α for many atoms in the Periodic Table has been computed by Schwarz (1972) by requiring that the total energy of the atom in its ground state calculated by the $X\alpha$ method is equal to the total energy of the atom calculated by the Hartree–Fock method. [It is well known (Slater 1968; 1974) that the Hartree–Fock energy of the atom is the optimum result within the one-electron approximation.] It turns out that for most atoms α estimated in this way lies in the region $2/3 < \alpha < 1$. The value α for a given atom is also used in calculating $V_{X\alpha}$ in the corresponding solid. [For a detailed discussion of the $X\alpha$ method the reader is referred to the book of Slater (1974).]

For nonmagnetic solids

$$n_\uparrow(\mathbf{r}) = n_\downarrow(\mathbf{r}) = \tfrac{1}{2}n(\mathbf{r}) \tag{3.6}$$

and the potential of Eq. (3.1) is the same for either spin. It is given by

$$V(\mathbf{r}) = V_c(\mathbf{r}) + V_{X\alpha}(\mathbf{r}) \tag{3.7}$$

where V_c is given by Eq. (3.2) and

$$V_{X\alpha}(\mathbf{r}) = -3\alpha e^2 \left[\frac{3}{8\pi} n(\mathbf{r}) \right]^{1/3} \tag{3.8}$$

Once the potential has been fixed the Schrödinger equation for the one-electron problem

$$\left[-\frac{\hbar^2}{2m} \nabla^2 + V(\mathbf{r}) \right] \phi_i(\mathbf{r}) = E_i \phi_i(\mathbf{r}) \tag{3.9}$$

can be solved to give the "eigenvalues" E_i and the corresponding orbitals ϕ_i. The subindex i stands for a set of quantum numbers including the spin one. We note, however, that E_i and $\phi_i(\mathbf{r})$ are independent of spin. The spin part of the orbital which multiplies $\phi_i(\mathbf{r})$ has been suppressed. The electron density is given by

$$n(\mathbf{r}) = \sum_i |\phi_i(r)|^2 n_i \tag{3.10}$$

where $n_i = 1$ if the ith orbital is occupied and $n_i = 0$ if this orbital is empty. When the many-electron system (solid) is in its ground state all orbitals with energy eigenvalue smaller than a certain value will be occupied and those with energy above this value will be empty. Since $V(\mathbf{r})$ is a function of $n(\mathbf{r})$, Eqs. (3.7)–(3.10) must, at least in principle, be solved self-consistently. A knowledge of the occupied orbitals is sufficient for a determination of the energy [see Eq. (3.17)] and other ground-state properties.

A second method, similar to the $X\alpha$ method but more systematic, for constructing an *effective* local one-electron potential is based on a theorem (Hohenberg and Kohn, 1964; Kohn and Sham, 1965) which states that the energy (similarly all other properties) of the ground state of the solid (or for that matter any system of N interacting electrons in an external field) is a unique functional of the density $n(\mathbf{r})$. Moreover, $n(\mathbf{r})$ can be obtained by formally solving a "one-electron" problem which takes the form of Eqs. (3.9) and (3.10) with $V(\mathbf{r})$ replaced by

$$V(\mathbf{r}) \rightarrow V_c(\mathbf{r}) + V_{xc}(n(\mathbf{r});\mathbf{r}) \tag{3.11}$$

where V_c is given by Eq. (3.2) and $V_{xc}(n(\mathbf{r});\mathbf{r})$ is an exchange-correlation operator which is a *functional* of $n(\mathbf{r})$. The theory does not say anything in rigorous terms on the physical meaning of the one-electron orbitals ϕ_i and the corresponding eigenvalues E_i. Since the form of the functional for $V_{xc}(n(\mathbf{r});\mathbf{r})$ in a real system of N interacting electrons is not known, one can proceed further only by making some approximations. In the simplest approximation, known as the local density approximation, the functional $V_{xc}(n(\mathbf{r});\mathbf{r})$ is replaced by the following function of $n(\mathbf{r})$ (we consider nonmagnetic materials):

$$V_{xc}(n(\mathbf{r})) = \left\{ \frac{d}{dn} \left[n\epsilon_{xc}(n) \right] \right\}_{n=n(\mathbf{r})} \tag{3.12}$$

$$\epsilon_{xc}(n) = \epsilon_x(n) + \epsilon_c(n) \tag{3.13}$$

where

$$\epsilon_x(n) \equiv -\frac{3}{2} e^2 \left(\frac{3n}{8\pi} \right)^{1/3} \tag{3.14}$$

is the exchange energy per electron for a *uniform* electron gas of density n, and $\epsilon_c(n)$ is the "correlation energy" per electron of the same gas. $\epsilon_c(n)$ can be defined as the difference between the exact energy (per electron) and the Hartree–Fock energy (per electron) of a uniform electron gas. Only approximate expressions are presently available for this quantity (see, e.g., Hedin and Lundqvist, 1971). Substituting Eq. (3.14) into Eq. (3.12) we find

$$V_{xc}(n(\mathbf{r})) = -2e^2 \left[\frac{3n(\mathbf{r})}{8\pi} \right]^{1/3} + \left\{ \frac{d}{dn} \left[n\epsilon_c(n) \right] \right\}_{n=n(\mathbf{r})} \tag{3.15}$$

Hence, in the local density approximation of the Kohn and Sham formalism, the one-electron potential in a solid is given by

$$V(\mathbf{r}) = V_c(\mathbf{r}) + V_{xc}(n(\mathbf{r})) \tag{3.16}$$

where V_c is given by Eq. (3.2) and V_{xc} by Eq. (3.15). The first term in Eq. (3.15) has the same physical origin as the $V_{X\alpha}$ term of the $X\alpha$ method and the two are identical for $\alpha = \frac{2}{3}$. We note that in deriving the above result by either method one makes the same assumption, namely, that $n(\mathbf{r})$ is slowly varying with position. The second term in Eq. (3.15), representing correlation, does not appear in the one-electron potential of the $X\alpha$ method. For this reason the Kohn and Sham formula [Eq. (3.16)] ought to be better than the corresponding $X\alpha$ formula [Eq. (3.7)]. This is certainly true in principle but not necessarily so in practice. By treating α as an adjustable parameter one produces a similar correction to the one-electron potential within the $X\alpha$ method.

We have already noted that, according to the Kohn and Sham (1965) formalism, the ground-state energy of the solid is determined from a knowledge of the occupied one-electron orbitals defined by Eq. (3.9). In the local density approximation of this formalism [Eq. (3.16)] the ground-state energy is given by

$$E = \sum_{i(\text{occ})} E_i - \int V(\mathbf{r})n(\mathbf{r}) \, d^3r + \int n(\mathbf{r})\epsilon_{xc}(n(\mathbf{r})) \, d^3r + E_{\text{es}} \tag{3.17}$$

where $\epsilon_{xc}(n)$ is given by Eq. (3.13), $V(\mathbf{r})$ by Eq. (3.16), and E_{es} is the total electrostatic energy of the system. The sum in Eq. (3.17) is over all occupied orbitals. The same expression [Eq. (3.17)] with ϵ_{xc} replaced by

$$\epsilon_{xc}(n) = \frac{3}{2}\alpha\epsilon_x(n) \tag{3.18}$$

gives the ground-state energy of the solid in the $X\alpha$ method. In this respect the $X\alpha$ method can be considered a special case of the Kohn and Sham formalism corresponding to the above choice (empirical) of $\epsilon_{xc}(n)$. Note that the semiempirical exchange correlation potential of the $X\alpha$ method [Eq. (3.8)] is obtained by substituting Eq. (3.18) into the more general equation for this potential given by Eq. (3.12).

In practice, we wish to know not only about the ground-state properties of the solid but also about its excited states, and for that purpose it is extremely useful to know what physical meaning, if any, one can attribute to the one-electron orbitals and eigenvalues that one obtains by solving the one-electron problem defined by Eqs. (3.9), (3.10), and (3.16). Consider an N-electron system in its ground state. It is described, in accordance with what we have already said, by N occupied one-electron orbitals denoted by ϕ_i ($i = 1, 2, \ldots, N$). Then consider an $(N - 1)$-electron system in a hypothetical state described by the $(N - 1)$ orbitals that enter into the description of the ground state of the above N-electron system and denote the missing orbital by ϕ_k (E_k need not be the maximum occupied eigenvalue). The question arises as to whether this hypothetical state represents a real situation, i.e., an excited state of the $(N - 1)$-electron system. That would imply that one can use essentially the same potential $V(\mathbf{r})$ to describe both the ground state of the N-electron system and the excited state of the $(N - 1)$-electron system, for otherwise one cannot obtain the same orbitals in the two cases. Obviously, this condition of "no relaxation" cannot be satisfied when N is a small number, as in the case of a light atom, since in this case removal of one electron will distort significantly the electronic charge distribution and in turn the potential field distribution. The case of a crystal where N is very large is quite different. If the orbital ϕ_k extends over the entire crystal, then removing an electron from this orbital will not affect $n(\mathbf{r})$ and hence $V(\mathbf{r})$ in the slightest and the above condition will be satisfied. In this case, it can be shown (Slater, 1974) that E_k is, to a good approximation, equal to the ionization energy associated with the removal of an electron from the kth orbital. By essentially the same arguments we reach the conclusion that if one electron is removed from an occupied orbital ϕ_i and placed into a (formerly) empty orbital ϕ_j (obviously $E_j > E_i$) both of which are extended in space, then $(E_j - E_i)$ will correspond to an actual excitation energy of the system. If either the initial or the final state is localized in space, the above will not be true, as "relaxation effects" are in that case important. We emphasize, however, that a generalization of the density functional method of Kohn and Sham (which includes the $X\alpha$ method as a special case), makes it possible to study "localized" excitations of the above type as well. This generalization is based on the concept of a "transition state" and of a potential depending on this transition state (Slater, 1974; Theophilou, 1979; Perdew and Zunger, 1981).

The one-electron orbitals corresponding to energies in the conduction bands of metals and in the conduction and valence bands of many semiconducting and insulator crystals are extended in space. It is because of this fact that energy-band-structure calculations of crystalline solids based on the independent-electron model turn out to be useful in describing not only the ground-state properties of the solid but also (at least approximately) processes involving excitation, heat absorption, electron transport, optical absorption, electron emission, etc. We should also point out that whereas the Hartree–Fock method, which treats the exchange interaction exactly but disregards correlation altogether, breaks down for metals (Hedin and Lundqvist, 1969) and for semiconductors (Bennet and Inkson, 1977), the methods described in the present section, which treat approximately exchange and correlation together, give results in good agreement with observations.

One must not assume, however, that one-electron orbitals extended in space correspond to true (stationary) states of one electron moving independently in the solid. Consider an electron in a metal at zero temperature and say that it occupies (at a given time) the orbital ϕ_j which has an energy eigenvalue $E_j > E_F$, where

E_F denotes as usual the Fermi level of the metal. If ϕ_j were a true one-electron state, its development with time would be given according to the quantum mechanical equation of motion by $\phi_j(r) \exp[-(i/\hbar)E_j t]$, from which follow all stationary properties of such states. This is not true in practice. Residual inter-action (correlation) with the other electrons in the crystal which has not been taken into account by the average field, represented by $V(\mathbf{r})$ in Eq. (3.9), will lead to a finite lifetime τ_j for this state. According to the uncertainty principle (see, e.g., Landau and Lifshitz, 1958), a finite lifetime implies an uncertainty (a spread) in the energy of the electron in the corresponding state given by $2\Gamma_j \simeq \hbar/\tau_j$. Accord-ingly, if we assume that Γ_j is relatively small we may describe, approximately, the time development of such a state by the following wave function:

$$\phi_j(t) = \phi_j(r) \exp[-(i/\hbar)(E_j - i\Gamma_j)t] \qquad (3.19)$$

which tells us that as the time goes on the electron has a finite probability of leaving the jth state. This occurs, for example, when the electron (in the jth state) collides with another electron in the crystal losing energy in the process. In such a collision the first electron "jumps" into a state of lower energy (we have assumed that $E_j > E_F$) and the second electron gains energy moving from an energy level below E_F to an energy level above E_F. It can be shown by very general consider-ations based on the conservation of energy and momentum in an electron–electron collision (see, e.g., Abrikosov, Gorkov, and Dzyaloshinski, 1963) that in metals and for energies near E_F,

$$\Gamma(E) \simeq \text{const } (E - E_F)^2 \qquad (3.20)$$

In Fig. 3.1 we show $\Gamma(E)$ for copper as calculated by Saccheti (1980). We note that a finite $\Gamma(E)$ for $E < E_F$ means that a hole, a vacant state created by remov-ing an electron from an occupied orbital with energy E, has a finite lifetime. The

FIGURE 3.1

Theoretical (open circles) width of one-electron energy levels in copper. The broken line is a parabola: $2\Gamma = 0.092(E - E_F)^2$ Ry. The full circles are experimental data by Thiry, Chandersis, Lecante, Guillot, Pin-chaux, and Petroff (1979). (From Sacchetti, 1980.)

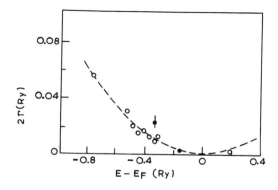

above-mentioned electron–electron collisions determine the lifetime of electrons and holes at relatively low excitation energies. An electron with energy 20 eV or so above the Fermi level (a similar argument applies to holes) can also interact with a collective (plasmon) excitation of the electrons in the metal (the energy of a plasmon is $E_p \simeq 20$ eV for most metals). This interaction is quite strong and for that reason $\Gamma(E)$ for $E > E_F + E_p$ is quite large, of the order of 5 eV or so. It is clear from the above discussion that nearly stationary electron (hole) states exist only in the immediate vicinity of the Fermi level. However, in spite of the fact that one-electron orbitals, corresponding to a real one-electron potential, cannot be interpreted as independent (stationary) one-electron states, their knowledge is extremely useful even at energies well removed from the Fermi level, provided such knowledge is properly interpreted (see also Section 4.5).

A calculation from first principles of the effect of inelastic collisions (electron–electron and electron–plasmon collisions) on an electron of energy E moving in a real crystal is a very difficult problem, and not much progress has been made in this direction. However, in many practical applications, what one wishes to know is the net effect of such collisions on the elastic channel of the motion of a primary electron, and this can be taken into account, on a semiempirical basis, by adding to the one-electron potential a (negative) imaginary component which acts as an absorber (a sink of electrons). [One can easily show that when an electron with energy E and wave vector **k** is incident upon a scatterer represented by a finite region (e.g., a sphere) within which the potential is complex, the outgoing flux, calculated in the usual manner (see any textbook on scattering theory) is *less* than the incident flux, and that the difference (the absorbed flux) is proportional to the imaginary component of the potential.] To a first approximation the imaginary part iV_{im} of the potential (in the bulk of the solid) is independent of position and we have $V_{im} = -\hbar/2\tau(E)$, where $\tau(E)$ is the lifetime of an electron in a state with energy E.

In what follows we assume, unless otherwise stated, that the potential is real. We do so in order to facilitate the interpretation of the formulas, and to be able to compare them (when relevant) with the results of the more traditional band-structure theories which are based on a real potential. We emphasize, however, that the derivation of the basic formulas, in particular the eigenvalue equation (3.48) and those relating to the scattering of the electron by a layer of atoms (Section 3.5), does *not* depend on the potential being real and therefore these formulas are equally valid for a complex potential.

An exact solution of the Schrödinger equation [Eq. (3.9)] for an electron moving in the three-dimensional potential field of a solid is practically impossible and one is forced to make some additional approximations. One such approximation, which turned out to be very successful in the calculation of the energy-band-structure of crystalline solids, is known as the "muffin-tin" approximation. According to this approximation each atom in the solid is surrounded by a sphere.

The spheres centered on different atoms do not overlap with each other. Within each sphere the potential is made spherically symmetric and the potential between the spheres is replaced by a constant. The muffin-tin potential can be made self-consistent as required by Eqs. (3.9), (3.10), and (3.15), but only within the limitation imposed by the form of this potential. In spite of the above limitation, the muffin-tin approximation has been used successfully not only in energy-band-structure calculations of metals (see, e.g., Moruzzi, Janak, and Williams, 1978) but, also, in calculations of the electronic properties of compounds, of clusters of atoms, and of molecules (Slater, 1974).

In many calculations the muffin-tin potential in the solid is not determined self-consistently but is, instead, derived from an electronic charge distribution, which is obtained by a superposition of the electronic charge distributions of the free atoms (or ions) that make up the solid, followed by spherical averaging within each muffin-tin sphere and volume averaging in the rest of the solid (Mattheiss, 1964). This way of constructing the potential works reasonably well in some cases (e.g., nonmagnetic metals), but it is less satisfactory in other cases (e.g., compounds of more than one element) where the possibility of charge transfer from one kind of atom in the solid to the other makes a self-consistent calculation necessary.

3.3. TWO- AND THREE-DIMENSIONAL LATTICES IN REAL AND RECIPROCAL SPACE

We can describe a crystalline solid as a stack of layers of atoms, the plane of each layer being parallel to a given crystallographic plane (which we assume to be parallel to the xy plane). We assume that the layers have a structure with two-dimensional (2D) periodicity described by the lattice.

$$\mathbf{R}_n^{(2)} = n_2\mathbf{a}_2 + n_3\mathbf{a}_3 \tag{3.21}$$

where $\mathbf{a}_2$ and $\mathbf{a}_3$ are two primitive vectors in the xy plane and n_2 and n_3 take all integer values between $-\infty$ and $+\infty$. As an example, we show in Fig. 3.2 the lattice sites of a body-centerd cubic (bcc) crystal viewed as a stack of atomic layers parallel to the (110) plane. In Fig. 3.2 the circles represent the lattice sites of a given layer and the squares those of the next layer along the positive z direction. The lattice sites of the following (third) layer have the same projection on the xy plane as those of the first layer. We note that in general there may be more than one atom per lattice site and therefore, more than one atom in the 2D unit (primitive) cell of the layer. In an infinite crystal (extending from $z = -\infty$ to $z = +\infty$) all layers parallel to a given crystallographic plane are identical with each other. Therefore, the ith layer along the positive z direction is obtained from the $(i - 1)$th layer by a simple translation described by a primitive vector $\mathbf{a}_1$ of the

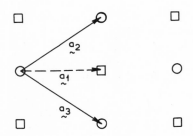

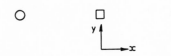

FIGURE 3.2

Lattice sites of two layers parallel to the (110) plane of a bcc crystal. The circles represent the lattice sites of a given layer and the squares those of the next layer along the positive z direction.

three-dimensional (3D) lattice corresponding to the crystal under consideration. Obviously, the three vectors $\mathbf{a}_1$, $\mathbf{a}_2$, and $\mathbf{a}_3$ constitute a set of primitive vectors for this lattice. The corresponding (3D) lattice vectors will be denoted by $\mathbf{R}_n^{(3)}$, i.e.,

$$\mathbf{R}_n^{(3)} = n_1 \mathbf{a}_1 + n_2 \mathbf{a}_2 + n_3 \mathbf{a}_3 \qquad (3.22)$$

where n_1, n_2, n_3 take all integer values from $-\infty$ to $+\infty$. We note that by construction the interlayer distance is equal to the z component of $\mathbf{a}_1$.

The 2D reciprocal lattice, corresponding to the 2D real lattice described by Eq. (3.21), is defined as follows:

$$\mathbf{g} = m_2 \mathbf{b}_2^{(2)} + m_3 \mathbf{b}_3^{(2)} \qquad (3.23)$$

where m_2 and m_3 take all integer values from $-\infty$ to $+\infty$ and $\mathbf{b}_2^{(2)}$ and $\mathbf{b}_3^{(2)}$ are two primitive vectors (in 2D reciprocal space) defined by Eqs. (3.24):

$$\mathbf{b}_i^{(2)} \cdot \mathbf{a}_j = 2\pi \, \delta_{ij} \qquad (3.24)$$

where $i,j = 2,3$ and δ_{ij} is the usual Kronecker delta ($\delta_{ij} = 1$ for $i = j$, $\delta_{ij} = 0$ for $i \neq j$). It follows from the above equations that the components of $\mathbf{b}_2$ and $\mathbf{b}_3$ along the x and y directions are

$$(b_{2x}^{(2)}, b_{2y}^{(2)}) = \frac{2\pi}{A_0} (a_{3y}, -a_{3x}) \qquad (3.25a)$$

$$(b_{3x}^{(2)}, b_{3y}^{(2)}) = \frac{2\pi}{A_0} (-a_{2y}, a_{2x}) \qquad (3.25b)$$

$$A_0 \equiv a_{2x}a_{3y} - a_{2y}a_{3x} \qquad (3.26)$$

We note that $|A_0|$ equals the area of the unit cell of the real 2D lattice. The 2D reciprocal lattice vectors have the following important property:

$$\exp(i\mathbf{g} \cdot \mathbf{R}_n^{(2)}) = 1 \tag{3.27}$$

which follows directly from Eqs. (3.23) and (3.24).

As an example, we show in Fig. 3.3 the reciprocal vectors $\mathbf{b}_2^{(2)}$ and $\mathbf{b}_3^{(2)}$ for the (110) crystallographic plane of a (bcc) lattice. They correspond to the following choice of primitive vectors for the 2D real lattice shown in Fig. 3.2 (a is the lattice constant):

$$\mathbf{a}_2 = \left(\frac{a}{\sqrt{2}}, \frac{a}{2} \right), \qquad \mathbf{a}_3 = \left(\frac{a}{\sqrt{2}}, -\frac{a}{2} \right) \tag{3.28}$$

The area within the hexagon corresponds to the first Brillouin zone of 2D reciprocal space for this particular crystallographic plane.

The first Brillouin zone, or surface Brillouin zone (SBZ), as it is most commonly known, corresponding to a given crystallographic plane (surface), denotes the area within a unit cell of the 2D reciprocal space and could be taken as the area of the parallelogram defined by $\mathbf{b}_2^{(2)}$ and $\mathbf{b}_3^{(2)}$. In practice it is more convenient to choose the SBZ in the following manner. We draw the $\mathbf{g}$ vectors from the origin (a site of the 2D reciprocal lattice) to all neighboring lattice sites. Then we draw straight lines through the midpoint and at a right angle to these vectors. The area enclosed by these lines constitutes a unit cell of the 2D reciprocal lattice and has

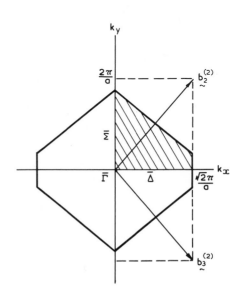

FIGURE 3.3

The SBZ of the (110) plane of a bcc lattice.

in addition the symmetry of the crystallographic plane under consideration. The SBZ of the bcc (110) plane, shown in Fig. 3.3 was constructed in the above manner. It has the rectangular symmetry appropriate to this plane. The physical significance of the SBZ will become apparent towards the end of this section.

The 3D reciprocal lattice corresponding to the real lattice described by Eq. (3.22) is given by

$$\mathbf{G} = m_1\mathbf{b}_1^{(3)} + m_2\mathbf{b}_2^{(3)} + m_3\mathbf{b}_3^{(3)} \tag{3.29}$$

where m_1, m_2, m_3 take all integer values from $-\infty$ to $+\infty$ and $\mathbf{b}_1^{(3)}$, $\mathbf{b}_2^{(3)}$, and $\mathbf{b}_3^{(3)}$ are three primitive vectors (in 3D reciprocal space) defined by Eqs. (3.30)

$$\mathbf{b}_i^{(3)} \cdot \mathbf{a}_j = 2\pi\,\delta_{ij} \tag{3.30}$$

It follows from the above equations that

$$\mathbf{b}_1^{(3)} = 2\pi\mathbf{a}_2 \times \mathbf{a}_3/\mathbf{a}_1 \cdot (\mathbf{a}_2 \times \mathbf{a}_3) \tag{3.31a}$$
$$\mathbf{b}_2^{(3)} = 2\pi\mathbf{a}_3 \times \mathbf{a}_1/\mathbf{a}_1 \cdot (\mathbf{a}_2 \times \mathbf{a}_3) \tag{3.31b}$$
$$\mathbf{b}_3^{(3)} = 2\pi\mathbf{a}_1 \times \mathbf{a}_2/\mathbf{a}_1 \cdot (\mathbf{a}_2 \times \mathbf{a}_3) \tag{3.31c}$$

The 3D reciprocal vectors have the property

$$\exp\,(i\mathbf{G} \cdot \mathbf{R}_n^{(3)}) = 1 \tag{3.32}$$

which follows from Eqs. (3.29) and (3.30). We see from Eq. (3.31a) that $\mathbf{b}_1^{(3)}$ is normal to the crystallographic plane under consideration (the xy plane). It is, also, easy to show that the projections of $\mathbf{b}_2^{(3)}$ and $\mathbf{b}_3^{(3)}$ on the xy plane coincide with $\mathbf{b}_2^{(2)}$ and $\mathbf{b}_3^{(2)}$, respectively. Hence the unit cell of the 2D reciprocal lattice (the parallelogram defined by $\mathbf{b}_2^{(2)}$ and $\mathbf{b}_3^{(2)}$) coincides with the projection on the xy plane of the unit cell of the 3D reciprocal lattice (the parallelepiped defined by $\mathbf{b}_1^{(3)}$, $\mathbf{b}_2^{(3)}$, and $\mathbf{b}_3^{(3)}$).

In the case of an infinite crystal $V(\mathbf{r})$ in Eq. (3.9) is periodic in space, i.e.,

$$V(\mathbf{r} + \mathbf{R}_n^{(3)}) = V(\mathbf{r}) \tag{3.33}$$

where $\mathbf{R}_n^{(3)}$ is given by Eq. (3.22). It follows from the above property of the potential (see, any textbook on solid-state physics) that the one-electron states [we refer to the orbitals denoted by ϕ_i in Eq. (3.9) as one-electron states or simply as electron states] must be of the Bloch type, i.e., $\phi_i \rightarrow \psi_{\mathbf{k}l}$, where

$$\psi_{\mathbf{k}l} = e^{i\mathbf{k}\cdot\mathbf{r}}u_{\mathbf{k}l}(\mathbf{r}) \tag{3.34}$$
$$u_{\mathbf{k}l}(\mathbf{r} + \mathbf{R}_n^{(3)}) = u_{\mathbf{k}l}(\mathbf{r}) \tag{3.35}$$

For a given wave vector **k** there exists a discrete set of such states with energy eigenvalues $E_l(\mathbf{k})$ belonging to different bands: $l = 1,2,3,\ldots$ It can be shown that all independent solutions of the above form are obtained by considering values of **k** within a single (arbitrarily chosen) unit cell of the 3D reciprocal lattice. For example, we need only consider values of **k** wtihin the parallelepiped defined by the vectors $\mathbf{b}_1^{(3)}$, $\mathbf{b}_2^{(3)}$, and $\mathbf{b}_3^{(3)}$. A more convenient (for our purposes) choice of this unit cell (sometimes referred to as the first Brillouin zone or the reduced **k** zone), which is permitted because of our specific choice of primitive vectors, is the following:

$$\mathbf{k} = (\mathbf{k}_\|,\mathbf{k}_z) \tag{3.36}$$

where $\mathbf{k}_\| = (k_x,k_y)$, the component of the wave vector parallel to the crystallographic plane under consideration, lies within the SBZ of this plane and k_z, the component normal to the crystallographic plane, lies in the region $0 \leq k_z < |\mathbf{b}_1^{(3)}|$ or, equivalently in the region $-\frac{1}{2}|\mathbf{b}_1^{(3)}| < k_z \leq \frac{1}{2}|\mathbf{b}_1^{(3)}|$.

We note that for a large but finite crystal (say a cube of side L), **k** (within the reduced zone) is a quasicontinuous variable given by

$$\mathbf{k} = (k_x,k_y,k_z) = \frac{2\pi}{L}(n_x,n_y,n_z) \tag{3.37}$$

where n_x,n_y,n_z are integers. There are N such points in the reduced **k** zone, where N is the number of unit cells in the crystal.

For the sake of clarity, we shall refer to the electron states defined by Eq. (3.34) as ordinary or propagating Bloch waves.

3.4. COMPLEX BAND STRUCTURE

We have seen that an infinite crystal can be viewed as a sequence (along the z axis) of identical atomic layers with a periodic structure in the xy plane. There is no loss of generality if we assume that these layers are "separated" from each other by a region of small thickness (a slot parallel to the xy plane) where the potential is constant denoted by V_0. In the final formulas describing such a system we can allow the thickness of these regions, shown by the shaded areas in Fig. 3.4, to go to zero. In the same figure we denote with a cross the center (a lattice point) of each layer. By construction the vector joining the center of the $(i - 1)$th layer to that of the ith layer is given by the lattice vector $\mathbf{a}_1$ introduced in Section 3.3. By construction the midpoint of the vector joining the above two centers lies in the region (slot) of constant potential. Its position, shown by a solid circle in Fig. 3.4,

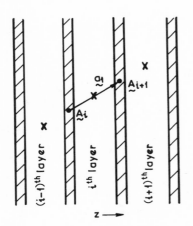

FIGURE 3.4

The shaded areas represent regions of constant potential between identical atomic layers. The crosses denote the centers of the respective layers. $A_{i+1} - A_i = a_1$.

will be denoted by A_i. We have $A_{i+1} - A_i = a_1$. For the moment we shall make no assumption as to the form of the potential within a layer other than its periodicity which is described by a 2D lattice (Eq. (3.21)].

We shall seek the solutions (Bloch waves) of the Schrödinger equation for an electron in the infinite crystal, described above, with a given energy E and with a given $k_{\parallel}$ within the SBZ of the crystallographic plane under consideration. In the region of constant potential between the $(i-1)$th and the ith layer, the most general solution of the Schrödinger equation which is compatible with the periodicity of the potential in the xy plane is given by

$$\psi = \sum_g \{U_{ig}^+ \exp[i K_g^+ \cdot (r - A_i)] + U_{ig}^- \exp[i K_g^- \cdot (r - A_i)]\} \quad (3.38)$$

where

$$K_g^\pm \equiv \left\{ k_{\parallel} + g, \pm \left[\frac{2m(E - V_0)}{\hbar^2} - (k_{\parallel} + g)^2 \right]^{1/2} \right\} \quad (3.39)$$

The same wave function in the region of constant potential between the ith and $(i+1)$th layer is given by

$$\psi = \sum_g \{U_{i+1\,g}^+ \exp[i K_g^+ \cdot (r - A_{i+1})]$$
$$+ U_{i+1\,g}^- \exp[i K_g^- \cdot (r - A_{i+1})]\} \quad (3.40)$$

and by similar expressions in every other slot of constant potential.

We call ψ a Bloch wave solution of the Schrödinger equation, or simply a Bloch wave, if

$$U^{\pm}_{i+1\ \mathbf{g}} = \exp\,(i\mathbf{k}\,\cdot\,\mathbf{a}_1)U^{\pm}_{i\ g} \qquad (3.41)$$

$$\mathbf{k} \equiv (\mathbf{k}_{\parallel},k_z) \qquad (3.42)$$

where, we emphasize, k_z may be a real or complex number.

It is obvious from Eq. (3.38) that for a given E we obtain all independent Bloch waves by considering values of $\mathbf{k}_{\parallel}$ within the SBZ only. For replacing $\mathbf{k}_{\parallel}$ by $\mathbf{k}_{\parallel} + \mathbf{g}$ in Eq. (3.38) renames the coefficients without changing the form of the wave function.

If k_z in Eq. (3.42) is real, the corresponding ψ represents a propagating Bloch wave. These are the ordinary Bloch waves, defined by Eq. (3.34) representing the one-electron states of the infinite crystal. If k_z is complex ψ represents an evanescent (Bloch) wave. It follows from Eq. (3.41) that the amplitude of an evanescent wave increases "exponentially" in the positive or negative z direction. Therefore, the evanescent waves do not represent one-electron states of the crystal. They are, nevertheless, extremely useful mathematical entities, as we shall see in the subsequent development of the theory of surface phenomena.

Since ψ and its derivative must be continuous throughout the crystal a relation must exist between the $U^{\pm}_{ig}$ coefficients in Eq. (3.38) and the $U^{\pm}_{i+1,\mathbf{g}}$ coefficients in Eq. (3.40) which is uniquely determined, for given E and $\mathbf{k}_{\parallel}$, by the potential field within the ith layer. We can express this relation in mathematical form as follows. We observe that $\exp\,[i\mathbf{K}^{+}_{\mathbf{g}}\,\cdot\,\mathbf{r}]$ represents a wave which propagates or decays in the positive z direction $\{$if $[2m(E\,-\,V_0)/\hbar^{2}\,-\,(\mathbf{k}_{\parallel}\,+\,\mathbf{g})^{2}] < 0$ the amplitude of the wave diminishes exponentially in the positive z direction$\}$. Similarly $\exp\,(i\mathbf{K}^{-}_{\mathbf{g}}\,\cdot\,\mathbf{r})$ represents a wave propagating or decaying in the negative z direction. Waves propagating or decaying in the positive z direction on the right of the ith layer $\{$the terms proportional to $\exp\,[i\mathbf{K}^{+}_{\mathbf{g}}\,\cdot\,(\mathbf{r}\,-\,\mathbf{A}_{i+1})]$ in Eq. (3.40)$\}$ can only derive from waves with the same E and $\mathbf{k}_{\parallel}$ (these quantities are conserved when the electron is scattered by the layer) which are incident on the ith layer from the left $\{$the terms proportional to $\exp\,[i\mathbf{K}^{+}_{\mathbf{g}}\,\cdot\,(\mathbf{r}\,-\,\mathbf{A}_i)]$ in Eq. (3.38)$\}$ which are partially transmitted through the layer, or from waves incident on the ith layer from the right $\{$the terms proportional to $\exp\,[i\mathbf{K}^{-}_{\mathbf{g}}\,\cdot\,(\mathbf{r}\,-\,\mathbf{A}_{i+1})]$ in Eq. (3.40)$\}$ which are partially reflected by this layer. Therefore, we have

$$U^{+}_{i+1\ \mathbf{g}} = \sum_{\mathbf{g}'} \{Q^{\mathrm{I}}_{\mathbf{g}\mathbf{g}'}U^{+}_{i\mathbf{g}'} + Q^{\mathrm{II}}_{\mathbf{g}\mathbf{g}'}U^{-}_{i+1\ \mathbf{g}'}\} \qquad (3.43)$$

By a similar argument we obtain

$$U^{-}_{i\mathbf{g}} = \sum_{\mathbf{g}'} \{Q^{\mathrm{III}}_{\mathbf{g}\mathbf{g}'}U^{+}_{i\ \mathbf{g}'} + Q^{\mathrm{IV}}_{\mathbf{g}\mathbf{g}'}U^{-}_{i+1\ \mathbf{g}'}\} \qquad (3.44)$$

Equations (3.43) and (3.44) define the scattering matrix elements $Q_{gg'}^{I}$, $Q_{gg'}^{II}$, $Q_{gg'}^{III}$, and $Q_{gg'}^{IV}$ of the layer. They are, of course, functions of E and $\mathbf{k}_{\parallel}$. Using Eq. (3.41) we can rewrite Eqs. (3.43) and (3.44) as follows:

$$\exp\left(i\mathbf{k}\cdot\mathbf{a}_1\right)U_{ig}^{+} = \sum_{g'}\{Q_{gg'}^{I}U_{ig'}^{+} + Q_{gg'}^{II}U_{i+1\,g'}^{-}\} \tag{3.45}$$

$$\exp\left(-i\mathbf{k}\cdot\mathbf{a}_1\right)U_{i+1,g}^{-} = \sum_{g'}\{Q_{gg'}^{III}U_{ig'}^{+} + Q_{gg'}^{IV}U_{i+1\,g'}^{-}\} \tag{3.46}$$

In practice a finite number, say n, of $\mathbf{g}$ vectors are included in the expansion of the wave function given by Eqs. (3.38) and (3.40). By ordering these vectors in some way, we can write Eqs. (3.45) and (3.46) in matrix form as follows:

$$\exp\left(i\mathbf{k}\cdot\mathbf{a}_1\right)\begin{bmatrix} \mathbf{I} & \mathbf{O} \\ \mathbf{Q}^{III} & \mathbf{Q}^{IV} \end{bmatrix}\begin{bmatrix} \mathbf{U}_i^{+} \\ \mathbf{U}_{i+1}^{-} \end{bmatrix} = \begin{bmatrix} \mathbf{Q}^{I} & \mathbf{Q}^{II} \\ \mathbf{O} & \mathbf{I} \end{bmatrix}\begin{bmatrix} \mathbf{U}_i^{+} \\ \mathbf{U}_{i+1}^{-} \end{bmatrix} \tag{3.47}$$

where the $\mathbf{Q}$ are $n \times n$ matrices, $\mathbf{I}$ and $\mathbf{O}$ are the unit and null $n \times n$ matrices, respectively, and $\mathbf{U}_i^{+}$ and $\mathbf{U}_{i+1}^{-}$ are n-component vectors. Using a bit of matrix algebra we finally obtain

$$\begin{bmatrix} \mathbf{Q}^{I} & \mathbf{Q}^{II} \\ -[\mathbf{Q}^{IV}]^{-1}\mathbf{Q}^{III}\mathbf{Q}^{I} & [\mathbf{Q}^{IV}]^{-1}[\mathbf{I} - \mathbf{Q}^{III}\mathbf{Q}^{II}] \end{bmatrix}\begin{bmatrix} \mathbf{U}_i^{+} \\ \mathbf{U}_{i+1}^{-} \end{bmatrix}$$
$$= \exp\left(i\mathbf{k}\cdot\mathbf{a}_1\right)\begin{bmatrix} \mathbf{U}_i^{+} \\ \mathbf{U}_{i+1}^{-} \end{bmatrix} \tag{3.48}$$

which represents a system of $2n$ linear homogeneous equations with $2n$ unknowns and has the form of an eigenvalue problem. For given E and $\mathbf{k}_{\parallel}$ there are $2n$ independent solutions, denoted by ψ_j, to the above system of equations corresponding to $2n$ (complex) values of k_z [within the reduced zone: $0 \le \mathrm{Re}(k_z) < |\mathbf{b}_1^{(3)}|$] which we denote by k_{jz} ($j = 1,2,\ldots,2n$). The periodicity of the energy band structure in the $\mathrm{Re}(k_z)$ direction follows directly from Eq. (3.48), which shows that the same solution ψ_j corresponds to values of $\mathbf{k}$ differing from

$$\mathbf{k}_j \equiv (\mathbf{k}_{\parallel}, k_{jz}) \tag{3.49}$$

by an integral multiple of the reciprocal vector $\mathbf{b}_1^{(3)}$, which according to Eq. (3.31a) is normal to the crystallographic plane under consideration (xy plane) and according to Eq. (3.32) satisfies the equation

$$\exp\left(i\mathbf{b}_1^{(3)} \cdot \mathbf{a}_1\right) = 1 \tag{3.50}$$

Equation (3.48) was first derived by McRae (1968) and has since played an important role in the theory of LEED (see, e.g., Pendry, 1974) and subsequently in many other calculations relating to surface phenomena.

For given $\mathbf{k}_\parallel$ the real energy lines in the complex plane defined by

$$k_{jz} = k_{jz}(E), \qquad -\infty < E < +\infty \tag{3.51}$$

constitute the complex band structure for the given crystallographic plane. The $k_{jz}(E)$ lines satisfy some very interesting and beautiful theorems which have been studied by a number of authors (see, e.g., Heine, 1963). For the purposes of the present volume it suffices to say that for given E and $\mathbf{k}_\parallel$ the solution of Eq. (3.48) provides us with $2n$ linearly independent Bloch waves ψ_j. Equally important is the fact that half of these solutions represent Bloch waves with their source at $z = -\infty$ (i.e., they propagate or "decay" in the positive z direction) and the other half represent Bloch waves which have their source at $z = +\infty$ (i.e., they propagate or decay in the negative z direction).

We emphasize that out of the $2n$ Bloch waves which one obtains by solving Eq. (3.48) only very few and in many instances none at all are propagating waves corresponding to real k_z. All the remaining solutions correspond to complex k_z and therefore represent evanescent waves (we assume a real potential; in a complex potential all waves are evanescent waves).

We note that the wave functions ψ_j, obtained in the above manner, are known (apart from a "normalization" constant) only in the regions (slots) of constant potential between the layers. In the region of constant potential between the $(i - 1)$th and the ith layer ψ_j is given by

$$\psi_j = \sum_g \left\{ U_{ig}^{j+} \exp\left[i\mathbf{K}_g^+ \cdot (\mathbf{r} - \mathbf{A}_i)\right] + U_{ig}^{j-} \exp\left[i\mathbf{K}_g^- \cdot (\mathbf{r} - \mathbf{A}_i)\right] \right\} \tag{3.52}$$

Obviously, it is sufficient to write down the $U^{j\pm}$ coefficients for the region between two particular layers (say the zeroth and the first layer). The corresponding coefficients for the region of constant potential between any two other layers is immediately obtained from the relation

$$U_{i+1\,g}^{j\pm} = \exp\left(i\mathbf{k}_j \cdot \mathbf{a}_1\right) U_{ig}^{j\pm} \tag{3.53}$$

In many applications, e.g., LEED and electron emission calculations, a knowledge of the wave function in the region of constant potential between the layers is sufficient. We note, however, that if the wave function is known in these regions it is always possible to obtain the wave function within the layers if that is required.

Finally we wish to remark that although in principle Eq. (3.48) can be used for the calculation of the complex band structure of any crystallographic plane, the difficulty of actually solving the problem on the computer will depend on the number of **g** vectors one has to include in Eq. (3.38) for an accurate description of the wave function. For layers with a relatively small 2D unit cell, no serious problem arises. For example, the low index plane of (bcc) and (fcc) metals can be handled without much difficulty. This is not the case for the high index planes of these metals which have a large unit cell. In such cases the number of **g** vectors that need to be included in the expansion of the wave function may be so large that the computation is practically impossible.

In some cases a plane of mirror symmetry exists parallel to the crystallographic plane under consideration. When this is the case the system of $2n$ equations described by Eq. (3.48) can be reduced to a system of n equations, which reduces considerably the computing time (Pendry, 1974).

So far, we have said nothing about the calculation of the Q matrix elements which describe the scattering of the electron by the layer potential. In the next section we shall see how this can be done for a layer of muffin-tin atoms.

3.5. SCATTERING BY A LAYER OF MUFFIN-TIN ATOMS

We shall consider the case when there is only one atom per unit cell of the layer. We can then assume that the centers (nuclei) of the atoms are located at the lattice points of a 2D lattice in the xy plane. We take the origin of coordinates at a lattice point (we refer to it as the center of the layer). Accordingly, the position vector of the nth atomic nucleus lies in the xy plane and it is given by

$$\mathbf{R}^{(n)} = \mathbf{R}_n^{(2)} \tag{3.54}$$

where $\mathbf{R}_n^{(2)}$ is a vector of the 2D lattice described by Eq. (3.21). Each atom is surrounded by a (muffin-tin) sphere. Outside these spheres, which do not overlap with each other, the potential energy field is constant and equal by definition to V_0. Inside each sphere the potential is spherically symmetric.

We treat the scattering of an electron by the above layer in two stages. First we present the formulas which describe the scattering of the electron by an individual muffin-tin atom. In the second stage, we take into account the multiple scattering of the electron within the layer.

3.5.1. SCATTERING BY AN INDIVIDUAL MUFFIN-TIN ATOM

In this case the potential energy field is given by

$$
\begin{aligned}
V &= v(r) \quad \text{for } r \le a_0 \\
&= V_0 \quad \text{for } r > a_0
\end{aligned}
\tag{3.55}
$$

where a_0 denotes the radius of the muffin-tin sphere. We have assumed that the muffin-tin atom under consideration is centered on the origin of the coordinates.

When an electron with momentum $\hbar\mathbf{q}$ and energy $E = (\hbar^2 q^2/2m + V_0)$ is incident on this potential, there is a probability that the electron will be scattered in a particular direction defined (in spherical coordinates) by the angular variables θ and ϕ (see Fig. 3.5). The corresponding mathematical problem, which is treated in practically every textbook on quantum mechanics, consists in finding the solution of the Schrödinger equation

$$
\left[-\frac{\hbar^2}{2m} \nabla^2 + V(r) \right] \psi(r) = E\psi(\mathbf{r})
\tag{3.56}
$$

which has the following asymptotic form:

$$
\psi \simeq e^{i\mathbf{q}\cdot\mathbf{r}} + f(\theta,\phi)\frac{e^{iqr}}{r}
\tag{3.57}
$$

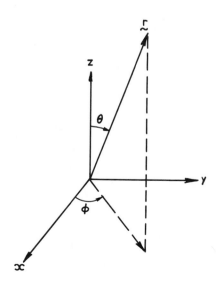

FIGURE 3.5

Angular variables θ and ϕ.

as $r \to \infty$. The first term in the above equation represents the incident wave and the second term the scattered wave. The incident wave can be expanded into a series of spherical waves, each of which has a definite total angular momentum $\hbar[l(l + 1)]^{1/2}$ and definite z component of angular momentum $m\hbar$. We have

$$e^{i\mathbf{q}\cdot\mathbf{r}} = \sum_{lm} 4\pi i^l (-1)^m Y_{l-m}(\Omega_\mathbf{q}) j_l(qr) Y_{lm}(\Omega) \qquad (3.58)$$

where $m = -l, -l + 1, \ldots, (l - 1), l$ and l takes all integer values from zero to infinity. $j_l(qr)$ denotes, as usual, a spherical Bessel function and Y_{lm} a spherical harmonic.† $\Omega \equiv (\theta, \phi)$, $\Omega_\mathbf{k} \equiv (\theta_\mathbf{k}, \phi_\mathbf{k})$.

We note that the spherical wave $j_l(qr) Y_{lm}(\Omega)$ is a solution of the Schrödinger equation for the case of constant potential, i.e., when $v(\mathbf{r}) = V_0$ in Eq. (3.55), representing an (incident) electron with energy E and angular momentum (lm). An independent solution of the Schrödinger equation (for the case of constant potential) for the same energy and angular momentum, valid for $r > 0$ is given by $h_l^{(1)}(qr) Y_{lm}(\Omega)$, where $h_l^{(1)}(qr)$ is the spherical Hankel function which has the following asymptotic form:

$$h_l^{(1)}(qr) \simeq \frac{(-i)^l e^{iqr}}{iqr} \qquad \text{as } r \to \infty \qquad (3.59)$$

Since angular momentum is conserved in a spherically symmetric potential field, each spherical wave in Eq. (3.58) scatters independently from all others and can, therefore, be treated separately. The solution of the Schrödinger equation [Eq. (3.56)] which describes the scattering of an electron, with energy E and angular momentum (l,m), by the potential $v(r)$ can be obtained as follows. From what we have already said it is obvious that, in the region of constant potential, i.e., outside the muffin-tin sphere the solution of the Schrödinger equation, which has the appropriate asymptotic form as $r \to \infty$, is given by

$$\psi_{lm} = j_l(qr) Y_{lm}(\Omega) + S_l h_l^{(1)}(qr) Y_{lm}(\Omega) \qquad (3.60a)$$

where $r \geq a_0$. The first term in the above equation represents the incident wave and the second term the scattered wave.

For $r \leq a_0$ the wave function has the form

$$\psi_{lm} = R_l(r) Y_{lm}(\Omega) \qquad (3.60b)$$

---

†For a definition and a brief summary of the properties of these and other special functions which appear in this volume, see, e.g., Pendry (1974).

Substituting Eq. (3.60b) into Eq. (3.56) and using standard properties of the spherical harmonics one finds (see, e.g., Landau and Lifshitz, 1958) that

$$\left\{ -\frac{1}{r^2}\frac{d}{dr}\left(r^2\frac{d}{dr}\right) + \frac{l(l+1)}{r^2} + \frac{2m}{\hbar^2}[v(r) - E]\right\} R_l(r) = 0 \quad (3.61)$$

from which $R_l(r)$ can be uniquely determined (apart from an unimportant multiplying constant) subject to the condition that $R_l(r)$ is finite everywhere within the sphere including the origin.

The wave function ψ_{lm} given by Eqs. (3.60) and its derivative must be continuous at $r = a_0$. Hence we find

$$S_l = \frac{L_l j_l(qa_0) - j_l'(qa_0)}{h_l^{(1)\prime}(qa_0) - L_l h_l^{(1)}(qa_0)} \quad (3.62)$$

where the prime denotes differentiation with respect to r and

$$L_l \equiv \left(\frac{1}{R_l}\frac{dR_l}{dr}\right)_{r=a_0} \quad (3.63)$$

One can easily show that

$$S_l = i \exp(i\delta_l) \sin\delta_l \quad (3.64)$$

where δ_l are the well-known scattering phase shifts as defined in most textbooks on scattering theory (see, e.g., Landau and Lifshitz, 1958). For our purposes, we can take Eq. (3.64) as the equation defining these phase shifts. Finally, the solution of the Schrödinger equation which describes the scattering of the plane wave Eq. (3.58) by the potential of Eq. (3.55) is given by

$$\psi = e^{i\mathbf{q}\cdot\mathbf{r}} + \sum_{lm} 4\pi i^l(-1)^m Y_{l-m}(\Omega_{\mathbf{q}}) S_l h_l^{(1)}(qr) Y_{lm}(\Omega) \quad (3.65)$$

for $r \geq a_0$. It is obvious from Eq. (3.59) that ψ, as $r \rightarrow \infty$, acquires the form of Eq. (3.57).

In problems relating to electron scattering by a "muffin-tin" atomic potential ($a_0 \simeq 1$–2 Å) only a few spherical waves corresponding to small values of angular momentum get scattered. The reason for this can be understood from the following semiclassical argument. Let an electron with linear momentum $\hbar\mathbf{q}$ be incident on the atom. If there were no interaction between the electron and the atom the electron would move, according to classical mechanics, along a straight line. If we denote by D the (normal) distance from the center of the atom to this straight line, the magnitude of the classical angular momentum of the electron with respect to the center of the atom will be given by $\hbar q D$. We may assume that this quantity is equal to one of the allowed values $\hbar [l(l+1)]^{1/2}$ (where l is an integer) of angular momentum. Using the correspondence principle of quantum mechanics (see, e.g., Leighton, 1959) we find that the average value of D for an electron of angular momentum $\hbar [l(l+1)]^{1/2}$ is given approximately by

$$D \simeq \frac{\hbar [l(l+1)]^{1/2}}{\hbar q} \simeq \frac{l}{q} \tag{3.66}$$

If we now switch on the interaction between the electron and the atomic potential field described by Eq. (3.55), it is obvious that the electron will "see" this potential only if $D \lesssim a_0$. Using Eq. (3.66) we find that only spherical waves with sufficiently small angular momentum will get scattered. In mathematical terms, this means that

$$S_l \simeq \delta_l \simeq 0 \qquad \text{if } l > l_{max} \simeq q a_0 \tag{3.67}$$

In practice, for, say, a transition metal atom and for energies relevant to energy band-structure calculations, $l_{max} \simeq 3$.

Finally, we should point out that the formulas given in this section remain valid when an imaginary component $iV_{im}(E)$ is added to the potential in Eq. (3.56). In this case, the phase shifts will, of course, become complex numbers.

3.5.2. MULTIPLE SCATTERING WITHIN THE LAYER

Let the wave $\exp(i\mathbf{K}_{\mathbf{g}}^+ \cdot \mathbf{r})$ be incident on the layer described at the beginning of Section 3.5 from the left. $\mathbf{K}_{\mathbf{g}}^+$ is the wave vector defined by Eq. (3.39). The corresponding scattered wave, denoted by $\psi_{sc}(\mathbf{r})$, has the form

$$\psi_{sc}(\mathbf{r}) = \sum_k \psi_{sc}^{(k)}(\mathbf{r}) \tag{3.68}$$

where $\psi_{sc}^{(k)}$ represents a scattered spherical wave centered on the kth atom and the sum is over all atoms in the layer. Outside the kth muffin-tin sphere, where the potential is constant (equal by definition to V_0), $\psi_{sc}^{(k)}$ is given by

$$\psi_{sc}^{(k)} = \sum_{lm} A_{lm}^{sc}(\mathbf{R}^{(k)}) h_l^{(1)}(K_0 r_k) Y_{lm}(\Omega_{\mathbf{r}_k}) \tag{3.69}$$

where $\mathbf{R}^{(k)}$ is the position vector of the kth atomic nucleus given by Eq. (3.54) and $\mathbf{r}_k$ and K_0 are defined by

$$\mathbf{r}_k = \mathbf{r} - \mathbf{R}^{(k)} \tag{3.70}$$

$$K_0 = K_{\mathbf{g}}^+ = +\left[\frac{2m}{\hbar^2}(E - V_0)\right]^{1/2} \tag{3.71}$$

In view of the 2D periodicity of the layer, $A_{lm}^{sc}(\mathbf{R}^{(k)})$ in Eq. (3.69) has the following form:

$$A_{lm}^{sc}(\mathbf{R}^{(k)}) = \exp(i\mathbf{k}_{\parallel} \cdot \mathbf{R}^{(k)}) A_{lm}^{sc}(\mathbf{K}_{\mathbf{g}}^+) \tag{3.72}$$

In writing the above equation we made use of Eq. (3.27) and of the fact that the position vectors $\mathbf{R}^{(k)}$ coincide with the vectors of the 2D lattice [see Eq. (3.54)].

The wave incident on the sth atom of the layer denoted by $\psi_{in}^{(s)}$ is given by

$$\psi_{in}^{(s)} = \exp(i\mathbf{K}_{\mathbf{g}}^+ \cdot \mathbf{r}) + \sum_{k(\neq s)} \psi_{sc}^{(k)} \tag{3.73}$$

The second term in Eq. (3.73) accounts for multiple scattering within the layer, i.e., it describes the incident flux on the sth atom of the layer that is due to scattered waves from the other atoms of the layer. The sum in Eq. (3.73) is over all atoms of the layer with the exception of the sth atom.

We proceed by expanding each term on the right-hand side of Eq. (3.73) into a series of spherical waves around the sth atom. The expansion of the first term is the same as in Eq. (3.58). We have

$$\exp(i\mathbf{K}_{\mathbf{g}}^+ \cdot \mathbf{r}) = \exp(i\mathbf{k}_{\parallel} \cdot \mathbf{R}^{(s)})\exp(i\mathbf{K}_{\mathbf{g}}^+ \cdot \mathbf{r}_s) \tag{3.74}$$

$$\exp(i\mathbf{K}_{\mathbf{g}}^+ \cdot \mathbf{r}_s) = \sum_{lm} A_{lm}^0(\mathbf{K}_{\mathbf{g}}^+)j_l(K_0 r_s)Y_{lm}(\Omega_{\mathbf{r}_s}) \tag{3.75}$$

$$A_{lm}^0(\mathbf{K}_{\mathbf{g}}^+) \equiv 4\pi i^l(-1)^m Y_{l-m}(\Omega_{\mathbf{K}_{\mathbf{g}}^+}) \tag{3.76}$$

We note that, in most textbooks $Y_{lm}(\theta,\phi)$ is defined by

$$Y_{l|m|}(\theta,\phi) = (-1)^{|m|}\left[\frac{2l+1}{4\pi}\frac{(l-|m|)!}{(l+|m|)!}\right]^{1/2} P_l^{|m|}(\cos\theta)\exp(i|m|\phi)$$

$$\tag{3.77a}$$

$$Y_{l-|m|}(\theta,\phi) = \left[\frac{2l+1}{4\pi}\frac{(l-|m|)!}{(l+|m|)!}\right]^{1/2} P_l^{|m|}(\cos\theta)\exp(-i|m|\phi) \tag{3.77b}$$

where $P_l^{|m|}(w)$ are the well-known associated Legendre functions, and w takes the real values between -1 and $+1$. We note, however, that $P_l^{|m|}(w)$ is an analytic function of w defined in the entire complex plane except for the two cuts shown in Fig. 3.6. Therefore, $Y_{l-|m|}\,(\Omega_{\mathbf{K}_{\mathbf{g}}^{\pm}})$ in Eq. (3.76) is uniquely defined even when the z component of $\mathbf{K}_{\mathbf{g}}^{+}$ is imaginary. In this case Eq. (3.77) remains valid but $\cos\theta_{\mathbf{K}_{\mathbf{g}}^{\pm}}$ must be replaced by

$$\cos\theta_{\mathbf{K}_{\mathbf{g}}^{\pm}} \to w = K_{\mathbf{g}z}^{\pm}/K_0 \tag{3.78}$$

where $K_{\mathbf{g}z}^{\pm}$ is the z component of $\mathbf{K}_{\mathbf{g}}^{\pm}$ as defined by Eq. (3.39) and K_0 is the "magnitude" of $\mathbf{K}_{\mathbf{g}}^{\pm}$ as defined by Eq. (3.71). The associated Legendre functions of complex argument can be calculated as easily as the associated Legendre functions of real argument, using the same formulas. We note, however, that relations of the type

$$Y_{lm}^{*}(\theta,\phi) = (-1)^m Y_{l,-m}(\theta,\phi) \tag{3.79}$$

which involve not analytic operations (e.g., complex conjugation) are not valid for complex arguments.

In order to expand the wave scattered from the kth atom, given by Eq. (3.69), around the sth atom we make use of the following mathematical identity (Nozawa, 1966; Beeby 1968):

$$h_l^{(1)}(K_0 r_k) Y_{lm}(\Omega_{\mathbf{r}_k}) = \sum_{l''m''} G_{lm;l''m''}(\mathbf{R}_{sk}) j_{l''}(K_0 r_s) Y_{l''m''}(\Omega_{\mathbf{r}_s}) \tag{3.80}$$

where

$$\mathbf{R}_{sk} \equiv \mathbf{R}^{(s)} - \mathbf{R}^{(k)} \tag{3.81}$$

$$G_{lm;l''m''}(\mathbf{R}_{sk}) = \sum_{l'm'} 4\pi(-1)^{(l-l'-l'')/2}(-1)^{m'+m''} h_{l'}^{(1)}(K_0 R_{sk})$$
$$\times\ Y_{l'-m'}(\Omega_{\mathbf{R}_{sk}}) B_{lm}(l'm';l''m'') \tag{3.82}$$

$$B_{lm}(l'm';l''m'') \equiv \int Y_{l'm'}(\Omega) Y_{lm}(\Omega) Y_{l''-m''}(\Omega)\ d\Omega \tag{3.83}$$

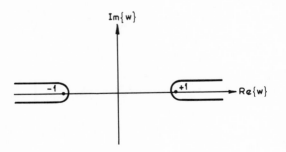

FIGURE 3.6

Domain of definition of $P_l^{|m|}(w)$.

Explicit formulas for the integrals over a product of three spherical harmonics can be found in many books (see, e.g., Slater, 1960; Pendry, 1974).

We are now in a position to replace the waves on the right-hand side of Eq. (3.73) by their corresponding expansions around the sth atom. We find

$$\psi_{\text{in}}^{(s)} = \exp{(i\mathbf{k}_{\parallel} \cdot \mathbf{R}^{(s)})} \sum_{l'm'} \left[A_{l'm'}^0(\mathbf{K_g^+}) + \sum_{lm} A_{lm}^{\text{sc}}(\mathbf{K_g^+}) Z_{lm;l'm'} \right]$$

$$\times \; j_{l'}(K_0 r_s) Y_{l'm'}(\Omega_{\mathbf{r}_s}) \tag{3.84}$$

where

$$Z_{lm;l'm'} \equiv \sum_{\mathbf{R}^{(k)}(\neq \mathbf{R}^{(s)})} \exp{(i\mathbf{k}_{\parallel} \cdot \mathbf{R}_{ks})} G_{lm;l'm'}(\mathbf{R}_{sk}) \tag{3.85}$$

Each spherical wave in Eq. (3.84) is scattered by the spherical potential of the sth atom in the manner described by Eq. (3.60). Therefore, the scattered wave from the sth atom is given by

$$\psi_{\text{sc}}^{(s)} = \exp{(i\mathbf{k}_{\parallel} \cdot \mathbf{R}^{(s)})} \sum_{l'm'} \left[A_{l'm'}^0(\mathbf{K_g^+}) + \sum_{lm} A_{lm}^{\text{sc}}(\mathbf{K_g^+}) Z_{lm;l'm'} \right]$$

$$\times \; S_{l'} h_{l'}^{(1)}(K_0 r_s) Y_{l'm'}(\Omega_{\mathbf{r}_s}) \tag{3.86}$$

Comparing Eq. (3.86) with Eq. (3.69) [taken together with Eq. (3.72)] we find that

$$A_{l'm'}^{\text{sc}}(\mathbf{K_g^+}) = S_{l'} A_{l'm'}^0(\mathbf{K_g^+}) + \sum_{lm} A_{lm}^{\text{sc}}(\mathbf{K_g^+}) Z_{lm;l'm'} S_{l'} \tag{3.87}$$

which we can solve with respect to $A_{lm}^{\text{sc}}(K_g^+)$ by straightforward matrix algebra. We obtain

$$A_{lm}^{\text{sc}}(\mathbf{K_g^+}) = \sum_{l'm'} A_{l'm'}^0(\mathbf{K_g^+})(I - SZ)_{l'm';lm}^{-1} S_l \tag{3.88}$$

where I is a unit matrix and S the diagonal matrix defined by

$$S_{lm;l'm'} = S_l \, \delta_{ll'} \, \delta_{mm'} \tag{3.89}$$

where δ_{ij} is the usual Kronecker delta. Using Eqs. (3.64) and (3.76), and introducing the matrix

$$
\begin{aligned}
X_{lm,l''m''} &= (SZ)_{lm;l''m''} \\
&= i \sin \delta_l \exp (i \delta_l) \sum_{\mathbf{R}^{(k)}(\neq \mathbf{R}^{(s)})} \exp (i\mathbf{k}_{\|} \cdot \mathbf{R}_{ks}) G_{lm;l''m''}(\mathbf{R}_{sk}) \quad (3.90)
\end{aligned}
$$

where $G_{lm;l''m''}$ is given by Eq. (3.82), we find that

$$
A_{lm}^{sc}(\mathbf{K_g^+}) = \sum_{l'm'} 4\pi i^{l'}(-1)^{m'} Y_{l'-m'}(\Omega_{\mathbf{K_g}}^+)(I - X)_{l'm';lm}^{-1} i \sin \delta_l \exp (i \delta_l) \quad (3.91)
$$

The scattered wave from the entire layer is obtained from Eqs. (3.68)–(3.72). We have

$$
\psi_{sc}(\mathbf{r}) = \sum_{\mathbf{R}^{(s)}} \exp (i\mathbf{k}_{\|} \cdot \mathbf{R}^{(s)}) \sum_{lm} A_{lm}^{sc}(\mathbf{K_g^+}) h_l^{(1)}(K_0 r_s) Y_{lm}(\Omega_{\mathbf{r}_s}) \quad (3.92)
$$

where $A_{lm}^{sc}(\mathbf{K_g^+})$ is given by Eq. (3.91). In order to proceed with the lattice sum in the above equation, we must transform the spherical waves around the individual atoms back into plane waves with every position vector referred to the origin of coordinates at the center of the layer. For this purpose we use Eq. (3.70) and the following mathematical theorem:

$$
h_l^{(1)}(K_0 r) Y_{lm}(\Omega) = \frac{(i)^{-l}}{2\pi K_0} \int\int_{-\infty}^{+\infty} \frac{d^2 q_{\|}}{(K_0^2 - q_{\|}^2)^{1/2}} Y_{lm}(\Omega_{\mathbf{q}\pm}) \exp (i\mathbf{q}^\pm \cdot \mathbf{r}) \quad (3.93)
$$

$$
\mathbf{q}^\pm \equiv (\mathbf{q}_{\|}, \pm (K_0^2 - q_{\|}^2)^{1/2}) \quad (3.94)
$$

where the $+(-)$ sign is used for $z > 0$ ($z < 0$), respectively, and $\mathbf{q}_{\|} = (q_x, q_y)$ as usual. We find

$$
\psi_{sc}(\mathbf{r}) = \frac{1}{2\pi K_0} \int\int_{-\infty}^{+\infty} d^2 q_{\|} \sum_{\mathbf{R}^{(s)}} \exp [i(\mathbf{k}_{\|} - \mathbf{q}_{\|}) \cdot \mathbf{R}^{(s)}]
$$
$$
\times \sum_{lm} \frac{(i)^{-l}}{(K_0^2 - q_{\|}^2)^{1/2}} A_{lm}^{sc}(\mathbf{K_g^+}) Y_{lm}(\Omega_{\mathbf{q}\pm}) \exp (i\mathbf{q}^\pm \cdot \mathbf{r}) \quad (3.95)
$$

where the $+(-)$ sign corresponds to $z > 0$ ($z < 0$), respectively. Because of the 2D periodicity of the layer, we have

$$
\sum_{\mathbf{R}^{(s)}} \exp [i(\mathbf{k}_{\|} - \mathbf{q}_{\|}) \cdot \mathbf{R}^{(s)}] = \frac{(2\pi)^2}{A_0} \sum_{\mathbf{g}'} \delta(\mathbf{k}_{\|} + \mathbf{g}' - \mathbf{q}_{\|}) \quad (3.96)
$$

where A_0 denotes the area of the corresponding 2D unit cell. Using the above identity and substituting for $A^{sc}_{lm}(\mathbf{K}_{\mathbf{g}}^+)$ from Eq. (3.91), we finally obtain

$$
\begin{aligned}
\psi_{sc}(\mathbf{r}) &= \sum_{\mathbf{g}'} M_{\mathbf{g}'\mathbf{g}}^+ \exp\left(i\mathbf{K}_{\mathbf{g}'}^+ \cdot \mathbf{r}\right) \qquad \text{for } z > 0 \\
&= \sum_{\mathbf{g}'} M_{\mathbf{g}'\mathbf{g}}^- \exp\left(i\mathbf{K}_{\mathbf{g}'}^- \cdot \mathbf{r}\right) \qquad \text{for } z < 0
\end{aligned}
\tag{3.97}
$$

where $\mathbf{K}_{\mathbf{g}}^-$ is defined by Eq. (3.39) and $M_{\mathbf{g}'\mathbf{g}}^\pm$ are given by the following formula:

$$
M_{\mathbf{g}'\mathbf{g}}^\pm = \frac{8\pi^2 i}{K_{\mathbf{g}'z}^+ A_0 K_0} \sum_{\substack{lm \\ l'm'}} \left[i^{l'}(-1)^{m'} Y_{l'-m'}(\Omega_{\mathbf{K}_{\mathbf{g}}^\pm}) \right]
$$

$$
\times (I - X)^{-1}_{l'm';lm} \left[i^{-l} Y_{lm}(\Omega_{\mathbf{K}_{\mathbf{g}}^\pm}) \right] \exp\left(i\delta_l\right) \sin \delta_l \tag{3.98}
$$

The above formula was initially derived by Beeby (1968) as part of a LEED theory.

The complete solution of the Schrödinger equation in the region of constant potential outside the muffin-tin spheres corresponding to a wave $\exp\left(i\mathbf{K}_{\mathbf{g}}^+ \cdot r\right)$ incident on the layer from the left is, therefore, given by

$$
\psi(\mathbf{r}) = \exp\left(i\mathbf{K}_{\mathbf{g}}^+ \cdot \mathbf{r}\right) + \psi_{sc}(\mathbf{r}) \tag{3.99}
$$

with ψ_{sc} given by Eq. (3.97).

It is obvious that the potential field of the layer under consideration is symmetric about the plane of the layer (xy plane). Therefore, the solution of the Schrödinger equation, denoted by $'\psi$, corresponding to a wave $\exp\left(i\mathbf{K}_{\mathbf{g}}^- \cdot r\right)$ incident on the layer from the right, can be obtained from Eq. (3.99) by the symmetry transformation $z \to (-z)$. We have

$$
'\psi(\mathbf{r}) = \exp\left(i\mathbf{K}_{\mathbf{g}}^- \cdot \mathbf{r}\right) + '\psi_{sc}(\mathbf{r}) \tag{3.100}
$$

$$
\begin{aligned}
'\psi_{sc}(\mathbf{r}) &= \sum_{\mathbf{g}'} M_{\mathbf{g}'\mathbf{g}}^+ \exp\left(i\mathbf{K}_{\mathbf{g}'}^- \cdot \mathbf{r}\right) \qquad \text{for } z < 0 \\
&= \sum_{\mathbf{g}'} M_{\mathbf{g}'\mathbf{g}}^- \exp\left(i\mathbf{K}_{\mathbf{g}'}^+ \cdot \mathbf{r}\right) \qquad \text{for } z > 0
\end{aligned}
\tag{3.101}
$$

where $M_{\mathbf{g}'\mathbf{g}}^\pm$ are given by Eq. (3.98).

We emphasize that the formulas derived in the present section are valid when the layer has one atom per unit cell. The extension of the theory to layers with more complicated unit cells is relatively straightforward (see, e.g., Pendry, 1974).

We are now in a position to write down explicit formulas for the Q matrix elements defined by Eqs. (3.43) and (3.44), for the case of a layer with a single atom per unit cell, in terms of the $M_{\mathbf{g}'\mathbf{g}}^\pm$ matrix elements of the layer given by Eq.

(3.98). We note that the coefficients $U_{ig}^{\pm}$ in Eqs. (3.43) and (3.44) refer to an origin ($\mathbf{A}_i$) displaced by $-\frac{1}{2}\mathbf{a}_1$ with respect to the center of the ith layer [see Eq. (3.38)], similarly $U_{i+1,g}^{\pm}$ refer to an origin ($\mathbf{A}_{i+1}$) displaced by $\frac{1}{2}\mathbf{a}_1$ with respect to the center of the ith layer [see Eq. (3.40)]. On the other hand the $M_{g'g}^{\pm}$ matrix elements in Eqs. (3.97)–(3.99) refer to an origin at the center of the ith layer. We note, also, that the first term in Eq. (3.43), similarly the second term in Eq. (3.44), represent the transmitted (incident + scattered) wave and not the scattered wave only. With the above in mind it is easy to show that

$$Q_{g'g}^{I} = (I + M^+)_{g'g} \exp\left[\tfrac{1}{2}i(\mathbf{K}_{g'}^+ \cdot \mathbf{a}_1 + \mathbf{K}_g^+ \cdot \mathbf{a}_1)\right] \tag{3.102a}$$

$$Q_{g'g}^{II} = M_{g'g}^- \exp\left[\tfrac{1}{2}i(\mathbf{K}_{g'}^+ \cdot \mathbf{a}_1 - \mathbf{K}_g^- \cdot \mathbf{a}_1)\right] \tag{3.102b}$$

$$Q_{g'g}^{III} = M_{g'g}^- \exp\left[\tfrac{1}{2}i(-\mathbf{K}_{g'}^- \cdot \mathbf{a}_1 + \mathbf{K}_g^+ \cdot \mathbf{a}_1)\right] \tag{3.102c}$$

$$Q_{g'g}^{IV} = (I + M^+)_{g'g} \exp\left[\tfrac{1}{2}i(-\mathbf{K}_{g'}^- \cdot a_1 - \mathbf{K}_g^- \cdot \mathbf{a}_1)\right] \tag{3.102d}$$

It would seem, at first sight, that in order to use the above formulas in Eq. (3.48), which determines the complex band structure of the infinite crystal associated with a given crystallographic plane, the diameter of the muffin-tin spheres must be smaller than the interlayer distance, for, otherwise, the midplane between two consecutive layers cuts the spheres of these two layers (as shown schematically in Fig. 3.7) and, therefore, cannot be identified with a "slot" of constant potential in the manner discussed in Section 3.4. The easiest way to convince oneself that the above is not a necessary condition for the application of Eqs. (3.102) to the complex band structure problem is to note that the properties of the muffin-tin atoms enter into Eqs. (3.102) only through the phase shifts δ_l. Hence, for the purpose of calculating the band structure (but not of the corresponding wave functions) we can replace the muffin-tin atoms by point atoms with the same phase shifts. A better way of looking at this problem, and one which allows us to interpret ψ, as defined by Eq. (3.38), as the true wave function (at least approximately) at the midplane between two layers, is based on the following observation. The scattered wave from a given layer is represented by a function which is analytic everywhere except at the atomic centers on the plane of this layer. This is obvious from Eq. (3.92). It follows from the above observation that the formulas given in the present section can be used to calculate the Bloch waves ψ_j in the region of constant potential between the planes of two consecutive layers of muffin-tin atoms even if the midplane between the two layers cuts the muffin-tin spheres of these layers as shown in Fig. 3.7. Moreover, in those cases where the section of the muffin-tin sphere cut by the midplane is relatively small (as is the case for the low-index planes of fcc and bcc metals), we may assume that the expression for ψ_j [Eq. (3.52)], with the coefficients determined in the above manner, will, for

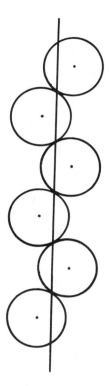

FIGURE 3.7

Cross section in the xz plane of two consecutive layers of muffin-tin atoms. The solid line shows the midplane between the layers.

reasons of continuity, be a good approximation to the wave function over the entire midplane.

3.5.3. PROPERTIES OF THE X MATRIX

It can be shown (Pendry, 1971; 1974) that, when all the atomic nuclei in a layer lie on a single plane (this is always the case when there is only one atom per unit cell) the matrix elements $X_{lm;l''m''}$, defined by Eq. (3.90) are zero, unless $(l + m)$ and $(l'' + m'')$ are either both even or both odd. Hence, **X** has the form shown in Fig. 3.8. Consequently the inversion of $(1 - X)$ required in Eq. (3.98) can be reduced to the more economical (computationally) problem of inverting two submatrices of smaller dimensions.

The most difficult part in the calculation of the **X** matrix is concerned with the evaluation of the lattice sum in Eq. (3.85). The corresponding series converges

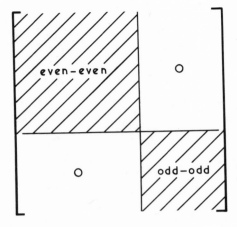

FIGURE 3.8

The matrix **X**: The matrix elements $X_{lm;l''m''}$ are different from zero only if $(l + m)$ and $(l'' \pm m'')$ are either both even or both odd.

absolutely only if a (negative) imaginary part is introduced in the crystal potential. In this case V_0 in Eq. (3.39) and in all subsequent formulas is given by

$$V_0 = V_{0r} + iV_{im}(E) \tag{3.103}$$

and the real phase shifts become complex. Otherwise all formulas remain valid. Since $V_{im} < 0$, K_0, defined by Eq. (3.71), has an imaginary component and $h_{l''}^{(1)}(K_0 R_{sk})$, which appears in $G_{lm;l''m''}$ goes to zero for sufficiently large values of R_{sk}. For energies near the Fermi level, the imaginary component of the potential is very small (see Fig. 3.1) and therefore, for such energies the lattice sum in Eq. (3.85) converges very slowly making an actual computation prohibitively long. A solution to this difficulty has been found by Kambe (1967a,b; 1968) who has shown that the sum over the 2D real lattice in Eq. (3.85) can be transformed, in the manner of Ewald (1921), into two sums, one over the real lattice and the other over the corresponding 2D reciprocal lattice, both of which converge quickly even when the imaginary part of the potential is equal to zero.

We rewrite Eq. (3.90) as follows:

$$X_{lm;l''m''} = -\frac{4\pi}{K_0} \exp(i\,\delta_l) \sin \delta_l \sum_{l'm'} (-1)^{m'} B_{lm}(l'm';l''m'')$$

$$\times (-1)^{(l''-l-l')/2} \left[\frac{iK_0\,\delta_{m'0}\,\delta_{l'0}}{(4\pi)^{1/2}} + D_{l'm'} \right] \tag{3.104}$$

where

$$D_{l'm'} = -\frac{iK_0\,\delta_{m'0}\,\delta_{l'0}}{(4\pi)^{1/2}} - iK_0 \sum_{\mathbf{R}^{(s)}(\neq\mathbf{R}^{(k)})} (-1)^{l'+m'} \exp\,(i\mathbf{k}_{\|}\cdot\mathbf{R}_{ks})$$
$$\times\, h_{l'}^{(1)}(K_0R_{sk})Y_{l'-m'}(\Omega_{\mathbf{R}_{sk}}) \qquad (3.105)$$

In writing Eq. (3.104) we made use of a well-known property of the B_{lm} coefficients, namely, that

$$B_{lm}(l'm';l''m'') = 0 \qquad \text{unless } l + l' + l'' = \text{even} \qquad (3.106)$$

Kambe (1967a,b; 1968) has shown that

$$D_{lm} = D_{lm}^{(1)} + D_{lm}^{(2)} + D_{lm}^{(3)} \qquad (3.107)$$

$$D_{lm}^{(1)} = -\frac{2^{-l}i^{1-m}}{A_0K_0}[(2l+1)(l+m)!(l-m)!]^{1/2}\sum_{\mathbf{g}}\exp\,[-im\phi(\mathbf{k}_{\|}+\mathbf{g})]$$
$$\times\sum_{n=0}^{(l-|m|)/2}\frac{(K_{\mathbf{g}z}^+/K_0)^{2n-1}(|\mathbf{k}_{\|}+\mathbf{g}|/K_0)^{l-2n}}{n![\tfrac{1}{2}(l-m-2n)]![\tfrac{1}{2}(l+m-2n)]!}$$
$$\times\,\Gamma[\tfrac{1}{2}(1-2n),\,e^{-i\pi}\,(K_{\mathbf{g}z}^+/K_0)^2\alpha] \qquad (3.108)$$

$$D_{lm}^{(2)} = -\frac{K_0 2^{-l}(-1)^{(l+m)/2}(-1)^l}{4\pi[\tfrac{1}{2}(l-m)]![\tfrac{1}{2}(l+m)]!}[(2l+1)(l-m)!(l+m)!]^{1/2}$$
$$\times\sum_{\mathbf{R}^{(n)}(\neq 0)}\exp\,[-\mathbf{k}_{\|}\cdot\mathbf{R}^{(n)} - im\phi(\mathbf{R}^{(n)})][\tfrac{1}{2}K_0|\mathbf{R}^{(n)}|]^l$$
$$\times\int_0^\alpha u^{-3/2-l}\exp\,[u-(K_0|\mathbf{R}^{(n)}|)/(4u)]\,du \qquad (3.109)$$

$$D_{lm}^{(3)} = -\frac{K_0\,\delta_{l0}\,\delta_{m0}}{2\pi}\left[2\int_0^{\alpha^{1/2}}\exp\,(t^2)\,dt - \alpha^{-1/2}\exp\,(\alpha)\right] \qquad (3.110)$$

where Γ is an incomplete gamma function, $\phi(\mathbf{k}_{\|}+\mathbf{g})$ and $\phi(\mathbf{R}^{(n)})$ are the azimuthal angles of $\mathbf{k}_{\|}+\mathbf{g}$ and $\mathbf{R}^{(n)}$, respectively. $\mathbf{R}^{(n)}$ is given by Eq. (3.54). α is a parameter chosen in such a way that the sums in real space [Eq. (3.109)] and in reciprocal space [Eq. (3.108)] converge equally well. According to Kambe (1967a,b; 1968)

$$\alpha = \left|\frac{A_0K_0^2}{4\pi}\right| \qquad (3.11)$$

A computer program for calculating the **X** matrix on the basis of the above formulas can be found in Pendry's (1974) book. In the same book the reader will find, also, a program for the evaluation of the $M_{\mathbf{g}\mathbf{g}}^\pm$ elements given by Eq. (3.98).

ELECTRON STATES IN A SEMI-INFINITE METAL

4.1. INTRODUCTION

A semi-infinite crystal extending from $z = -\infty$ to $z = 0$ (we shall always assume that this is the case every time we refer to a semi-infinite crystal) can be viewed as a stack of atomic layers parallel to the given crystallographic plane (xy plane) but, in contrast to the infinite crystal considered in the previous chapter, we cannot assume, without further justification, that *all* layers are identical with each other. Obviously, the layers which lie away from the surface, deep into the solid, will be unaffected by the presence of the surface, but the top few layers are bound to be different from the bulk layers in more than one way. In the first instance the relative position of the atomic nuclei at the surface may be different from those in the bulk. For metals where a layer consists of a single plane of atoms (one atom per unit surface cell) this difference usually consists of a simple dilation away from or contraction towards the bulk of the metal of the top one or two layers. The analysis of relevant LEED experiments (see, e.g., Pendry, 1974; Van Hove and Tong, 1979) shows that the interlayer spacing changes by 10% or less for the first pair of layers, the change is much less for the second pair (second and third layers) and practically nonexistent further into the solid. There are cases, however [e.g., tungsten (100) below 350 K and possibly at higher temperatures], where lattice reconstruction does occur leading to a top (surface) layer with a different 2D unit cell (Debe and King, 1977; 1979; Felter, Barker, and Estrup, 1977). Reconstruction is much more common on semiconductor surfaces (see, e.g., Harrison, 1976).

In the present chapter we shall be concerned exclusively with clean metal surfaces. We shall disregard the possibility of reconstruction at the surface and the relatively small dilation or contraction of the top layers, assuming a constant interlayer distance throughout the semi-infinite crystal. Now, even in the absence of lattice reconstruction, the potential the electron sees at the metal–vacuum interface is bound to be different from that in the bulk because the electronic charge distribution is different at the surface. In principle, therefore, this charge distribution and the corresponding one-electron potential must be calculated self-consistently. So far, self-consistent calculations of the potential at the metal–vacuum interface of a *semi-infinite* metal have been performed for Na (100) (Appelbaum and Hamann, 1972b) and for Cu (111) (Appelbaum and Hamann, 1978). More recently, self-consistent calculations have been performed for *very thin films* (infinite in the xy plane) of transition metals consisting of a few (usually less than ten layers) parallel to a given crystallographic plane (see, e.g., Kerker, Ho, and Cohen, 1978; Louie, 1979; Louie, 1981; Posternak, Krakauer, Freeman, and Keolling, 1980). The potential and the electronic charge distribution at the metal–vacuum interface of such films is practically the same with that of a semi-infinite crystal. These self-consistent calculations have shown that any deviation from the bulk potential is limited to the top one or two layers of the metal, and that on the vacuum side of the interface the potential reduces to a function depending only on the z coordinate at a distance of about 2 Å from the center of the top atomic layer. The above observations suggest that for, at least, a semiquantitative discussion of the electron states of a semi-infinite metal one can approximate the one electron potential as follows. One may assume that the potential inside the metal right up to the metal vacuum interface (at $z = 0$) is given, to a reasonable degree of approximation, by the bulk potential and that on the vacuum side of the interface ($z > 0$) the potential is a function of the z-coordinate only. The relative simplicity of the above model allows a mathematical description of the electronic states of the semi-infinite crystal (see this chapter) and of electron emission phenomena (see following chapters) which is both elegant and physically transparent. We note that in many cases the results of surface states calculations based on this model are not significantly different from those obtained from more realistic potentials (see, e.g., Caruthers, Kleinman, and Alldredge, 1974), although this is not always the case (see, e.g., Caruthers and Kleinman, 1975; and Section 4.2.2). We note also that calculations of the intensity of angle-resolved photo emission current from a number of transition metal surfaces based on this model give results in good quantitative agreement with experiment (Pendry and Hopkinson, 1978; Moore and Pendry, 1978; Durham and Kar, 1981). Finally, the above model has been used, so far exclusively, in field emission calculations (Chapter 5), thermionic emission calculations (Chapter 9) and in secondary electron emission calculations (Chapter 10).

4.2. BOUND ONE-ELECTRON STATES

We assume that the one-electron potential energy field is described, to a good degree of approximation, by

$$V(\mathbf{r}) = \sum_n v(|\mathbf{r} - \mathbf{R}_n|) \qquad \text{for } z < 0 \qquad (4.1a)$$

$$= V_s(z) \qquad \text{for } z > 0 \qquad (4.1b)$$

The metal–vacuum interface (the plane $z = 0$) is taken a distance $(d/2 + D_s)$ from the center of the top atomic layer; d is the interlayer distance and D_s an adjustable parameter $(D_s < d/2)$. $v(|\mathbf{r} - \mathbf{R}n|)$ denote nonoverlapping spherical (muffin-tin) potentials centered on the atomic nuclei in the solid. These are grouped into identical layers, parallel to and periodic in the xy plane, in the manner described in Chapter 3. We have taken the constant potential between the muffin-tin spheres as the zero of energy and we have assumed that $V_s(z)$, the surface barrier, is a function of the z coordinate only. An example of such a barrier, which we use in later applications, is

$$V_s = E_F + \phi - \frac{e^2}{4(z + z_0)} \qquad (4.2)$$

E_F and ϕ are (respectively) the Fermi level and the work function, and z_0 is an adjustable parameter. The one-dimensional analog of the potential field described by Eqs. (4.1) and (4.2) is shown schematically in Fig. 4.1. Throughout the present chapter we assume, unless otherwise stated, that the interface is located half an interlayer distance above the center of the top layer, i.e., that $D_s = 0$ in Fig. 4.1, and that the origin of coordinates lies at the midpoint of the vector joining the center of the top (surface) layer of the truncated solid with the center of the would-be subsequent layer if the crystal were an infinite one. In Sections 4.2 and 4.3 we shall be concerned with electron states bound within the semi-infinite metal. The energy of the electron in such a state lies below the vacuum level and therefore the corresponding wave function decays exponentially into the vacuum region ($z > 0$). We have two kinds of such states. We refer to the first kind as bulk states. The wave function of such a state extends over the entire semi-infinite crystal. The second kind are the so-called surface states. These are localized states, i.e., the wave function of such a state decays "exponentially" on both sides of the metal–vacuum interface.

FIGURE 4.1

One-dimensional analog of the one-electron potential near the metal-vacuum interface. The potential wells correspond to planes of muffin-tin atoms. The potential depends only on the z coordinate for $z > 0$. The two curves (a) and (b) show (schematically) the barrier described by Eq. (4.2) for two different values of z_0.

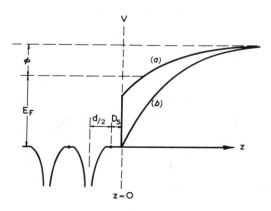

4.2.1. BULK STATES

These are intimately connected with the electron states of the corresponding *infinite* crystal. We have seen that for a given energy and a given wave vector $\mathbf{k}_{\|}$ = (k_x, k_y) within the SBZ of the 2D reciprocal lattice corresponding to the surface under consideration, one can always find, in the manner described in Chapter 3, a set of (say $2n$) Bloch waves (solutions of the Schrödinger equation for the infinite crystal) which includes all propagating waves that may exist for the given energy and $\mathbf{k}_{\|}$, plus a number of evanescent waves which decay either in the positive or the negative z direction (normal to the surface). For each propagating Bloch wave, denoted by ψ^B, which has its source at $z = -\infty$, there corresponds as we shall see, an electron state of the semi-infinite crystal, which we shall denote by Ψ^B. The index B which characterizes both states stands for the following set of quantum numbers: $B = (E^B, \mathbf{k}_{\|}, \alpha)$, where E^B is the energy of the Bth state and α is a band index. We note that for each $\mathbf{k}_{\|}$ there exist regions of energy (energy gaps) where no propagating Bloch wave exists and, therefore, no bulk (extended) state of the semi-infinite crystal exists either. On the other hand for a given $\mathbf{k}_{\|}$ and at a given energy there may be more than one Bloch waves propagating in the positive direction, with corresponding bulk states in the semi-infinite crystal. These are distinguished by the band-index α. (We use Greek letters to designate these bands). We note that the notation employed here is slightly different from that which is common in the literature on the energy-band-structure of infinite crystals. The situation is clarified by reference to Fig. 4.2. The solid lines in this figure represent dispersion curves in the traditional form

$$E_A = E_A(\mathbf{k}_{\|}; k_z) \tag{4.3}$$

where for a given band A (we use capital letters to designate these bands) and for fixed $\mathbf{k}_{\|}$, one plots the energy versus k_z in the reduced zone of the latter [see comments following Eq. (3.36)]. For this hypothetical crystal we see that for $\mathbf{k}_{\|} = 0$ and $E = E_1$ there exists only one Bloch wave propagating in the positive z direction ($\partial E/\partial k_z > 0$). This is represented by a circle on the figure. It belongs to the band α which is represented by a broken curve. There is an energy gap between the maximum of band A and the minimum of the next band where no propagating Bloch wave exists. The solid line denoted by CC′ stands for a doubly degenerate band, so that for $E = E_2$, for example, there are four Bloch waves propagating in the positive z direction. They belong to the bands (broken curves) denoted by β, γ, $\mu\mu'$ ($\mu\mu'$ being doubly degenerate). Usually, for given E and $\mathbf{k}_{\|}$, states belonging to different α bands have a different z component of the wave vector, given by (see Section 3.4)

$$k_{\alpha z} = k_{\alpha z}(E; \mathbf{k}_{\|}) \tag{4.4}$$

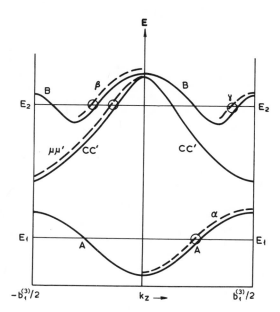

FIGURE 4.2

Relation of α-bands (broken lines) to A-bands (solid lines) for a fixed value of $\mathbf{k}_\|$ (say $\mathbf{k}_\| = 0$).

but at symmetry points of the Brillouin zone one often obtains states belonging to different bands having the same k_z component. For example in Fig. 4.2 we have $k_{\mu z}(E;\mathbf{k}_\| = 0) = k_{\mu'z}(E;\mathbf{k}_\| = 0)$. It is obvious that in general an α band coincides with a particular section of an A band and that the part of an A band which corresponds to Bloch waves propagating in the positive z direction can be split into sections so that each section coincides with a particular α band. For example, the β and γ bands coincide with two different sections of the band B. Over the region of the (E,k_z) space where an A band coincides with an α band we obtain

$$\frac{\partial k_{\alpha z}}{\partial E} = \left(\frac{\partial E_A}{\partial k_z}\right)^{-1} \tag{4.5}$$

since $k_{\alpha z}(E)$ given by Eq. (4.4) is the inverse function of $E_A(k_z)$ given by Eq. (4.3).

We construct the one-electron state Ψ^B of the semi-infinite crystal corresponding to the Bloch wave ψ^B as follows (Modinos and Nicolaou, 1976). For $z < 0$ (inside the crystal) we write

$$\Psi^B = \frac{1}{(2V)^{1/2}} \psi^B + \frac{1}{(2V)^{1/2}} \sum_{j=1}^{n} C_j^B \psi_j \tag{4.6}$$

C_j^B are constants, V denotes the volume of the semi-infinite and $2V$ that of the infinite crystal ($V \to \infty$ in the final formulas; it has been introduced here for normalization purposes). Like ψ^B, the ψ_j are solutions of the Schrödinger equation for the infinite crystal for the given energy ($E = E^B$) and wave vector $\mathbf{k}_{\parallel}$. We emphasize that the sum over j includes *only* the Bloch waves (n in number) which propagate or decay in the negative z direction. We note that every term in Eq. (4.6) and therefore Ψ^B satisfies the Schrödinger equation for the *semi-infinite crystal*

$$-\frac{\hbar^2}{2m} \nabla^2 \Psi^B + V(\mathbf{r})\Psi^B = E^B \Psi^B(\mathbf{r}) \tag{4.7}$$

in the half-space $z < 0$, since $V(\mathbf{r})$, according to our adopted model [Eq. (4.1)], is identical with that of the infinite crystal for $z < 0$. Moreover, Ψ^B remains finite as $z \to -\infty$.

On the vacuum side ($z > 0$) of the interface Ψ^B can be written in the following form:

$$\Psi^B = \frac{1}{(2V)^{1/2}} \sum_{\mathbf{g}} D_{\mathbf{g}}^B \Phi_{\mathbf{k}_{\parallel}+\mathbf{g}}^0(\mathbf{r}) \tag{4.8}$$

where the sum over $\mathbf{g}$ includes the same (n in number) $\mathbf{g}$ vectors which enter in the plane wave expansion, described by Eq. (3.52), of the Bloch waves ψ^B and ψ_j which appear in Eq. (4.6). The $D_{\mathbf{g}}^B$ are constants to be determined. Putting

$$\Phi_{\mathbf{q}_{\parallel}}^0(\mathbf{r}) = e^{i\mathbf{q}_{\parallel}\cdot\mathbf{r}_{\parallel}} \phi_{\mathbf{q}_{\parallel}}^0(z) \tag{4.9}$$

where $\mathbf{q}_{\parallel} \equiv \mathbf{k}_{\parallel} + \mathbf{g}$, and using the fact that in our model the potential depends only on z on the vacuum side of the interface, we find that each term in Eq. (4.8) satisfies the Schrödinger equation [Eq. (4.7)] for $z \geq 0$, provided $\phi_{\mathbf{q}_{\parallel}}^0(z)$ satisfies the following equation:

$$\left[-\frac{\hbar^2}{2m}\frac{d^2}{dz^2} + V_s(z)\right]\phi_{\mathbf{q}_{\parallel}}^0(z) = \left(E^B - \frac{\hbar^2 q_{\parallel}^2}{2m}\right)\phi_{\mathbf{q}_{\parallel}}^0(z) \tag{4.10}$$

$\phi_{\mathbf{q}_{\parallel}}^0$ must also satisfy the boundary condition,

$$\phi_{\mathbf{q}_{\parallel}}^0(z) \to 0 \qquad \text{as } z \to \infty \tag{4.11}$$

which characterizes an electron state bound within the semi-infinite crystal. Equations (4.10) and (4.11) determine, apart from an unimportant multiplicative constant, $\phi_{\mathbf{q}_{\parallel}}^0(z)$. It is easy to calculate $\phi_{\mathbf{q}_{\parallel}}^0(z)$, at least numerically, for any surface barrier $V_s(z)$.

The $2n$ coefficients, C_j^B and D_g^B, in Eqs. (4.6) and (4.8) are determined by the requirement that Ψ^B and $\partial \Psi^B / \partial z$ are continuous at the interface ($z = 0$). In the region of constant potential at the interface ψ^B and ψ_j are given according to Eq. (3.52), by

$$\psi^B = \sum_g [U_{1g}^{B+} \exp(i\mathbf{K}_g^+ \cdot \mathbf{r}) + U_{1g}^{B-} \exp(i\mathbf{K}_g^- \cdot \mathbf{r})] \qquad (4.12)$$

$$\psi_j = \sum_g [U_{1g}^{j+} \exp(i\mathbf{K}_g^+ \cdot \mathbf{r}) + U_{1g}^{j-} \exp(i\mathbf{K}_g^- \cdot \mathbf{r})] \qquad (4.13)$$

The notation follows from Eq. (3.52) and the way we have chosen the origin [see comments following Eq. (4.2); in enumerating the layers of the infinite crystal we take the top (surface) layer of the truncated solid as the zeroth layer, the immediate next layer in the positive z direction as the first layer, and so on]. Substituting Eqs. (4.12) and (4.13) into Eq. (4.6) and equating the resulting expressions for $\Psi^B(z = 0-)$ and $(\partial \Psi^B / \partial z)(z = 0-)$ with the corresponding expressions for $\Psi^B(z = 0+)$ and $(\partial \Psi^B / \partial z)(z = 0+)$ obtained from Eqs. (4.8) and (4.9), we find

$$D_g^B \phi_{\mathbf{k}_\parallel + \mathbf{g}}^0 (0) - \sum_{j=1}^{n} C_j^B (U_{1g}^{j+} + U_{1g}^{j-}) = U_{1g}^{B+} + U_{1g}^{B-} \qquad (4.14a)$$

$$D_g^B \dot{\phi}_{\mathbf{k}_\parallel + \mathbf{g}}^0 (0) - iK_{gz}^+ \sum_{j=1}^{n} C_j^B (U_{1g}^{j+} - U_{1g}^{j-}) = iK_{gz}^+ (U_{1g}^{B+} - U_{1g}^{B-}) \qquad (4.14b)$$

for every one of the n $\mathbf{g}$ vectors included in the expansion of the wave functions. The dot over ϕ denotes a derivation with respect to z, i.e., $\dot{\phi}^0 \equiv (d\phi^0 / dz)$. Equations (4.14) represent a system of $2n$ inhomogeneous equations which can be solved by standard numerical techniques to provide the values for the $2n$ coefficients, C_j^B and D_g^B. The wave function Ψ^B obtained in the above manner is, of course, not normalized unless ψ^B is normalized. A convenient way of normalizing ψ^B starts from the following general equation (see, e.g., Mott and Jones, 1958):

$$\frac{\hbar}{2mi} \iiint_{2V} \left(\psi_{\mathbf{k}A}^* \frac{\partial}{\partial z} \psi_{\mathbf{k}A} - \psi_{\mathbf{k}A} \frac{\partial}{\partial z} \psi_{\mathbf{k}A}^* \right) d^3r = \frac{1}{\hbar} \frac{\partial E_A}{\partial k_z} \qquad (4.15)$$

which is valid for any propagating Bloch wave $\psi_{\mathbf{k}A}$ with reduced wave vector $\mathbf{k}$ and energy $E = E_A(\mathbf{k}_\parallel, k_z)$ given by Eq. (4.3), and normalized as follows:

$$\iiint_{2V} |\psi_{\mathbf{k}A}(\mathbf{r})|^2 \, d^3r = 1 \qquad (4.16)$$

where $2V$ is the volume of the infinite crystal. We note that the integral

$$\frac{\hbar}{2mi} \int_{-L}^{L} \int_{-L}^{L} \left(\psi_{kA}^* \frac{\partial}{\partial z} \psi_{kA} - \psi_{kA} \frac{\partial}{\partial z} \psi_{kA}^* \right) dx\, dy$$

represents the probability of the electron passing in a unit of time through a plane described by $z = $ const and for a stationary state this is independent of z. Putting $\psi_{kA} = (2V)^{-1/2} \psi^B$, evaluating the above integral at $z = 0$ and substituting into Eq. (4.15) we obtain

$$\frac{1}{L^2} \left(\frac{\hbar}{2mi} \right) \int_{-L}^{L} \int_{-L}^{L} \left(\psi^{B*} \frac{\partial}{\partial z} \psi^B - \psi^B \frac{\partial}{\partial z} \psi^{B*} \right)_{z=0} dx\, dy = \frac{1}{\hbar} \frac{\partial E_A}{\partial k_z} \quad (4.17)$$

where L^2 is the area of the crystal parallel to the surface under consideration ($L \to \infty$ in the final formulas). In terms of the coefficients $U_{1g}^{B\pm}$ which determine ψ^B at the interface [Eq. (4.12)] the above normalization condition becomes

$$\frac{\hbar}{m} \sum_{g} [(\,|\,U_{1g}^{B+}|^2 - |\,U_{1g}^{B-}|^2)\mathrm{Re}(K_{gz}^+) - 2$$

$$\times \mathrm{Im}(U_{1g}^{B-*} U_{1g}^{B+})\mathrm{Im}(K_{gz}^+)] = \frac{1}{\hbar} \left(\frac{\partial k_{az}}{\partial E} \right)^{-1} \quad (4.18)$$

Note that the group velocity on the right-hand side of Eq. (4.17) has been replaced by its equal according to Eq. (4.5). We shall assume throughout this volume that, unless otherwise stated, ψ^B is normalized in the above manner. The normalization of Ψ^B follows automatically from that of ψ^B in accordance with a general theorem (see, e.g., Roman, 1965) which states that the set of scattering states (scattering state = incident + scattered wave) corresponding to a set of orthonormal incident waves (states) is, itself, an orthonormal set of states. In the present case ψ^B are the "incident" and Ψ^B the corresponding scattering states. It is worth noting that since scattering at the metal–vacuum interface does not allow the electron to move into the vacuum region, Ψ^B extends over the half space ($z < 0$; volume V) whereas ψ^B extends over all space (volume $2V$).

In many instances we need to know the sum over all bulk states with energy between E and $E + dE$ of a quantity

$$F(B) \equiv \int \Psi^{B*}(\mathbf{r}) \hat{F} \Psi^B(\mathbf{r})\, d^3 r$$

where $\hat{F}$ is a spin-independent one-electron operator. Such a sum can be transformed into an integral in the usual manner [see, e.g., our derivation of Eq. (1.4b)] as follows:

$$2 \sum_{\substack{B \\ E < E^B < E+dE}} F(B) = \frac{2(2V)dE}{(2\pi)^3} \iint_{SBZ} d^2 k_{\parallel}$$

$$\times \left[\sum_{\alpha} \left(\frac{\partial k_{\alpha z}}{\partial E} \right) F(E^B, \mathbf{k}_{\parallel}, \alpha) \right]_{E^B = E} \quad (4.19)$$

We emphasize that our model potential [Eq. (4.1)] does not depend on spin and, therefore, we have two independent electron states (corresponding to spin up and spin down) with the same spatial wave function Ψ^B. The resulting (spin) degeneracy is accounted for in Eq. (4.19) by the multiplicative factor of 2, which appears on both sides of this equation. The sum on the right-hand side of Eq. (4.19) takes into account the fact that there may be more than one bulk state (corresponding to different α bands) for given E and $\mathbf{k}_{\parallel}$. Obviously, over those sections of the SBZ where none such state exists (for the given energy) the integrand in Eq. (4.19) should be replaced by zero.

4.2.2. SURFACE STATES AND RESONANCES

For a given $\mathbf{k}_{\parallel}$ in the SBZ of a given crystallographic plane (surface) there exist regions of energy where no propagating Bloch wave of the corresponding infinite crystal exists and, therefore no bulk state of the semi-infinite crystal exists either. This is demonstrated in Fig. 4.3, taken from a paper by Tomasek and Pick (1978), which shows a projection of the bulk band structure (PBS) of tungsten on the SBZ of the (100) plane, for $\mathbf{k}_{\parallel}$ points along the symmetry lines of the latter (shown in Fig. 4.4). Areas within the solid lines correspond to bands of states (with symmetry $\overline{\Delta}_1$, $\overline{Y}_1$, and $\overline{\Sigma}_1$) which are symmetrical under the symmetry operation of the respective symmetry direction. For example, along the $\overline{\Gamma X}$ direction these states (their wave functions) are symmetrical (even) under the operation $y \rightarrow -y$. Areas within the dotted lines correspond to bands of states (with symmetry $\overline{\Delta}_2$, $\overline{Y}_2$, $\overline{\Sigma}_2$) which are antisymmetrical (odd) under the symmetry operation of the respective symmetry direction. For example, the states along the $\overline{\Gamma X}$ symmetry line are antisymmetrical (odd) under the operation $y \rightarrow -y$. An energy region where no bulk state exists (for the given $\mathbf{k}_{\parallel}$) is referred to as an (absolute) energy gap. An energy region where bulk states of a given symmetry do not exist while states of another symmetry exist is referred to as a symmetry or resonance energy gap. Obviously, a resonance gap can only exist at symmetry points or along

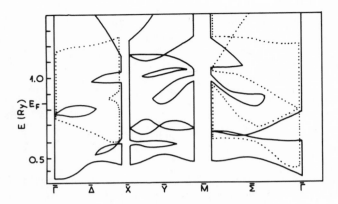

FIGURE 4.3

Projection of the bulk energy band structure of tungsten on the SBZ of the (100) plane. Areas within solid (broken) lines correspond to symmetrical (antisymmetrical) states with respect to the symmetry operation of the respective symmetry direction. (From Tomasek and Pick, 1978.)

symmetry lines of the SBZ. In the case of W(100), we see from Fig. 4.3 that resonance gaps exist along the $\overline{\Gamma X}$ and $\overline{\Gamma M}$ directions. Along the $\overline{MX}$ direction we have only absolute gaps. Additional resonance gaps exist at the symmetry points $\overline{\Gamma}, \overline{M}$, and $\overline{X}$ (these are not shown in Fig. 4.3).

For given $\mathbf{k}_\parallel$ and for an energy E within an energy gap (let it be an absolute energy gap), the $2n$ Bloch waves which one obtains by solving the complex-band-structure problem [Eq. (3.48)] are all evanescent waves, half of them decaying in the positive z direction and the other half in the negative z direction. An electron state in the semi-infinite crystal described by the model potential of Eq. (4.1), if

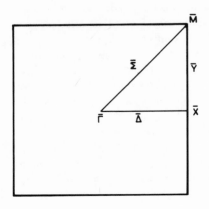

FIGURE 4.4

The SBZ of the (100) plane of a bcc crystal.

such exists for the given E and $\mathbf{k}_{\|}$, must have the following form inside the crystal ($z \leq 0$):

$$\Psi^S = \frac{1}{\sqrt{A}} \sum_{j=1}^{n} C_j^S \psi_j \qquad (4.20a)$$

where the surface area, $A = L^2$, has been introduced for normalization purposes, C_j^S are constants to be determined and ψ_j are the n evanescent Bloch waves which decay "exponentially" in the negative z direction, i.e., Ψ^S vanishes away from the surface. For $z \geq 0$, i.e., on the vacuum side of the interface, we have

$$\Psi^S = \frac{1}{\sqrt{A}} \sum_{\mathbf{g}} D_{\mathbf{g}}^S \Phi_{\mathbf{k}_{\|}+\mathbf{g}}^0(\mathbf{r}) \qquad (4.20b)$$

where $\Phi_{\mathbf{k}_{\|}+\mathbf{g}}^0(\mathbf{r})$ are the functions defined by Eqs. (4.9)–(4.11), $D_{\mathbf{g}}^S$ are constants to be determined and the sum includes the n $\mathbf{g}$ vectors which enter in the plane wave expansion [Eq. (3.52)] of the Bloch waves ψ_j. The wave function Ψ^S, described by Eq. (4.20), satisfies the Schrödinger equation for the given potential [Eq. (4.1)] on either side of the metal–vacuum interface and behaves well ($\Psi^S \rightarrow 0$) as $z \rightarrow \pm\infty$. However, for it to constitute a physical solution of the Schrödinger equation (an eigenstate of the electron in the potential fields of the semi-infinite crystal), it and its derivative must be continuous at the interface (at $z = 0$). Following the same steps as in the derivation of Eqs. (4.14a and b), we find that the above requirement implies that

$$D_{\mathbf{g}}^S \phi_{\mathbf{k}_{\|}+\mathbf{g}}^0(0) - \sum_{j=1}^{n} C_j^S (U_{1\mathbf{g}}^{j+} + U_{1\mathbf{g}}^{j-}) = 0 \qquad (4.21a)$$

$$D_{\mathbf{g}}^S \dot{\phi}_{\mathbf{k}_{\|}+\mathbf{g}}^0(0) - iK_{\mathbf{g}z}^+ \sum_{j=1}^{n} C_j^S (U_{1\mathbf{g}}^{j+} - U_{1\mathbf{g}}^{j-}) = 0 \qquad (4.21b)$$

for every one of the n $\mathbf{g}$ vectors included in the expansion of the wave functions. Equations (4.21) can be put into a matrix form

$$\wedge (\mathbf{k}_{\|}; E)\mathbf{y} = 0 \qquad (4.21c)$$

where $\mathbf{y}$ is a column vector of $2n$ unknowns ($D_{\mathbf{g}}^S$ and C_j^S) and $\wedge$ is a $(2n) \times (2n)$ matrix. A solution of the above system of equations exists only if the determinant of $\wedge$ vanishes. Hence, we conclude that if

$$\det \Lambda(E; \mathbf{k}_{\|}) = 0 \qquad (4.22)$$

at some value of E within an energy gap (for the given $\mathbf{k}_{\parallel}$), then a *surface state* exists at this energy. It is called so, because the wave function of such a state, given by Eq. (4.20), is localized at the surface, decaying "exponentially" on either side of it. We shall assume that Ψ^S is normalized as follows:

$$\iiint |\Psi^S(\mathbf{r})|^2 \, d^3r = 1 \qquad (4.23)$$

where the integral in the xy plane extends over the area A of the surface. Of course, $|\Psi^S|^2$ has the periodicity of the surface lattice so that the integral over A reduces to an integral over the surface unit cell. The surface states obtained in the above manner, i.e., on the basis of the model potential defined by Eq. (4.1), are sometimes called Shockley surface states (SS states). Shockley (1939) proved that in the case of the one-dimensional crystal (shown schematically in Fig. 4.1), which is the one-dimensional analog of the potential defined by Eq. (4.1), the existence or otherwise of a surface state in a given energy gap is completely determined by the nature of that gap. He showed in particular that a hybridizational energy gap (HG gap) always contains one surface state and that a non-HG gap cannot contain a surface state. (The nature of an HG gap is described in one of the following paragraphs.) Of course, the exact position of the energy level of an SS state does depend on the shape of the surface barrier on the vacuum side of the interface. It is reasonable to expect that while a small deviation from the Shockley model potential in the region of the top one or two layers may affect the position of an SS state, it will not destroy it altogether. In three dimensions, the establishment of rigorous criteria which will tell us with a minimum of computation whether a surface state exists in a given gap, without us having to search for possible roots of Eq. (4.22) over the entire gap, appears to be very difficult. On the other hand, certain empirical rules have been established, on the basis of simple model calculations, and the validity of these rules has been confirmed, in some cases, by extensive numerical calculations. The interested reader will find an extensive list of references and a general discussion of these rules in relation to SS states on transition metal surfaces (of particular importance in relation to electron emission) in a review article by Tomasek and Pick (1978). In brief, these rules say that a SS state can exist *only* in an HG gap and that in such a gap (corresponding to a given $\mathbf{k}_{\parallel}$) there may be 0, 1, or 2 SS states. Which of the above alternatives takes place in a particular gap cannot be determined without detailed numerical calculations except in some special cases, e.g., when the surface has both a mirror plane parallel to the surface and a center of symmetry and the gap itself is narrow (Pendry and Gurman, 1975).

The nature of a HG gap is demonstrated in Fig. 4.5. The bands shown in Fig. 4.5a correspond to (hypothetical) nonhybridized $s-p$ and d bands along the Δ-symmetry line [normal the (100) plane] of a typical (bcc) transition metal like

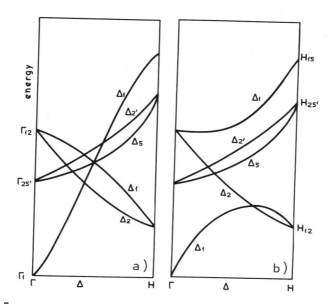

FIGURE 4.5

The interaction between the two Δ_1-bands shown in (a) opens up the HG (resonance) gap shown in (b).

tungsten or molybdemum. The s–p band (the Δ_1 curve from Γ_1 to H_{15}) interacts (hybridizes) with the d band of the same symmetry (Δ_1 curve from Γ_{12} to H_{12}) to produce the actual band structure shown in Fig. 4.5b. The two Δ_1 bands in Fig. 4.5b are separated by a gap which is a HG gap, in the sense that it would not have been there if the hybridization interaction did not exist or if the nonhybridized s–p and d bands did not cross. We note that this gap is not an absolute gap but a symmetry (resonance) gap as defined at the beginning of this section. It exists only at the center of the SBZ of the (100) plane (point $\bar{\Gamma}$ in Fig. 4.4). In deriving Eq. (4.22) as the condition for the existence of a surface state, we have assumed an absolute energy gap, but a surface state can exist in a symmetry gap as well. In the present case among the n Bloch waves ψ_j appearing in the formal expansion of the wave function, given by Eq. (4.20), there is at least one Bloch wave which propagates in the negative z direction. Let us assume for the sake of clarity that there exists only one such wave and let us denote it by $\psi_{j'}$ and the corresponding coefficient in Eq. (4.20) by $C_{j'}^S$. It is possible that Eq. (4.22) is satisfied at some value of the energy within the symmetry gap, and that $C_{j'}^S = 0$ at the same time, in which case we obtain at this energy a (true) surface state whose wave function Ψ^S is localized on the surface decaying exponentially on either side of it.

The fact that $C_{j'}^S = 0$ implies, in general, that Ψ^S has a different symmetry from that of $\psi_{j'}$ and therefore the two cannot mix. This follows from a general rule of quantum mechanics (see, e.g., Slater, 1972) which states that the different components of the wave function (of a stationary electron state) must have the same symmetry. This general rule, taken together with the PBS shown in Fig. 4.3, tells us, for example, that a (true) surface state of $\overline{\Delta}_1$ (even) symmetry can exist in an HG even band gap (no bulk states with $\overline{\Delta}_1$ symmetry exist in this gap) along the $\overline{\Gamma X}$ symmetry line, because it will not mix with the bulk states of odd symmetry which exist at the same energy and wavevector $\mathbf{k}_{\|}$.

The empirical rules mentioned above in relation to the existence of SS states apply equally well to SS states in symmetry energy gaps. It is important to remember, however, that a symmetry gap can exist only at symmetry points or along symmetry lines of the SBZ zone. When we move off the symmetry line to a general point of the SBZ, symmetry arguments do not apply, and therefore, if a propagating Bloch wave exists at the given energy the corresponding coefficient in Eq. (4.20a) is, in general, different from zero. Therefore, at a *general* point of the SBZ, surface states can exist only in absolute energy gaps. The question arises as to what happens to a surface state in a symmetry gap when we move an infinitesimal amount away from the symmetry point onto a general point of the SBZ. It cannot simply disappear. We find, in practice, that the wave functions of bulk states in the vicinity of such a point peak at the surface in a manner reminiscent of a true surface state for energies within a narrow region: $E_r(\mathbf{k}_{\|}) - \tfrac{1}{2}\Gamma$ $(\mathbf{k}_{\|}) < E < E_r(\mathbf{k}_{\|}) + \tfrac{1}{2}\Gamma(\mathbf{k}_{\|})$. We describe the above situation by saying that at the given $\mathbf{k}_{\|}$ a surface resonance of width Γ exists at the energy $E_r(\mathbf{k}_{\|})$. The nature of the wave function near a surface resonance is demonstrated in Fig. 4.6. The solid curve shows the wave function of a bulk state averaged over the xy plane (parallel to the surface) versus z (the distance from the surface), at an energy and at a value of $\mathbf{k}_{\|}$ where a surface resonance exists. The broken curve shows the same quantity for an ordinary bulk state with the same energy but different value of $\mathbf{k}_{\|}$ (away from the resonance). [The calculation (Nicolaou and Modinos, 1976) based on a model potential for $W(100)$, was done only for values of z halfway between consecutive layers (crosses and circles on the figure) where the potential is constant according to the model described by Eq. (4.1). The straight lines joining these points have no physical significance.] Note that the wave function at a surface resonance reduces to that of an ordinary bulk state away from the surface; it does not go to zero as is the case with a true surface state. For given $\mathbf{k}_{\|}$ the magnitude of the wave function at the surface is maximum at the center $E_r(\mathbf{k}_{\|})$ of the resonance and diminishes rapidly away from it, reaching half its maximum value at $E = E_r(\mathbf{k}_{\|}) \pm \tfrac{1}{2}\Gamma(\mathbf{k}_{\|})$, where Γ is by definition the width of the resonance. The probability density in the surface region integrated over the width of the resonance is approximately equal to that which one would obtain from a true surface state with a well-defined energy level at the given $\mathbf{k}_{\|}$. In that respect a *narrow* surface resonance is physically indistinguishable from a true surface state. It is

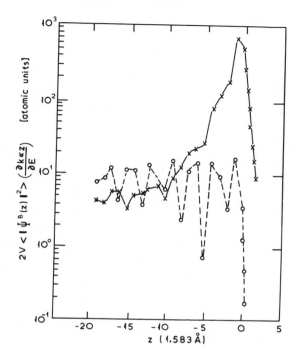

FIGURE 4.6

The solid line shows the wave function of a bulk state (averaged over a surface unit cell) at an energy and at a value of $\mathbf{k}_\parallel$ where a surface resonance exists. The broken line shows the same quantity for an ordinary bulk state. The wave functions decrease exponentially on the vacuum side of the interface ($z > 0$).

also clear that although a band of surface resonances is more likely to exist in the vicinity of a symmetry point or line in the SBZ, it can, in principle, exist in any region of the SBZ. The energy $E_r(\mathbf{k})$ and width $\Gamma(\mathbf{k})$ of the resonance within a band are smooth functions of $\mathbf{k}_\parallel$, with $\Gamma(\mathbf{k}_\parallel)$ increasing and the magnitude of the wave function at the surface decreasing as $\mathbf{k}_\parallel$ approaches the "boundary" of the band. [It is clear that the "boundary" in $\mathbf{k}_\parallel$ space of a band of surface resonances is blurred and not a well-defined line].

Another way of looking at a surface resonance is the following. The energy level of a (true) surface state coincides with a singularity, a pole, on the real energy axis of the Green's function defined by Eq. (5.41). We have seen that such a state can exist at a symmetry point of the SBZ. Obviously, moving away from that point by a small amount onto a general point of the SBZ does not remove the singularity but pushes it off the real axis. In that respect, a surface resonance can be looked upon as a virtual surface state, a nonstationary state, with complex energy $E_r(\mathbf{k}_{\parallel})$ $- (i/2)\Gamma(\mathbf{k}_{\parallel})$, and therefore a finite lifetime given by $\tau \simeq \hbar/\Gamma$. Finally, we must emphasize that surface resonances are not independent one-electron states of the semi-infinite crystal, and ought not to be counted as such. For example, their contribution to the local density of states at the surface is automatically taken into account when one calculates the contribution to the above quantity of the bulk states defined in the previous section.

In the last few years a considerable number of detailed numerical calculations have been performed of surface states and resonances on particular metal surfaces. Some of these calculations have been based on the Shockley model [Eq. (4.1)] or on slight variations of this potential (see Gurman and Pendry, 1973; Gurman, 1976a; 1976b; Inglesfield, 1978; Noguera, Spanjaard, and Jepsen, 1978; Spanjaard, Jepsen, and Marcus, 1979), while others have been based on more elaborate self-consistent calculations of the electron states of *thin* metal films (see, e.g., Kerker *et al.*, 1978; Posternak *et al.*, 1980). At the same time, considerable progress has been made in the experimental study of surface states and resonances. [A summary of theoretical and experimental results relating to surface states can be found in the review article of Inglesfield and Holland (1981).] With the aid of angle-resolved photoemission it is now possible to establish experimentally the dispersion of a band of surface states/resonances along any particular direction in the SBZ. The symmetry of the surface state along a symmetry direction or at a symmetry point of the SBZ can also be established from such measurements. [Excellent review articles are available on photoemission spectroscopy (Feuerbacher and Willis, 1976; Feuerbacher, Fitton, and Willis, 1978; Eastman and Himpsel, 1981; Inglesfield and Holland, 1981; Plummer and Eberhardt, 1982.] Considerable effort has been devoted, in particular, to the experimental documentation of the properties of the surface states/resonances on tungsten (100) (Willis, Feuerbacher, and Fitton, 1976; Weng, Plummer, and Gustafson, 1978; Campuzano, Inglesfield, King, and Somerton, 1981; Holmes and Gustafson, 1981). Some of the experimental results for W(100) are summarized in Fig. 4.7. In the same figure we give the results of some theoretical calculations for the sake of comparison. Figure 4.7a shows the dispersion of the surface states/resonances along the $\overline{\Gamma X}$ symmetry line (see Fig. 4.4). The open circles, the squares, and the closed circles in Fig. 4.7a are the results of angle-resolved photoemission measurements by Weng *et al.* (1978). The amplitude (at the surface) of the state associated with the first high-energy band (open circles) diminishes as we move away from $\overline{\Gamma}$. We note that this is a true surface state only at $\overline{\Gamma}$ where it exists in an HB $\overline{\Delta}_1$ symmetry gap (see Fig. 4.5), becoming a surface resonance for $k_{\parallel} > 0$. It has an even $(\overline{\Delta}_1)$ symmetry along the $\overline{\Gamma X}$ line. The second high-energy band (squares) does not exist at $\overline{\Gamma}$. The amplitude of the corresponding wave function (at the surface) is maximum at about $k_{\parallel} \simeq 0.3 \text{ Å}^{-1}$. Its symmetry has not been identified experimentally. The low-energy band (closed circles) at about 4.5 eV below E_F has a $\overline{\Delta}_1$ symmetry. It is now known that this band extends from $\overline{\Gamma}$ to $\overline{X}$ with very little dispersion

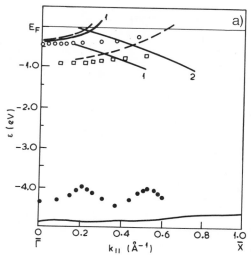

FIGURE 4.7

Surface states/resonances on W(100). (a) along the $\overline{\Gamma X}$ symmetry line. (b) along the $\overline{\Gamma M}$ symmetry line. The open circles, the squares, the crosses and the closed circles are the results of angle-resolved photoemission measurements by Weng *et al.* (1978) and by Holmes and Gustafson (1981). The solid lines are surface bands calculated self-consistently by Posternak *et al.* (1980). The states marked 1 and 2 have (respectively) even and odd parity with respect to the appropriate mirror plane. The broken lines have been calculated (unpublished results by the author) using the model potential defined by Eqs. (4.1) and (4.2). Also shown in (b) are the even gap (shaded area) and the odd gap (area between solid lines marked by dashes) from Grise *et al.* (1979).

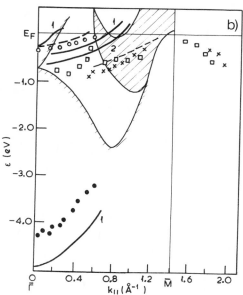

(Campuzzano *et al.*, 1981). The experimental curves along the $\overline{\Gamma M}$ direction, shown in Fig. 4.7b, have been obtained by Weng *et al.* (1978) and by Holmes and Gustafson (1981). We have the high-energy band centered at $\overline{\Gamma}$ (open circles). The amplitude (at the surface) of the corresponding wave function [of even ($\overline{\Sigma}_1$) symmetry along $\overline{\Gamma M}$] diminishes as we move away from $\overline{\Gamma}$. Then we have a *double*

band with an even component (squares) and an odd component (crosses). The odd state exists entirely within an odd-band-gap, but the even state crosses the even-band-gap at approximately $k_{\parallel} \simeq 0.7$ Å^{-1} leading to a perturbation in the dispersion. (The two gaps are shown in Fig. 4.7b. Note that the even gap is, in fact, an absolute gap.) Finally, we have a low-lying band of even symmetry (closed circles). Along the $\overline{\Gamma M}$ direction this band shows considerable dispersion in contrast to the situation along the $\overline{\Gamma X}$ direction.

The solid lines in Fig. 4.7 show the bands of surface states/resonances as calculated self-consistently by Posternak *et al.* (1980). [Although the calculation was done for a seven-layer film one may assume (in view of the localized nature of these states) that the surface states and resonances are the same as for the corresponding semi-infinite crystal.] The mark 1 and 2 (respectively) even and odd parity with respect to the appropriate mirror plane [the (100) plane along the $\overline{\Gamma X}$ direction and the (110) plane along the $\overline{\Gamma M}$ direction]. We see that there is reasonably good agreement between theory and experiment as far as the high-energy band (open circles) is concerned, although the calculated dispersion differs significantly from the experimental one especially along the $\overline{\Gamma M}$ direction. There is very poor agreement between the theory and the second high-energy band along the $\overline{\Gamma X}$ direction. [The fact that the calculation predicts an odd state along this direction is somewhat surprising because there does not exist an odd-band-gap at this energy (see, e.g., Fig. 4.3).] The situation is different along the $\overline{\Gamma M}$ direction. There we have an HG even-band-gap (an absolute gap) and a much larger HG odd-band-gap in the energy region under consideration, and a double band (with an even and an odd component) is to be expected. Such a double band is found both experimentally (squares and crosses) and theoretically. The calculated dispersion and the splitting between the odd and the even states is, however, significantly different from the corresponding experimental values. As far as the low-energy band (closed circles) is concerned, the self-consistent calculation gives correctly the dispersion but not the absolute energy, which differs from the experimental value by approximately 0.6 eV. We note that like the first high-energy band this is a true surface state only at $\overline{\Gamma}$ where it exists in the HG $\overline{\Delta}_1$ symmetry gap.

The broken curves in Fig. 4.7 have been calculated (unpublished results by the author) using the (Shockley) model potential defined by Eqs. (4.1) and (4.2). [The muffin-tin potential of the center layer of the seven-layer film of Posternak *et al.* (1980) has been used throughout the crystal right up to the metal–vacuum interface (I am indebted to N. Kar for providing me with the scattering phase shifts for this potential).] The values of the other parameters used in the calculation were $E_F = 0.47$ Hartree (12.79 eV), $\phi = 4.5$ eV, $D_s = 0.397$ Å, $z_0 = 0.278$ Å. Along the $\overline{\Gamma X}$ direction this model produces reasonably well the high-energy bands (open circles and squares) but it fails to produce the low-lying band (closed circles). Along the $\overline{\Gamma M}$ direction the agreement between this model calculation and the experiment is good as far as the high-energy band (open circles) and the even component of the double band (squares) are concerned. We cannot tell with certainty whether the odd state is there. Either such a state does not exist (for this model) or it peaks around the second (or third) atomic layer and we missed it because of the way the calculation was done {the surface states/resonances are recognized as peaks (which are sensitive to the choice of surface barrier) in the average surface density of states at $z = 0$, which is given by $-(1/\pi)\text{Im}$ $\langle \chi_{q_\parallel} | G(E) | \chi_{q_\parallel} \rangle$ [see comments following Eq. (5.43)]}. The low-energy band (closed circles) is certainly not there. The failure of this model [Eqs. (4.1) and (4.2)] to produce the low-lying state is, probably, due to the neglect of the transverse variation of the potential in the region between the muffin-tin spheres of the top layer and the interface plane at $z = 0$. It is worth pointing out that earlier calculations (Modinos and Nicolaou, 1976; Kar and Soven, 1976b; Smith and Mattheiss, 1976) based on this model but using a different bulk muffin-tin potential [that of Mattheiss (1965)] failed to predict the high-energy band centered around $\bar{\Gamma}$ (open circles in Fig. 4.7), although the bulk energy-band-structure obtained with the Mattheiss muffin-tin potential is not significantly different from that obtained with the potential of Posternak *et al.*

It is clear from Fig. 4.7 that for a full description of the surface states/resonances on W(100) one must go beyond the model potential defined by Eqs. (4.1) and (4.2). This may be true for other, relatively open surfaces, as well (Caruthers and Kleinman, 1975). It is also clear from Fig. 4.7 that, although self-consistency has improved the agreement between theory and the experimental results, the situation is not entirely satisfactory. This may be, at least partly, due to the fact that spin–orbit coupling, which may be significant for tungsten (see Section 5.4.1), has been neglected in the calculation. [We note, however, that relativistic corrections other than spin–orbit interaction (i.e., kinematic effects) have been taken into account in the calculation of Posternak *et al.* (1980).]

Apart from surface states/resonances of the Shockley type which appear only in hybridizational gaps and are most cases rather insensitive to the details of the potential in the surface region, there may exist Tamm surface states (Tamm, 1932). These are states pulled off the bulk bands when the potential in the (top) surface layer is significantly distorted. Self-consistent calculations (Gay, Smith, and Arlinghaus, 1979; Smith, Gay, and Arlinghaus, 1980) and angle-resolved photoemission measurements (Heimann, Hermanson, Miosga, and Neddermeyer, 1979) have shown, for example, that such states exist on the (100) plane of copper.

Finally, we would like to mention that the surface states and resonances we have described so far, whether they are Shockley states or Tamm states, are described by wave functions whose center of gravity lies in the immediate neighborhood of the metal–vacuum interface and in most cases on the metal side of it (see, e.g., Fig. 4.6). Apart from such states there may be "barrier-induced" surface states whose wave function lies almost entirely on the vacuum side of the interface.

These states owe their existence to the Coulombic tail of the surface barrier, $[V_s(z) \rightarrow \text{const} - e^2/4z \text{ as } z \rightarrow \infty]$ which traps electrons between the interface and $z = +\infty$, in a manner analogous to that which gives rise to the Rydberg states in an atom. Obviously, such states can only exist for energies very near the vacuum level (within 0.1 eV or so from it), and they are likely to appear as a series of such states as in the case of the atomic Rydberg states. So far no attention has been paid to *bound* Rydberg states in contrast to the closely related (Rydberg type) surface resonances at energies above the vacuum level (see Section 4.4.3).

4.3. THE LOCAL DENSITY OF STATES

The local density of states (LDS) is defined by

$$\rho(E,\mathbf{r}) = 2 \sum_M |\Psi^M(\mathbf{r})|^2 \, \delta(E - E^M) \tag{4.24}$$

where $\delta(x)$ is the Dirac δ function, M stands for the set of quantum numbers (apart from spin) which characterize the Mth state, E^M denotes the energy of this state, and the factor of 2 takes account of spin degeneracy. We can write

$$\rho(E,\mathbf{r}) = \rho_b(E,\mathbf{r}) + \rho_s(E,\mathbf{r}) \tag{4.25}$$

where $\rho_b(E,\mathbf{r})$ is the contribution to ρ from the bulk (extended) states and ρ_s the contribution to ρ from surface states. Using Eq. (4.19) we find that

$$\rho_b(E,\mathbf{r}) = \frac{(2V)}{4\pi^3} \iint_{\text{SBZ}} d^2k_\parallel \left[\sum_\alpha \left(\frac{\partial k_{\alpha z}}{\partial E} \right) |\Psi^B(\mathbf{r})|^2 \right]_{E^B = E} \tag{4.26}$$

where $B = (E_B, \mathbf{k}_\parallel, \alpha)$ characterizes the Bth bulk state and the sum over α takes into account the fact that for given E and $\mathbf{k}_\parallel$, there may be none (in which case the integrand vanishes), one, or more than one such states. It is worth noting that if $(\partial k_{\alpha z}/\partial E)$ is replaced by its equal according to Eq. (4.18), the quantity $\{(\partial k_\alpha/\partial E) \Psi^B(\mathbf{r})|^2\}_{E^B = E}$ becomes independent of the normalization factor in Ψ^B.

The density of states averaged over a bulk unit cell *deep inside the crystal* is given by

$$\frac{dN}{dE} = \frac{1}{2\pi^3} \iint_{\text{SBZ}} d^2k_\parallel \sum_\alpha \left(\frac{\partial k_{\alpha z}}{\partial E} \right) \tag{4.27}$$

which follows from Eq. (4.26), since

$$\iiint_V |\Psi^B(\mathbf{r})|^2 d^3r = 1 \qquad (4.28)$$

where V is the volume of the truncated crystal [see comments following Eq. (4.18)]. We note that the factor of 2 in $(2V)$ on the right of Eq. (4.26) [with a consequent factor of 2 in Eq. (4.27)] compensates for the fact that the α-bands correspond to those states of the infinite crystal (half of the total number of states at the given energy) which represent waves propagating in the positive z direction.

The contribution to the local density of states from the surface states is

$$\rho_s(E,\mathbf{r}) = 2 \sum_S \delta(E - E^S) |\Psi^S(\mathbf{r})|^2$$

$$= \frac{A}{2\pi^2} \iint_{SBZ} \sum_\mu \delta(E - E_\mu(\mathbf{k}_\parallel)) |\Psi^S(\mathbf{r})|^2 d^2k_\parallel \qquad (4.29)$$

where $S \equiv (\mathbf{k}_\parallel, \mu)$ are the quantum numbers which characterize the surface state, μ being a (surface) band index. The energy of the Sth state is denoted by $E^S = E_\mu(\mathbf{k}_\parallel)$ and Ψ^S is assumed normalized according to Eq. (4.23). Putting $\mathbf{k}_\parallel = (2\pi/L)(n_x, n_y)$, where n_x, n_y are integers and $L = \sqrt{A}$, the sum over $\mathbf{k}_\parallel$ has been transformed into an integral in the usual manner. The factor of 2 takes into account spin degeneracy.

Sometimes $\rho(E,\mathbf{r})$ is averaged over a given layer to give a "layer density of states." The results of such a calculation by Posternak et al. (1980) for tungsten (100) are shown in Fig. 4.8. Although the above calculation refers to a thin metal film made up of seven layers, the LDS at the center layer of such a film is practically identical with that of a bulk layer in a semi-infinite crystal. The reader will note the marked difference between the density of states of the top surface layer and that of the center layer (bulk density of states). It is also worth noting that the density of states of the first layer in from the surface looks more like that of the center layer (bulk) rather than that of the surface layer.

An equally useful concept, especially in relation to the analysis of field emission measurements, is the "average density of states at z," which we define as follows:

$$\langle \rho(E,z) \rangle = \frac{1}{A_0} \iint_{A_0} \rho(E;\mathbf{r}_\parallel,z) d^2\mathbf{r}_\parallel \qquad (4.30)$$

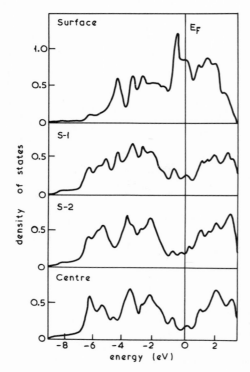

FIGURE 4.8

Theoretical layer density of states for a seven layer tungsten (100) film. S-1 and S-2 are (respectively) the first and second layer in from the surface. (From Posternak *et al.*, 1980).

where A_0 denotes the area of the surface unit cell [$\rho(E,\mathbf{r})$ is periodic in the xy plane]. We have

$$\langle \rho(E,z) \rangle = \langle \rho_b(E,z) \rangle + \langle \rho_s(E,z) \rangle \tag{4.31}$$

where $\langle \rho_b \rangle$ and $\langle \rho_s \rangle$ are the contributions to $\langle \rho \rangle$ from the bulk states and from the surface states, respectively. They are given by

$$\langle \rho_b(E,z) \rangle = \frac{(2V)}{4\pi^3} \iint\limits_{SBZ} d^2 k_{\parallel} \left[\sum_{\alpha} \left(\frac{\partial k_{\alpha z}}{\partial E} \right) \langle |\Psi^B(z)|^2 \rangle \right]_{E^B = E} \tag{4.32}$$

$$\langle \rho_s(E,z) = \frac{A}{2\pi^2} \iint\limits_{SBZ} d^2 k_{\parallel} \sum_{\mu} \delta(E - E_{\mu}(k_{\parallel})) \langle |\Psi^S(z)|^2 \rangle \tag{4.33}$$

where

$$\langle |\Psi^M(z)|^2 \rangle \equiv \frac{1}{A_0} \iint\limits_{A_0} |\Psi^M(\mathbf{r}_{\parallel},z)|^2 \, d^2 r_{\parallel} \tag{4.34}$$

4.4. SCATTERING STATES

These are one-electron states whose energy lies above the vacuum level and therefore are described by wave functions which extend over the entire space. We shall assume for the sake of simplicity that the surface barrier defined by Eq. (4.1b) is constant beyond a certain distance from the surface, i.e.,

$$V_s(z) = V_\infty \qquad \text{for } z > z_\infty \tag{4.35}$$

where z_∞ may be very large but finite. V_∞ defines the vacuum level and

$$\mathbf{E} = E - V_\infty \tag{4.36}$$

is the energy of the electron measured from this level. We have scattering states whose source lies inside the metal at $z = -\infty$ and scattering states whose source lies outside the metal at $z = +\infty$.

4.4.1. SCATTERING STATES WITH SOURCE AT $z = -\infty$

These can be constructed, for the model potential described by Eq. (4.1), in exactly the same manner as the bulk states defined by Eqs. (4.6) and (4.8). Let ψ^B represent a propagating Bloch wave of the infinite crystal with its source at $z = -\infty$ and with energy $\mathbf{E}_B > 0$. One can easily show that a corresponding state of the semi-infinite crystal exists, given by

$$
\begin{aligned}
\Psi^B &= \frac{1}{(2V)^{1/2}} \psi^B + \frac{1}{(2V)^{1/2}} \sum_{j=1}^{n} C_j^B \psi_j \qquad \text{for } z < 0 \\
&= \frac{1}{(2V)^{1/2}} \sum_{\mathbf{g}} D_{\mathbf{g}}^B \exp\left[i(\mathbf{k}_\| + \mathbf{g}) \cdot \mathbf{r}_\|\right] \phi_{\mathbf{k}_\|+\mathbf{g}}^+(z) \qquad \text{for } z > 0
\end{aligned}
\tag{4.37}
$$

where $B = (\mathbf{E}_B, \mathbf{k}_\|, \alpha)$ is the set of quantum numbers which describe both states (Ψ^B and ψ^B) in exactly the same way as for the bound bulk states (Section 4.2.1). The rest of the notation also has the same meaning as before, the only difference arising in relation to $\phi_{\mathbf{k}_\|+\mathbf{g}}^+(z)$, which replaces $\phi_{\mathbf{k}_\|+\mathbf{g}}^0(z)$ in Eq. (4.9). ϕ^+ satisfies the same differential equation [Eq. (4.10)] but a different boundary condition, given by

$$\phi_{\mathbf{q}_\|}^+(z) = \exp\left[i(2m\mathbf{E}_B/\hbar^2 - q_\|^2)^{1/2}\right] \qquad \text{for } z > z_\infty \tag{4.38}$$

Therefore, when $\mathbf{E}_B > \hbar^2 q_\|^2 / 2m$, the above expression represents a wave propagating away from the metal–vacuum interface towards $z \to +\infty$. We see that, for Ψ^B to be a scattering state extending over the entire space we must have $E_B > \hbar^2 k_\parallel^2/2m$ and $D_0^B \neq 0$. The $2n$ coefficients (C_j^B and D_g^B) in Eq. (4.37) are determined in exactly the same manner as for the bulk states [Eq. (4.6) and (4.8)] and so is the normalization of Ψ^B.

4.4.2. SCATTERING STATES WITH SOURCE AT $z = +\infty$ (LEED STATES)

Let $\exp(iq_\parallel \cdot r_\parallel)\phi_{q_\parallel}^-(z)$ represent an electronic wave, of energy E and wave vector parallel to the surface $q_\parallel$ (not necessarily within the SBZ), incident on the metal surface from the right (we assume that $E - \hbar^2 q_\parallel^2/2m > 0$). $\phi_{q_\parallel}^-(z)$ satisfies Eq. (4.10) and has the asymptotic form

$$\phi_{q_\parallel}^-(z) = \exp\left[-i(2mE/\hbar^2 - q_\parallel^2)^{1/2}z\right] \qquad \text{for } z > z_\infty \qquad (4.39)$$

Incident upon the metal surface the above wave will be partly "reflected" and partly transmitted into the metal. Because of the periodicity of the potential field parallel to the metal surface, the "reflected" wave will consist of a specular and a number of diffracted beams, and, similarly, the "transmitted" wave will consist of a superposition of Bloch waves, denoted by ψ_j in Eq. (4.41), of the same energy and the same *reduced* wave vector $k_\parallel$. We note that it is always possible to write

$$q_\parallel = k_\parallel + g' \qquad (4.40)$$

where $k_\parallel$ lies within the SBZ and g' is a particular 2D reciprocal vector of the surface under consideration. The complete wave function is given by

$$\Psi_{E,q_\parallel} = \frac{1}{(2V)^{1/2}} \exp(iq_\parallel \cdot r_\parallel)\phi_{q_\parallel}^-(z) \qquad (4.41a)$$

$$+ \frac{1}{(2V)^{1/2}} \sum_g D_{q_\parallel+g,q_\parallel} \exp\left[i(q_\parallel + g) \cdot r_\parallel\right] \phi_{q_\parallel+g}^+(z) \qquad z > 0$$

$$= \frac{1}{(2V)^{1/2}} \sum_{j=1}^n C_j \psi_j, \qquad z < 0 \qquad (4.41b)$$

In practice the sum over g includes n such vectors. The term $g = 0$ corresponds to the specularly reflected beam and the rest ($g \neq 0$) correspond to various diffracted beams. The n Bloch waves ψ_j appearing in Eq. (4.41) propagate or decay in the negative z direction. The $2n$ coefficients $D_{q_\parallel+g,q_\parallel}$ and C_j are determined by the continuity of the wave function and its derivatives at the interface (the plane $z = 0$) in the manner described in Section 4.2.1. The determination of the amplitudes $D_{q_\parallel+g,q_\parallel}$ of the diffracted beams constitutes the central problem of the theory of low energy electron diffraction (LEED) and as such it has been studied extensively (see, e.g., Pendry, 1974; Van Hove and Tong, 1979). The normalization factor $(2V)^{-1/2}$ in Eq. (4.41) is, of course, entirely unimportant as far as the LEED problem is concerned.

4.4.3. SURFACE RESONANCES (E > 0)

Within an energy gap corresponding to a given $\mathbf{k}_{\|}$ no propagating Bloch wave of the infinite crystal [ψ^B in Eq. (4.37)] exists, and therefore no scattering state Ψ^B with its source at $z = -\infty$ exists either. Let us further assume that over a region of energy within this gap $E - (\hbar^2/2m)(\mathbf{k}_{\|} + \mathbf{g})^2 < 0$, except for $\mathbf{g} = 0$. One can easily see that within the latter region a surface state may exist under certain conditions. The wave function of the surface state, if such exists, will have the form

$$
\begin{aligned}
\Psi^S &= \sum_j C_j^S \psi_j \qquad \text{for } z < 0 \\
&= \sum_{\mathbf{g}} D_{\mathbf{g}}^S \exp\left(i[\mathbf{k}_{\|} + \mathbf{g}] \cdot \mathbf{r}_{\|}\right) \phi_{\mathbf{k}_{\|}+\mathbf{g}}^+ (z) \qquad \text{for } z > 0
\end{aligned}
\tag{4.42}
$$

Continuity of the wave function and its derivative at $z = 0$ leads to a system of $2n$ homogeneous equations, so that a solution is possible only if the determinant of the coefficients in this system of equations vanishes. Suppose now that a real value of the energy exists for which this actually happens while at the same time D_0^S [in Eq. (4.42)] vanishes. The corresponding state is a surface state because all the remaining terms (corresponding to $\mathbf{g} \neq 0$) go to zero exponentially on the vacuum side of the interface and all ψ_j, being evanescent waves, go "exponentially" to zero on the metal side of the interface. Even when a truly localized state cannot be sustained a surface resonance may exist, in the sense that a scattering state $\Psi_{E,\mathbf{k}_{\|}}$, as defined by Eq. (4.41), may peak in the surface region within a narrow region of energy corresponding to the width of the resonance. The one-dimensional analog of such a resonance is shown schematically in Fig. 4.9b. Figure 4.9a shows the penetration of the wave function into the metal away from the resonance. In this example we have assumed the existence of an (absolute) energy gap in the band structure so that ψ_j [in equation (4.42)] are all evanescent waves. However, it will not be difficult for the reader to convince himself that surface resonances may exist in a resonance gap or indeed at a general point of the SBZ for exactly the same reasons that make resonances possible at energies below the vacuum level. A surface resonance of this type is shown schematically in Fig. 4.9c. It is also worth noting that a surface resonance may exist even when more than one beam (corresponding to different $\mathbf{g}$ vectors) can escape from the metal.

Surface resonances at energies above the vacuum level may be crystal induced

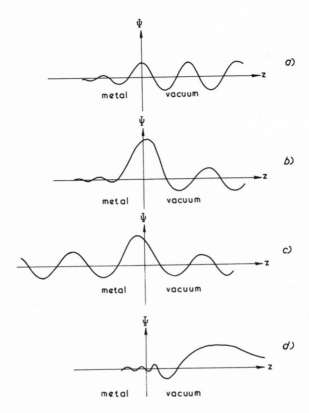

FIGURE 4.9

Schematic description of scattering states. (a) Ordinary scattering state, energy within a band-gap. (b) A surface resonance, energy within a band-gap. (c) A surface resonance, allowed-energy-region of the band structure. (d) A barrier-induced surface resonance.

in which case the wave function peaks near the metal–vacuum interface (Figs. 4.9b and 4.9c), or barrier induced, in which case the wave function peaks on the vacuum side of the interface as shown (schematically) in Fig. 4.9d. Because of the Coulombic tail of the potential, barrier induced surface resonances form a Rydberg series, similar to the Rydberg series of states in an atom. It is worth noting that because the wave function of a Rydberg state lies almost entirely on the vacuum side of the metal–vacuum interface, where the imaginary part of the potential is practically zero, their lifetime is very large. It is not, however, infinite because the wave function contains "outgoing" beams that can escape from the metal besides the "trapped" beams which are responsible for localization and the existence of the Rydberg state in the first instance. Only when the coupling between the trapped beams and the outgoing beams is sufficiently small, so that the broadening of the Rydberg levels is smaller than their separation (of the order of 0.04 eV), can the Rydberg resonances be resolved into a series of such states (Echenique and Pendry, 1978). Surface barrier resonances have been observed in LEED experiments on a number of surfaces (see, e.g., Jennings, 1978; 1979; Baribeau and Carette, 1981; Thurgate and Jennings, 1982). A review of theoretical and experimental studies on surface resonances has been written by McRae (1979).

4.5. LIFETIME EFFECTS

In the preceding discussion of the electron states of a semi-infinite crystal we have assumed that the one-electron potential, described by Eq. (4.1), is real. The corresponding one-electron states, described in Sections 4.2–4.4, represent stationary one-electron states in this potential field. We know, however, that the excited states of the crystal, which involve electrons (similarly holes) with energy away from the Fermi level, cannot be described, simply, in terms of electrons occupying well-defined (stationary) one-electron states because of the inelastic collisions, which we described in Section 3.2, and which lead to a broadening of the one-electron energy levels and a finite lifetime of the corresponding states. For example, electrons in metals with energy around the vacuum level ($E - E_F \simeq 4$ eV for a clean surface), which is the region of interest in thermionic emission, are scattered inelastically to a degree which is, already, by no means negligible and therefore cannot be neglected in a quantitative theory of thermionic emission (see Chapter 9). The same applies, of course, at the higher electron energies associated with photoemission and secondary electron emission. For this reason rigorous theories of electron emission, with the exception of field electron emission, must be constructed without any reference to stationary one-electron states, simply because such states do not exist. It turns out (see Chapters 9 and 10) that, under certain conditions, the central problem in thermionic and secondary electron emission can be reduced to that of calculating (as in LEED theory) the elastic scattering of an electron by a complex potential field (representing the semi-infinite metal)

$$V(\mathbf{r}) = V_r(\mathbf{r}) + iV_{im}(E,\mathbf{r}) \tag{4.43}$$

where the real part is given, say, by Eq. (4.1) and the imaginary part takes into account the effect of inelastic collisions on the elastic channel (electrons are removed from the elastic channel because of inelastic scattering). Sometimes, e.g., in LEED calculations at higher energies, above 50 eV or so, it is assumed that

$$V_{im}(E,\mathbf{r}) = V_{im}(E) \quad \text{for } z < 0$$
$$= 0 \quad \text{for } z < 0 \tag{4.44}$$

At lower electron energies (below 20 eV or so) it is necessary (see, e.g., Chapter 9) to take into account inelastic scattering on the vacuum side of the interface as well. {The use of a complex potential in the solution of a practical problem can be readily demonstrated in the case of LEED by reference to Eq. (4.41). The equation remains perfectly valid when the real potential inside the metal is replaced by the complex potential [Eqs. (4.43) and (4.44)]. Now of course, $\Psi_{E,q_{\parallel}}$ does not represent a stationary state. This, however, is irrelevant to the LEED problem, which can be stated as follows: Given an incident electron beam [described by the first term in Eq. (4.41a)], what are the values of the amplitudes of the diffracted beams [denoted by D_g in Eq. (4.41a)]? We note that the Bloch waves ψ_j are obtained in exactly the same way (described in Chapter 3) whether the potential is real or complex. Of course, when the potential is complex, *all* waves are evanescent waves.}

The question arises: If the excited states described in this chapter (corresponding to a real potential) do not correspond to actual one-electron states in a real crystal why bother with them at all? The answer is, of course, partly historical. Quite a lot of effort has gone into the calculation of the energy-band-structure of metals and other crystals on the basis of a real one-electron potential. These calculations are, of course, perfectly valid in relation to ground-state properties such as the shape of the Fermi surface in **k** space of a metal, and they provide a firm foundation for at least a semiquantitative understanding of various transport properties (see, e.g., Harrison, 1970). At the same time, band-structure calculations usually provide us with a lot of information about the empty bands (corresponding to excited one-electron states in the given potential). Traditionally optical absorption by metals and other crystals has been analyzed, at least on a semiquantitative level, in terms of one-electron transitions from an occupied to an empty state (see, e.g., Harrison, 1970). More recently, measured energy and angular distributions of photoemitted electrons (see, e.g., Eastman and Himpsel, 1981) and of secondary electrons (Chapter 10) have also been interpreted on a semiquantitative level in terms of calculated band structures of one-electron levels. It seems that lifetime effects modify but do not destroy the one-electron picture. We can best describe this in terms of the density of (one-electron) states as defined by Eq. (4.24). We have seen that a finite lifetime is by the uncertainty principle associated with a broadening of the one-electron level and that this can be taken into account by adding an imaginary part to the real one-electron potential. Accordingly, a sharp (in energy) state (in the infinite crystal) is replaced by a normalized Lorenzian curve with full width at half maximum equal to $2\Gamma = -2V_{im}(E)$, where $V_{im}(E)$ is the imaginary component of the potential in the bulk of the metal, which in turn leads to a modified local density of states given by

$$\bar{\rho}(E,\mathbf{r}) = \frac{1}{\pi} \int \frac{\Gamma}{(E' - E)^2 + \Gamma^2} \rho(E',\mathbf{r}) \, dE' \qquad (4.45)$$

where $\rho(E,\mathbf{r})$ is the local density of states in the absence of inelastic collisions (real potential). A similar formula is obtained for the density of states associated with a given $\mathbf{k}$ (for an infinite crystal) or a given $\mathbf{k}_{\parallel}$ in a semi-infinite crystal. It is clear that sharp edges between allowed regions of energy and gaps will be smoothed out, and so will sharp peaks in the density of states, but they will not necessarily disappear.

FIELD EMISSION SPECTROSCOPY OF METALS

5.1. INTRODUCTION

A theory of field emission which takes into account the energy-band-structure of the metal emitter was initially proposed by Itskovich (1966, 1967). The next significant step was made by Politzer and Cutler (1972). Their calculation of the spin-polarized field emission current from ferromagnetic nickel (100) constitutes the first serious attempt to include in a quantitative calculation of this kind the energy band structure of the metal and the nature of the corresponding electronic wave functions (Bloch waves). A critical review of theoretical work relating to field emission up to the year 1972 can be found in the article of Gadzuk and Plummer (1973). Further progress was made when it was realized (Nicolaou and Modinos, 1975; Modinos and Nicolaou, 1976) that a method can be established, using the mathematical apparatus developed by LEED theorists, which allows one to calculate the energy distribution of the field-emitted electrons (the current is simply the integral of this quantity) accurately and efficiently, at least for some models of the metal–vacuum interface. The theory to be presented here is based on the above-mentioned papers and on subsequent work by the author (Modinos, 1976). The following review articles are also directly relevant to the subject matter of this chapter: Plummer (1975); Soven, Plummer, and Kar (1976); Feuchtwang and Cutler (1976); Modinos (1978a).

5.2. ENERGY DISTRIBUTION OF THE EMITTED ELECTRONS. BASIC FORMULAS

We assume that the one-electron potential, when an electric field appropriate to a field emission experiment is applied to the surface, is given by Eq. (4.1) with $V_s(z)$ replaced by

$$V_F(z) = V_s(z) - eFz \qquad (5.1)$$

where F denotes the magnitude of the field. Equation (5.1) implies that the field lines terminate on the metal–vacuum interface at $z = 0$. [We note that this plane is also the "image plane," i.e., $V_s(z) \simeq E_F + \phi - e^2/4z$ as $z \to \infty$.] This is of course an approximation. In reality the screening charge is not distributed on a plane but extends over a finite thickness, of the order of ½Å or so. Its distribution in space and the corresponding variation (field-induced) of the surface potential near the interface must in principle be calculated self-consistently. However, explicit numerical calculations of the field-induced variation in the potential, based on a jellium model of the surface (see, e.g., Theophilou and Modinos, 1972; Lang and Kohn, 1973), have shown that for a metal and for applied fields commonly used in field emission experiments ($F \lesssim 0.5$ F/Å), the difference between the actual potential and that given by Eq. (5.1) is rather small and that no significant error is introduced into the calculation of the energy distribution of the emitted electrons by neglecting it. In any case, introducing a small field-induced correction to the potential is not meaningful when the *zero-field* potential itself has not been evaluated self-consistently.

The wave function of a one-electron state of the semi-infinite crystal at the metal–vacuum interface (at $z = 0$) where the potential is constant (the region of constant potential at $z = 0$ can be infinitesimally small) has the following general form:

$$\Psi^M = \sum_g [A_g^{M+} \exp(i\mathbf{K}_g^+ \cdot \mathbf{r}) + A_g^{M-} \exp(i\mathbf{K}_g^- \cdot \mathbf{r})] \tag{5.2}$$

where $\mathbf{K}_g^{\pm}$ are the vectors defined by Eq. (3.39), and M denotes as usual the set of quantum numbers, which includes the energy E^M and $\mathbf{k}_{\parallel}$ but not the spin, which characterize this state. Because (by assumption) the surface barrier (the potential on the vacuum side of the interface) depends only on z, when a plane wave is scattered by it, its transverse momentum is conserved. Therefore A_g^{M-} in Eq. (5.2) is related to A_g^{M+} in the same equation as follows:

$$A_g^{M-} = R_s(E^M, \mathbf{k}_{\parallel} + g) A_g^{M+} \tag{5.3}$$

The reflection matrix element, $R_s(E, \mathbf{k}_{\parallel} + g)$, depends only on the surface barrier and can be evaluated numerically, if need be, without much difficulty. When no external field is applied to the surface the wave function Ψ^M decays exponentially on the vacuum side of the interface (only bound states are of interest in field emission) and has the following general form:

$$\Psi^M = \sum_g D_g^M \Phi_{\mathbf{k}_{\parallel}+g}^0(\mathbf{r}), \qquad z > 0 \tag{5.4}$$

where $\Phi_{\mathbf{k}_{\parallel}+g}^0$ is the function defined by Eqs. (4.9)–(4.11).

Now, for temperatures and applied fields within the field emission region as defined by Eqs. (1.55) and (1.56) (see, also, Fig. 1.4), the electrons which contribute to the emitted (tunneling) current have energies well below the top of the surface barrier described by Eq. (5.1). At these energies the reflection matrix element $R_s(E,\mathbf{k}_\| + \mathbf{g})$, when the field is on, is not significantly different from its zero-field value for it is essentially determined by that region of the barrier (near the interface) where the field term is small. [We note that the tunneling probability at these energies for typical values of the parameters ($\Phi \simeq 4.5$ eV; $F \simeq 0.4$ V/Å) is very small, of the order of 10^{-6} or so]. In what follows we assume that $R_s(E,\mathbf{k}_\| + \mathbf{g})$ and therefore $A_\mathbf{g}^{M\pm}$ in Eq. (5.2) are practically independent of the applied field. In other words the electronic wave functions and the surface density of states in the relevant energy region are not in any significant way affected by the application of the field. [We note, however, that, if need be, the dependence of $R_s(E,\mathbf{k}_\| + \mathbf{g})$ on the field can be taken into account without much difficulty.] On the other hand, when the electric field is applied to the surface the asymptotic form of the wave function as $z \to \infty$ is very much different from its zero field expression [Eq. (5.4)]. We now have

$$\Psi^M = \sum_\mathbf{g} B_\mathbf{g}^M \Phi_{\mathbf{k}_\|+\mathbf{g}}(\mathbf{r}) \tag{5.5}$$

$$\Phi_{q\|}(\mathbf{r}) = \exp(i\mathbf{q}_\| \cdot \mathbf{r}_\|)\phi_{q\|}(z) \tag{5.6}$$

$$\phi_{q\|}(z) = \frac{1}{[\lambda_{q\|}(z)]^{1/2}} \exp\left[i \int_{bq\|}^z \lambda_{q\|}(z)\,dz\right], \qquad z \gg b_{q\|} \tag{5.7}$$

$$\lambda_{q\|}(z) = \left|\frac{2m}{\hbar^2}[E^M - V_F(z)] - q_\|^2\right|^{1/2} \tag{5.8}$$

$b_{q\|}$ in Eq. (5.7) denotes the larger of the two roots (classical turning points) of the equation $\lambda_{q\|}(z) = 0$. The asymptotic expression given by Eq. (5.7) is the WKB solution of the appropriate Schrödinger equation in the region $z \gg b_{q\|}$ (see Section 1.2), and represents an electronic wave propagating in the positive z direction. Obviously

$$B_\mathbf{g}^M = T_s(E^M,\mathbf{k}_\| + \mathbf{g})A_\mathbf{g}^{M+} \tag{5.9}$$

where $A_\mathbf{g}^{M+}$ are the coefficients appearing in Eq. (5.2) and $T_s(E,\mathbf{k}_\| + \mathbf{g})$ is a transmission matrix element, which depends only on the surface barrier $V_F(z)$. We can obtain $T_s(E,\mathbf{k}_\| + \mathbf{g})$ for any surface barrier, which depends only on z, by numerically integrating the appropriate Schrödinger equation in the manner described in Section 1.2. The contribution to the (tunneling) current density from the state Ψ^M, assuming that this state is occupied at the given temperature, is given by

$$j_M = \lim_{A\to\infty}\left[\frac{1}{A}\iint_A \left(\frac{\hbar}{2mi}\right)\left(\Psi^{M*}\frac{\partial}{\partial z}\psi^M - \Psi^M\frac{\partial}{\partial z}\Psi^{M*}\right)_{z\to\infty} d^2r_\|\right] \tag{5.10}$$

where A is the area of the emitting surface. We assume that in practice this area is much larger than the surface unit cell (see Section 2.4). Substituting Eq. (5.5) into Eq. (5.10) and using Eqs. (5.3) and (5.9) we obtain

$$j_M = \frac{\hbar}{m} \sum_{\mathbf{g}} |A_{\mathbf{g}}^{M+} + A_{\mathbf{g}}^{M-}|^2 \frac{|T_s(E^M, \mathbf{k}_\parallel + \mathbf{g})|^2}{|1 + R_s(E^M, \mathbf{k}_\parallel + \mathbf{g})|^2} \qquad (5.11)$$

The total energy distribution of the field-emitted electrons is given by

$$j(E) = 2f(E) \sum_M j_M \, \delta(E - E^M) \qquad (5.12)$$

where $f(E)$ is the Fermi–Dirac distribution function [Eq. (1.5)] and the factor of 2 takes into account spin degeneracy. We emphasize that Eq. (5.11) is valid quite generally subject only to our assumption that the metal–vacuum interface (the plane $z = 0$) can be chosen in such a way that the surface barrier (the potential for $z > 0$) depends only on z. It is important to note in this respect that, so far, no reference has been made to the form of the potential inside the metal ($z < 0$) other than its 2D periodicity parallel to the surface. Our assumption that a slot of constant potential exists at $z = 0-$ is no restriction of generality.

In order to clarify the relation between the energy distribution of the emitted electron and the surface density of states we shall now apply the above formulas to the model potential defined at the beginning of Section 4.2 [Eqs. (4.1); $D_s = 0$]. We have seen how the coefficients $A_{\mathbf{g}}^{M\pm}$ appearing in Eq. (5.11) can be calculated for this model. For a bulk state [$M = B = (E^B, \mathbf{k}_\parallel, \alpha)$] these are given, according to Eqs. (4.6), (4.12), and (4.13) by

$$A_{\mathbf{g}}^{M\pm} = (2V)^{-1/2} A_{\mathbf{g}}^{B\pm} \qquad (5.13)$$

$$A_{\mathbf{g}}^{B\pm} = (U_{1\mathbf{g}}^{B\pm} + \sum_j C_j^B U_{1\mathbf{g}}^{j\pm}) \qquad (5.14)$$

and for a surface state [$M = S = (\mathbf{k}_\parallel, \mu)$], according to Eqs. (4.13) and (4.20), by

$$A_{\mathbf{g}}^{M\pm} = A^{-1/2} A_{\mathbf{g}}^{S\pm} \qquad (5.15)$$

$$A_{\mathbf{g}}^{S\pm} = \sum_j C_j^S U_{1\mathbf{g}}^{j\pm} \qquad (5.16)$$

It is assumed that the wave functions Ψ^B and Ψ^S are normalized in the manner described in Sections 4.2.1 and 4.2.2, respectively. Therefore, $A_{\mathbf{g}}^{B\pm}$ are dimensionless quantities and $A_{\mathbf{g}}^{S\pm}$ have a dimension of (length)$^{-1/2}$. Substituting the above expressions in Eq. (5.11), and transforming the sum over bulk and surface states in Eq. (5.12) into integrals [as in Eqs. (4.19) and (4.29)], we obtain

$$j(E) = j_b(E) + j_s(E)$$

$$j_b(E) = \frac{\hbar f(E)}{4\pi^3 m} \iint_{SBZ} d^2 k_{\parallel} \left[\sum_{\alpha} \left(\frac{\partial k_{\alpha z}}{\partial E} \right) \sum_{\mathbf{g}} |A_{\mathbf{g}}^{B+} + A_{\mathbf{g}}^{B-}|^2 \right]_{E^B = E}$$

$$\times \frac{|T_s(E, \mathbf{k}_{\parallel} + \mathbf{g})|^2}{|1 + R_s(E, \mathbf{k}_{\parallel} + \mathbf{g})|^2}$$ (5.18)

$$j_s(E) = \frac{\hbar f(E)}{2\pi^2 m} \iint_{SBZ} d^2 k_{\parallel} \sum_{\mu} \delta(E - E_{\mu}(\mathbf{k}_{\parallel})) \sum_{\mathbf{g}}$$

$$\times |A_{\mathbf{g}}^{S+} + A_{\mathbf{g}}^{S-}|^2 \frac{|T_s(E, \mathbf{k}_{\parallel} + \mathbf{g})|^2}{|1 + R_s(E, \mathbf{k}_{\parallel} + \mathbf{g})|^2}$$ (5.19)

j_b and j_s are the contributions to the total energy distribution of the emitted electrons from bulk and (true) surface states, respectively. The sum over α in Eq. (5.18) has the same meaning as in Eq. (4.19). It is important to remember that for given E there are sections of the SBZ over which no bulk state exists and that there the integrand in Eq. (5.18) vanishes identically.

We emphasize that in our derivation of Eq. (5.19) we have assumed, although this was not explicitly stated, that when an electron tunnels out from a surface state it is *automatically* replaced by another electron from the bulk metal, so that at any moment the occupation of a surface state, like that of the bulk states, is determined by the Fermi–Dirac distribution function. This is a reasonable approximation for metal surfaces where the lifetime of a "hole" in the surface band is generally much smaller than the lifetime, with respect to tunneling, of an electron in the same band (Gadzuk, 1972). This is not necessarily the case in other situations, e.g., in field emission from semiconductor surfaces (see Section 8.6).

It turns out that for typical values of the work function ($\phi \simeq 4.5$ eV) and of the applied field ($F \simeq 0.4$ V/Å), the transmission matrix element $T_s(E, \mathbf{k}_{\parallel} + \mathbf{g})$ for $\mathbf{g} \neq 0$ is so small, that one can drop the corresponding terms in Eqs. (5.18) and (5.19). One can demonstrate this quite easily for the case of the triangular barrier, defined by

$$V_F(z) = E_F + \phi - eFz, \qquad z > 0$$
$$= 0, \qquad z < 0$$ (5.20)

The solution of the Schrödinger equation for the above barrier, which represents a wave propagating in the positive z direction as $z \to \infty$, can be obtained analytically in terms of the well-known Airy functions (see, e.g., Gadzuk and Plummer, 1973). It turns out that for electron energies in the vicinity of the Fermi level,

$$T_s(E,\mathbf{k}_\| + \mathbf{g}) \simeq \exp\left(\frac{\pi i}{4}\right) \sqrt{\eta}\,[1 + R_s(E,\mathbf{k}_\| + \mathbf{g})] \tag{5.21}$$

$$\times \exp\left\{-\frac{2}{3}\left(\frac{2m}{\hbar^2 e^2 F^2}\right)^{1/2}\left[\phi - \epsilon + \frac{\hbar^2}{2m}(\mathbf{k}_\| + \mathbf{g})^2\right]^{3/2}\right\}$$

$$R_s(E,\mathbf{k}_\| + \mathbf{g}) \simeq \frac{\eta + i\zeta}{\eta - i\zeta} \tag{5.22}$$

$$\eta = [(2m/\hbar^2)(\phi - \epsilon) + (\mathbf{k}_\| + \mathbf{g})^2]^{1/2} \tag{5.23}$$

$$\zeta = [(2m/\hbar^2)(E_F + \epsilon) - (\mathbf{k}_\| + \mathbf{g})^2]^{1/2} \tag{5.24}$$

where $\epsilon = E - E_F$. It is obvious from Eq. (5.21) that for the triangular barrier the transmission probability, which is proportional to $|T_s(E,\mathbf{k}_\| + \mathbf{g})|^2$, diminishes rapidly as the magnitude of $(\mathbf{k}_\| + \mathbf{g})$ increases. The above conclusion remains valid for any reasonable shape of the surface barrier. We may, therefore, conclude that practically the whole of the tunneling current comes from states with $\mathbf{k}_\|$ in the central region of the SBZ and that only the $\mathbf{g} = 0$ components of the corresponding wave functions contribute to this current. We have

$$j_b(E) = \frac{\hbar f(E)}{4\pi^3 m} \iint_{\substack{\text{central}\\\text{SBZ}}} d^2 k_\| \left[\sum_\alpha \left(\frac{\partial k_{\alpha z}}{\partial E}\right)|A_0^{B+} + A_0^{B-}|^2\right]_{E^B = E}$$

$$\times \frac{|T_s(E,\mathbf{k}_\|)|^2}{|1 + R_s(E,\mathbf{k}_\|)|^2} \tag{5.25}$$

$$j_s(E) = \frac{\hbar f(E)}{2\pi^2 m} \iint_{\substack{\text{central}\\\text{SBZ}}} d^2 k_\| \sum_\mu \delta(E - E_\mu(\mathbf{k}_\|))|A_0^{S+} + A_0^{S-}|^2$$

$$\times \frac{|T_s(E,\mathbf{k}_\|)|^2}{|1 + R_s(E,\mathbf{k}_\|)|^2} \tag{5.26}$$

In practice, the limits of the $\mathbf{k}_\|$ integration may vary slightly from one case to the other but it is reasonable to assume that in most cases the "central SBZ" coincides with $0 \le |\mathbf{k}_\|| \le 0.5$ Å^{-1} or so. We note that the SBZ of the low-index planes of transition metals, which are most commonly used in field emission experiments, extends well beyond this "central region." It is only in such cases that one may replace Eqs. (5.18) and (5.19) by Eqs. (5.25) and (5.26), respectively.

One can easily show that for the model under consideration the average density of states at the interface ($z = 0$) as defined by Eqs. (4.30) and (4.31) is given by

$$\langle \rho(E,z = 0) \rangle = \langle \rho_b(E,z = 0) \rangle + \langle \rho_s(E,z = 0) \rangle \qquad (5.27)$$

$$\langle \rho_b(E,z = 0) \rangle = \frac{1}{4\pi^3} \iint_{\text{SBZ}} d^2k_{\|} \left[\sum_{\alpha} \left(\frac{\partial k_{\alpha z}}{\partial E} \right) \sum_{g} |A_g^{B+} + A_g^{B-}|^2 \right]_{E^B = E} \qquad (5.28)$$

$$\langle \rho_s(E,z = 0) \rangle = \frac{1}{2\pi^2} \iint_{\text{SBZ}} d^2k_{\|} \sum_{\mu} \delta(E - E_\mu(\mathbf{k}_{\|})) \sum_{g} |A_g^{S+} + A_g^{S-}|^2 \qquad (5.29)$$

At this stage we would like to introduce an additional quantity, a partial density of states, which we define as follows:

$$\langle \rho(E,z = 0) \rangle_{c;0} = \langle \rho_b(E,z = 0) \rangle_{c;0} + \langle \rho_s(E,z = 0) \rangle_{c;0} \qquad (5.30)$$

$$\langle \rho_b(E,z = 0) \rangle_{c;0} = \frac{1}{4\pi^3} \iint_{\substack{\text{central} \\ \text{SBZ}}} d^2k_{\|} \left[\sum_{\alpha} \left(\frac{\partial k_{\alpha z}}{\partial E} \right) |A_0^{B+} + A_0^{B-}|^2 \right]_{E^B = E} \qquad (5.31)$$

$$\langle \rho_s(E,z = 0) \rangle_{c;0} = \frac{1}{2\pi^2} \iint_{\substack{\text{central} \\ \text{SBZ}}} d^2k_{\|} \sum_{\mu} \delta(E - E_\mu(\mathbf{k}_{\|})) |A_0^{S+} + A_0^{S-}|^2 \qquad (5.32)$$

$\langle \rho(E,z = 0) \rangle_{c;0}$ represents the contribution to $\langle \rho(E,z = 0) \rangle$ from the $\mathbf{g} = 0$ component of the wave function of all states with $\mathbf{k}_{\|}$ in the *central* SBZ.

One can easily see by comparing Eqs. (5.25) and (5.26) with Eqs. (5.31) and (5.32) that, any structure in $j(E)$ or, more specifically, in the enhancement factor

$$R(E) = \frac{j(E)}{j_0(E)} \qquad (5.33)$$

as defined by Eq. (1.93), implies a corresponding structure in $\langle \rho(E,z = 0) \rangle_{c;0}$. We note that $j_0(E)$, which we can write in terms of $T_s^0(E,\mathbf{k}_{\|})$ (calculated for an image barrier in the WKB approximation) as follows:

$$j_0(E) = \frac{f(E)}{4\pi^3\hbar} \iint_{\substack{\text{central} \\ \text{SBZ}}} d^2k_{\|} \frac{|T_s^0(E,\mathbf{k}_{\|})|^2}{[(2m/\hbar^2)(E - V_0) - k_{\|}^2]^{1/2}} \qquad (5.34)$$

is a smooth function of the energy (see, e.g., Fig. 1.8a). The above is demonstrated quite clearly in Fig. 5.1 which shows $R(E)$ (solid line) and $\langle \rho(E,z = 0) \rangle_{c;0}$ (broken line) for a model tungsten (100) emitter (Modinos, 1978; Nicolaou and Mod-

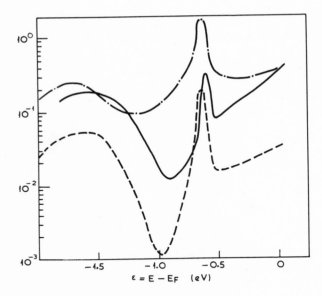

FIGURE 5.1

The enhancement factor R (solid line); $\langle\rho(E,z = 0)\rangle_{c;0}$ as defined by Eq. (5.30) in 10^{-2} eV^{-1} Å^{-3} (broken line); $\langle\rho(E, z = 0)\rangle_c$ as defined by Eq. (5.35) in 10^{-2} eV^{-1} Å^{-3} (dash–dot line) for a model tungsten (100) emitter.

inos, 1976).† We see from this figure that the structure in $R(E)$ (solid line) parallels very closely that of $\langle\rho(E,z = 0)\rangle$ (broken line). The dash-dot line in the same figure represents the contribution $_{c;0}$ ⟩ to $\langle\rho(E,z = 0)\rangle$, as defined by Eq. (5.27), from the states within the central region of the SBZ, i.e.,

$$\langle\rho(E,z = 0)\rangle_c = \frac{1}{4\pi^3} \iint_{\substack{\text{central}\\\text{SBZ}}} d^2k_\parallel \left[\sum_\alpha \left(\frac{\partial k_{\alpha z}}{\partial E}\right) \sum_g |A_g^{B+} + A_g^{B-}|^2 \right]_{E^B = E}$$

$$+ \frac{1}{2\pi^2} \iint_{\substack{\text{central}\\\text{SBZ}}} d^2k_\parallel \sum_\mu \delta(E - E_\mu(\mathbf{k}_\parallel)) \sum_g |A_g^{S+} + A_g^{S-}|^2 \qquad (5.35)$$

We note that $\langle\rho(E,z = 0)\rangle_c$ has the same, or almost the same, dependence on energy as $\langle\rho(E,z = 0)\rangle_{c;0}$ and $R(E)$, the main features of all three curves being a sharp peak at $\epsilon = E - E_F \simeq -0.65$ and a broad peak at $\epsilon \simeq -1.5$ eV.

---

†The fact that the calculated $R(E)$ for this model of W(100) does not agree well with the experimental $R(E)$ for W(100) (see Section 5.4.1) does not invalidate our conclusion concerning the general relation between $R(E)$ and $\langle\rho(E,z = 0)\rangle_{c;0}$.

5.2.1. A GREEN'S FUNCTION METHOD FOR CALCULATING THE TOTAL ENERGY DISTRIBUTION OF THE EMITTED ELECTRONS

Equations (5.25) and (5.26) are very useful not least because they make the physics of the field emission process transparent, but it must be realized that the computation of the quantitites entering these equations, namely $(\partial k_{\alpha z}/\partial E)|A_0^{B+} + A_0^{B-}|^2$ in Eq. (5.25) and $E_\mu(\mathbf{k}_\parallel)$ and $|A_0^{S+} + A_0^{S-}|^2$ in Eq. (5.26) is not an easy task. We have given explicit formulas for the evaluation of these quantities only for a specific model potential [that described by Eqs. (4.1) and (5.1)]. Even for this potential the computation of the energy distribution of the emitted electrons on the basis of Eqs. (5.25) and (5.26) may be prohibitively long. One must first evaluate the complex band structure and Bloch waves of the infinite crystal in the manner described in Sections 3.4 and 3.5. Then the relevant electron states of the semi-infinite crystal must be constructed in the manner described in Section 4.2. The computation of the bulk states of the semi-infinite crystal, and therefore of $(\partial k_{\alpha z}/\partial E)|A_0^{B+} + A_0^{B-}|^2$, is relatively straightforward once the Bloch waves of the infinite crystal have been computed, but this is not the case for surface states. If there are absolute gaps in the central SBZ one must search for possible surface states in these gaps (at least over the energy region that matters in field emission). If such states exist the corresponding wave functions must then be normalized according to Eq. (4.23) in order to determine the coefficients $A_0^{S\pm}$ which enter Eq. (5.26). This makes a long and laborious calculation longer and more difficult. In this section we describe an alternative method of calculation (Modinos, 1976) which avoids some of these difficulties. This method applies equally well when the potential (of the muffin-tin type) in the top layer (or layers) is different from that of the bulk layers. The formalism which we describe in this and the following two sections applies when the potential in the interface region [between the muffin-tin atoms of the top layer and the interface (the plane $z = 0$), which is taken at a distance $\frac{1}{2}d + D_s$ from the center of the top atomic layer; see comments following Eq. (4.1)] is constant and equal to the interstitial potential in the bulk of the metal. Although the approximation of constant potential between the muffin-tin atoms in the bulk of the metal appears to be satisfactory (Moruzzi, Janak, and Williams, 1978) a constant potential in the interface region may not be a good approximation at least for some (relatively open) surfaces. In principle, one can treat the region of the interface as an additional layer (within the doubling layer scheme to be described in Section 5.3), provided an efficient way of calculating the scattering matrix elements of such a layer can be established. This has not been done as yet but it is certainly not an impossible problem. We note that our assumption that the potential depends only on z for $z > 0$ does not constitute an approximation when the plane $z = 0$ is placed at a reasonable distance from the center of the top layer. [It is worth noting, however, that the "image plane" in a self-consistent evaluation of the potential may not coincide with the interface plane ($z = 0$), in which case the asymptotic form of the barrier may be different from that in Eq. (5.1); e.g., the field term will be given by $-eF(z - z_{im})$ for large z if the image plane occurs at $z = z_{im}$.]

We have already pointed out that Eqs. (5.11) and (5.12) are valid quite generally [see comments following Eq. (5.12)]. We note that $|A_g^{M+} + A_g^{M-}|^2$ in Eq. (5.11) is equal to

$$|A_g^{M+} + A_g^{M-}|^2 = (1/A)\langle\chi_{k_\parallel+g}|\Psi^M\rangle\langle\Psi^M|\chi_{k_\parallel+g}\rangle \tag{5.36}$$

where $\chi_{q_\parallel}(\mathbf{r})$ is defined as follows:

$$\chi_{q_\parallel}(\mathbf{r}) = \frac{1}{\sqrt{A}}\exp(i\mathbf{q}_\parallel \cdot \mathbf{r}_\parallel)\,\delta(z) \tag{5.37}$$

A denotes the area of the surface, and

$$\langle f|g\rangle = \int f^*(\mathbf{r})g(\mathbf{r})\,d^3r$$

as usual. Equation (5.36) follows from the assumed form [Eq. (5.2)] of the wave function at $z = 0$. (Since an infinitesimal slot at $z = 0$ where the potential is constant can be assumed without any loss of generality, the wave function can always be written in that form.) Substituting Eq. (5.36) into Eq. (5.11) and using the resulting expression for j_M in Eq. (5.12) we obtain

$$j(E) = 2f(E)\left(\frac{\hbar}{m}\right)\frac{1}{A}\sum_{k'_\parallel}\sum_g\sum_M \delta_{k_\parallel k'_\parallel}\,\delta(E - E^M)$$

$$\times \langle\chi_{k'_\parallel+g}|\Psi^M\rangle\langle\Psi^M|\chi_{k'_\parallel+g}\rangle \frac{|T_s(E,k'_\parallel + g)|^2}{|1 + R_s(E,k'_\parallel + g)|^2} \tag{5.38}$$

where $\mathbf{k}_\parallel$ and $\mathbf{k}'_\parallel$ lie in the SBZ and

$$\delta_{k_\parallel k'_\parallel} = 1 \qquad \text{for } \mathbf{k}_\parallel = \mathbf{k}'_\parallel \tag{5.39}$$
$$\qquad\quad = 0 \qquad \text{for } \mathbf{k}_\parallel \neq \mathbf{k}'_\parallel$$

We remember that M, the set of quantum numbers which characterize the Mth state, includes the reduced wave vector $\mathbf{k}_\parallel$ of that state. It follows that $\langle\chi_{k'_\parallel+g}|\Psi^M\rangle = 0$ when $\mathbf{k}'_\parallel \neq \mathbf{k}_\parallel$ and that, therefore, $\delta_{k_\parallel k'_\parallel}$ in Eq. (5.38) can be omitted. Using the identity (see, e.g., Roman, 1965)

$$\frac{1}{E - E^M + i\bar{\epsilon}} = P \frac{1}{E - E^M} - i\pi \, \delta(E - E^M) \tag{5.40}$$

where as usual $\bar{\epsilon}$ denotes a small (infinitesimal) positive energy and "$P \ldots$" means "Principal part of $\ldots$," the Green's function defined by

$$G(E; \mathbf{r}, \mathbf{r}') = \sum_M \frac{\Psi^M(\mathbf{r}) \Psi^{M*}(\mathbf{r}')}{E - E^M + i\bar{\epsilon}} \tag{5.41}$$

and replacing the double sum $A^{-1} \Sigma_{\mathbf{k}'_\parallel} \Sigma_\mathbf{g}$ in Eq. (5.38) by the double integral

$$(4\pi^2)^{-1} \iint_{-\infty} d^2 q_\parallel$$

we finally obtain

$$j(E) = - \left(\frac{\hbar}{2m\pi^3} \right) f(E) \iint_{-\infty}^{+\infty} d^2 q_\parallel \, \mathrm{Im} \langle \chi_{\mathbf{q}\parallel} | G(E) | \chi_{\mathbf{q}\parallel} \rangle \frac{|T_s(E, \mathbf{q}_\parallel)|^2}{|1 + R_s(E, \mathbf{q}_\parallel)|^2} \tag{5.42}$$

where "$\mathrm{Im} \langle \cdots \rangle$" means "imaginary part of $\langle \cdots \rangle$," and

$$\langle \chi_{\mathbf{q}\parallel} | G(E) | \chi_{\mathbf{q}\parallel} \rangle \equiv \iint \chi_{\mathbf{q}\parallel}^*(\mathbf{r}) G(E, \mathbf{r}, \mathbf{r}') \chi_{\mathbf{q}\parallel}(\mathbf{r}') \, d^3 r \, d^3 r' \tag{5.43}$$

$T_s(E, \mathbf{q}_\parallel)$ and $R_s(E, \mathbf{q}_\parallel)$ can easily be evaluated numerically for any surface barrier, hence the problem of evaluating the energy distribution of the emitted electrons is reduced to that of evaluating the matrix element defined by Eq. (5.43). One can think of the quantity $-(1/\pi) f(E) \, \mathrm{Im} \langle \chi_{\mathbf{q}\parallel} | G(E) | \chi_{\mathbf{q}\parallel} \rangle$ as the average density of electrons at the interface ($z = 0$) with energy E and transverse momentum $\hbar \mathbf{q}_\parallel$. We note that the integral in Eq. (5.42) is over all $\mathbf{q}_\parallel$ space. In practice, however, only the central region of the SBZ contributes significantly to this integral for the reasons already explained in the previous section [see comments following Eq. (5.24)]. The $\mathbf{q}_\parallel$ integration is further reduced if one makes use of the symmetry of the crystal.

We note that within the present formalism emission from surface states is treated on the same footing as emission from bulk states. The contribution from surface states, if such exist, is automatically taken into account in the evaluation of the Green's function G at the metal–vacuum interface. We have of course assumed that the occupation of all states under conditions appropriate to a field emission experiment is the same as for equilibrium conditions [see comments following Eq. (5.19)].

5.2.2. CALCULATION OF $\langle \chi_{q_{\parallel}} | G(E) | \chi_{q_{\parallel}} \rangle$

In this section we derive a formula for the above matrix element which is valid for a potential field of the following general form:

$$V(\mathbf{r}) = V_0 + V'(\mathbf{r}) \tag{5.44}$$

$$\begin{aligned} V'(\mathbf{r}) &= V_s(z), \quad z > 0 \\ &= V_c(\mathbf{r}), \quad z < 0 \end{aligned} \tag{5.45}$$

where V_0 is a constant. We assume that the potential on the right of the interface (the plane $z = 0$) depends only on z. The potential to the left of the interface has a 2D periodicity in the xy plane, but, at present, no other restriction is imposed on it. We also assume that a region exists around $z = 0$ (it can be a slot of infinitesimal thickness) where $V_c = V_s = 0$.

The Green's function $G_0(E,\mathbf{r},\mathbf{r}')$, as defined by Eq. (5.41), for an electron in the constant potential field V_0 can be written in the following form:

$$\begin{aligned} G_0(E;\mathbf{r},\mathbf{r}') = &-\frac{i}{A} \sum_{q_{\parallel}} \frac{m}{\hbar^2(K_0^2 - q_{\parallel}^2)^{1/2}} \exp\{i[\mathbf{q}_{\parallel} \cdot (\mathbf{r}_{\parallel} - \mathbf{r}'_{\parallel}) \\ &+ (K_0^2 - q_{\parallel}^2)^{1/2}|z - z'|]\} \end{aligned} \tag{5.46}$$

where K_0 is given by Eq. (3.71). The Green's function for an electron in the potential field $V(\mathbf{r})$ is given by†

$$\begin{aligned} G(E;\mathbf{r},\mathbf{r}') = &\ G_0(E;\mathbf{r},\mathbf{r}') + \int G_0(E;\mathbf{r},\mathbf{r}'')V'(\mathbf{r}'')G_0(E;\mathbf{r}'',\mathbf{r}')\, d^3\mathbf{r}'' \\ &+ \int G_0(E;\mathbf{r},\mathbf{r}''')V'(\mathbf{r}''')G_0(E;\mathbf{r}''',\mathbf{r}'')V'(\mathbf{r}'')G_0(E;\mathbf{r}'',\mathbf{r}')\, d^3r'''d^3r' + \cdots \end{aligned} \tag{5.47}$$

†Eq. (5.47) can be proved as follows. From the definition of the Green's function [Eq. (5.41)], it follows that

$$(E - H + i\bar{\epsilon})G(E;\mathbf{r},\mathbf{r}') = \delta(\mathbf{r} - \mathbf{r}') \tag{i}$$
$$(E - H_0 + i\epsilon)G_0(E;\mathbf{r}',\mathbf{r}) = \delta(\mathbf{r} - \mathbf{r}') \tag{ii}$$

where $\delta(\mathbf{r} - \mathbf{r}')$ is the Dirac delta function, and

$$H_0 = -\frac{\hbar^2}{2m}\nabla^2 + V_0$$
$$H = -\frac{\hbar^2}{2m}\nabla^2 + V(r)$$

Operating with $(E - H_0 + i\epsilon)$ on Eq. (5.47) and using Eq. (ii) we see that G, as defined by Eq. (5.47), satisfies Eq. (i) and that it has the correct asymptotic behavior (the same as G_0) as $r \to \infty$.

It follows from the definition of $\chi_{\mathbf{q}_\parallel}$ that in order to calculate the matrix element defined by Eq. (5.43) we need to know

$$G|\chi_{\mathbf{q}_\parallel}\rangle = \int G(E;\mathbf{r},\mathbf{r}')\chi_{\mathbf{q}_\parallel}(\mathbf{r}')\,d^3r' \qquad (5.48)$$

for $\mathbf{r}$ on the interface ($z = 0$). We have, from Eq. (5.47),

$$G|\chi_{\mathbf{q}_\parallel}\rangle = \int G_0(E;\mathbf{r},\mathbf{r}')\chi_{\mathbf{q}_\parallel}(\mathbf{r}')\,d^3r' + \int G_0(E;\mathbf{r},\mathbf{r}''')T(\mathbf{r}''',\mathbf{r}'')$$
$$\times G_0(E,\mathbf{r}'',\mathbf{r}')\,\chi_{\mathbf{q}_\parallel}(\mathbf{r}')\,d^3r''\,d^3r''\,d^3r' \qquad (5.49)$$

where

$$T(\mathbf{r},\mathbf{r}') = V'(\mathbf{r})\,\delta(\mathbf{r}-\mathbf{r}') + V'(\mathbf{r})G_0(E;\mathbf{r},\mathbf{r}')V'(\mathbf{r}') + \int V'(\mathbf{r})G_0$$
$$\times (E;\mathbf{r},\mathbf{r}'')V'(\mathbf{r}'')G_0(E;\mathbf{r}'',\mathbf{r}')V'(\mathbf{r}')\,d^3r'' + \cdots \qquad (5.50)$$

In what follows we represent an expansion of the above form as follows:

$$T = V' + V'G_0V' + V'G_0V'G_0V' + \cdots \qquad (5.51)$$

One can easily show that with G_0 given by Eq. (5.46)

$$\int G_0(E;\mathbf{r},\mathbf{r}')\chi_{\mathbf{q}_\parallel}(\mathbf{r}')\,d^3r' = \Phi^-_{\mathbf{q}_\parallel}(\mathbf{r}) + \Phi^+_{\mathbf{q}_\parallel}(\mathbf{r}) \qquad (5.52)$$

where

$$\Phi^-_{\mathbf{q}_\parallel}(\mathbf{r}) = \alpha_{\mathbf{q}_\parallel}\frac{\exp(i\mathbf{q}_-\cdot\mathbf{r})}{\sqrt{A}}\,\Theta(-z) \qquad (5.53)$$

$$\Phi^+_{\mathbf{q}_\parallel}(\mathbf{r}) = \alpha_{\mathbf{q}_\parallel}\frac{\exp(i\mathbf{q}_+\cdot\mathbf{r})}{\sqrt{A}}\,\Theta(z) \qquad (5.54)$$

$$\mathbf{q}_\pm = \{\mathbf{q}_\parallel, \pm q_z\} \qquad (5.55)$$

$$q_z = \left[\frac{2m}{\hbar^2}(E-V_0) - \mathbf{q}_\parallel^2\right]^{1/2} \qquad (5.56)$$

$$\alpha_{\mathbf{q}_\parallel} = -\left(\frac{im}{\hbar^2}\right)\frac{1}{q_z} \qquad (5.57)$$

$$\Theta(z) = 1, \qquad z > 0$$
$$= 0, \qquad z < 0 \qquad (5.58)$$

Substituting Eq. (5.52) into Eq. (5.49) and using the fact that the potential $V'(\mathbf{r})$ consists of two nonoverlapping potentials $V_c(\mathbf{r})$ and $V_s(z)$ [see Eq. (5.45)], we obtain

$$G|\chi_{\mathbf{q}_\parallel}\rangle = (1 + G_0T_c + G_0T_sG_0T_c + G_0T_cG_0T_sG_0T_c + \cdots)\Phi^-_{\mathbf{q}_\parallel}$$
$$+ (1 + G_0T_s + G_0T_cG_0T_s + G_0T_sG_0T_cG_0T_s + \cdots)\Phi^+_{\mathbf{q}_\parallel} \qquad (5.59)$$

The notation in the above equation is made obvious by the following example:

$$G_0 T_c \Phi_{\mathbf{q}_\parallel}^- = \int G_0(E;\mathbf{r},\mathbf{r}'') \, T_c(\mathbf{r}'',\mathbf{r}') \Phi_{\mathbf{q}_\parallel}^-(\mathbf{r}') \, d^3r'' \, d^3r' \qquad (5.60)$$

The matrices T_c and T_s, defined as follows:

$$T_c = V_c + V_c G_0 V_c + V_c G_0 V_c G_0 V_c + \cdots \qquad (5.61)$$

$$T_s = V_s + V_s G_0 V_s + V_s G_0 V_s G_0 V_s + \cdots \qquad (5.62)$$

describe to all orders the scattering of the electron by V_c (the potential field to the left of the interface) and by V_s (the potential field to the right of the interface), respectively. Obviously, $T_c(\mathbf{r},\mathbf{r}') = 0$ if $z \geq 0$ or $z' \geq 0$, and $T_s(\mathbf{r},\mathbf{r}') = 0$ if $z \leq 0$ or $z' \leq 0$. The first term in Eq. (5.59), $\Phi_{\mathbf{q}_\parallel}^-$, represents a wave propagating in the negative z direction. The second term, written explicitly in Eq. (5.60), represents, in the region $z \geq 0$, a reflected wave propagating in the positive z direction. This follows directly from the form of G_0 [Eq. (5.46)] and the fact that V_c and T_c are identically zero for $z \geq 0$. Because V_c is periodic parallel to the interface (let $\{\mathbf{g}\}$ be the corresponding set of 2D reciprocal vectors), we obtain

$$G_0 T_c \Phi_{\mathbf{q}_\parallel}^- = \frac{\alpha_{\mathbf{q}_\parallel}}{\sqrt{A}} \sum_{\mathbf{g}} \exp(i\mathbf{K}_\mathbf{g}^+ \cdot \mathbf{r}) R_{\mathbf{g}0}^c, \qquad z \geq 0 \qquad (5.63)$$

where

$$K_\mathbf{g}^\pm = \left\{ \mathbf{q}_\parallel + \mathbf{g}, \pm \left[\frac{2m}{\hbar^2}(E - V_0) - (\mathbf{q}_\parallel + \mathbf{g})^2 \right]^{1/2} \right\} \qquad (5.64)$$

Note that in contrast to $\mathbf{k}_\parallel$ in Eq. (3.39), $\mathbf{q}_\parallel$ in Eq. (5.64) is not restricted within the SBZ of the two-dimensional reciprocal lattice. The matrix elements $R_{\mathbf{g}'\mathbf{g}}^{c}(E,\mathbf{q}_\parallel)$ are defined quite generally by the following equation:

$$\int G_0(E;\mathbf{r},\mathbf{r}'') T_c(\mathbf{r}'',\mathbf{r}') \exp(i\mathbf{K}_\mathbf{g}^- \cdot \mathbf{r}') \, d^3r'' \, d^3r'$$
$$= \sum_{\mathbf{g}} R_{\mathbf{g}\mathbf{g}'}^c \exp(i\mathbf{K}_\mathbf{g}^+ \cdot \mathbf{r}), \qquad z \geq 0 \quad (5.65)$$

The physical meaning of $R_{\mathbf{g}\mathbf{g}'}^c$ is obvious. When an incident plane wave $\exp(i\mathbf{K}_\mathbf{g}^- \cdot \mathbf{r}')$ is scattered by the periodic field to the left of the interface a number of diffracted beams are generated corresponding to different $\mathbf{g}$ vectors. $R_{\mathbf{g}\mathbf{g}'}^c$ is the amplitude of the $\mathbf{g}$th beam. In what follows we shall assume that the $\mathbf{g}$ vectors are enumerated in a given order and the $R^c_{\mathbf{g}\,\mathbf{g}'}$ matrix elements will be collectively denoted by the matrix $\mathbf{R}^c$. We note that $\mathbf{R}^c$ is a function of E and $\mathbf{q}_{\parallel}$ although this dependence will not be explicitly denoted, and we also note that $\mathbf{R}^c$ needs to be evaluated only in the first (SBZ) of $\mathbf{q}_{\parallel}$. In other zones the indices $\mathbf{g}$ and $\mathbf{g}'$ should merely be rearranged.

We can describe reflection by the surface barrier in a similar manner. A plane wave $\exp(i\mathbf{K}^+_{\mathbf{g}'} \cdot \mathbf{r})$ incident on the barrier $V_s(z)$ [Eq. (5.45)] from the left will give rise to a reflected wave described, for given E and $\mathbf{q}_{\parallel}$, by

$$\int G_0(E;\mathbf{r},\mathbf{r}'')\,T_s(\mathbf{r}'',\mathbf{r}')\,\exp(i\mathbf{K}^+_{\mathbf{g}'} \cdot \mathbf{r}')\,d^3r''\,d^3r'$$
$$= \sum_{\mathbf{g}} R^s_{\mathbf{g}\mathbf{g}'} \exp(i\mathbf{K}^-_{\mathbf{g}} \cdot \mathbf{r}), \qquad z \le 0 \qquad (5.66)$$

where the $\mathbf{K}^{\pm}_{\mathbf{g}}$ are given by Eq. (5.64). Because the surface barrier is a function of z only, we obtain

$$R^s_{\mathbf{g}\mathbf{g}'} = R_s(E,\mathbf{q}_{\parallel} + \mathbf{g})\,\delta_{\mathbf{g}\mathbf{g}'} \qquad (5.67)$$

i.e., $\mathbf{R}^s$ is a diagonal matrix. $R_s(E,\mathbf{q}_{\parallel} + \mathbf{g})$ are identical (when $\mathbf{q}_{\parallel}$ lies in the SBZ) with the matrix elements introduced by Eq. (5.3).

In the region of the interface ($z \simeq 0$) *where the potential is constant* [see comments following Eq. (5.45)] each term in Eq. (5.59) is a plane wave. The terms proportional to $\Phi^-_{\mathbf{q}_{\parallel}}$ can be interpreted as follows. The first term is an "incident" wave propagating to the left; when scattered by the crystal it gives rise to a reflected wave (the second term); when this is scattered by the surface barrier it gives rise to the third term, which, in turn, is scattered by the crystal, giving rise to the fourth term, and so on. A similar interpretation applies to the terms proportional to $\Phi^+_{\mathbf{q}_{\parallel}}$.

Using Eqs. (5.65) and (5.66) and the rules of matrix multiplication we find that, in the region of constant potential at the interface ($z \simeq 0$), Eq. (5.59) can be written as follows:

$$\begin{aligned}
G|\chi_{\mathbf{q}_{\parallel}}\rangle &= \Phi^-_{\mathbf{q}_{\parallel}} + \sum_{\mathbf{g},\mathbf{g}'} \frac{\exp(i\mathbf{K}^+_{\mathbf{g}} \cdot \mathbf{r})}{\sqrt{A}}\,R^c_{\mathbf{g}\mathbf{g}'}(\mathbf{I} - \mathbf{R}^s\mathbf{R}^c)^{-1}_{\mathbf{g}'\mathbf{o}}\alpha_{\mathbf{q}_{\parallel}} \\
&+ \sum_{\mathbf{g},\mathbf{g}',\mathbf{g}''} \frac{\exp(i\mathbf{K}^-_{\mathbf{g}} \cdot \mathbf{r})}{\sqrt{A}}\,R^s_{\mathbf{g}\mathbf{g}''}(\mathbf{I} - \mathbf{R}^c\,\mathbf{R}^s)^{-1}_{\mathbf{g}''\mathbf{g}}R^c_{\mathbf{g}'\mathbf{o}}\alpha_{\mathbf{q}_{\parallel}} + \Phi^+_{\mathbf{q}_{\parallel}} \\
&+ \sum_{\mathbf{g},\mathbf{g}'} \frac{\exp(i\mathbf{K}^-_{\mathbf{g}} \cdot \mathbf{r})}{\sqrt{A}}\,R^s_{\mathbf{g}\mathbf{g}'}(\mathbf{I} - \mathbf{R}^c\mathbf{R}^s)^{-1}_{\mathbf{g}'\mathbf{o}}\alpha_{\mathbf{q}_{\parallel}} \\
&+ \sum_{\mathbf{g},\mathbf{g}''} \frac{\exp(i\mathbf{K}^+_{\mathbf{g}} \cdot \mathbf{r})}{\sqrt{A}}\,R^c_{\mathbf{g}\mathbf{g}''}(\mathbf{I} - \mathbf{R}^s\mathbf{R}^c)^{-1}_{\mathbf{g}''\mathbf{g}}R^s_{\mathbf{g}'\mathbf{o}}\alpha_{\mathbf{q}_{\parallel}}
\end{aligned} \qquad (5.68)$$

where I is the unit matrix and $\Phi_{\mathbf{q}_\parallel}^{\pm}$ are defined by Eqs. (5.53) and (5.54). Substituting Eq. (5.68) into Eq. (5.43) and performing the final integration with respect to $\mathbf{r}$, we obtain

$$\langle \chi_{\mathbf{q}_\parallel} | G(E) | \chi_{\mathbf{q}_\parallel} \rangle = \alpha_{q_\parallel} \{ 1 + [1 + 2R_s(E, \mathbf{q}_\parallel)]$$
$$\times \sum_{\mathbf{g}} [I - R^c R^s]_{\overline{0}\mathbf{g}}^{-1} R_{\mathbf{g}0}^c + R_s(E, \mathbf{q}_\parallel)[I - R^c R^s]_{\overline{0}\overline{0}}^{-1} \} \quad (5.69)$$

where $R_s(E, \mathbf{q}_\parallel)$ is the diagonal matrix element defined by Eq. (5.67). This matrix can be evaluated numerically without difficulty. Therefore, the problem of evaluating the matrix element $\langle \chi_{\mathbf{q}_\parallel} | G(E) | \chi_{\mathbf{q}_\parallel} \rangle$ is reduced to that of evaluating the reflection matrix $R^c(E, \mathbf{q}_\parallel)$. Because this is formally the same with the matrix one meets in the LEED problem, its evaluation can be carried out, under certain conditions, in exactly the same manner. In the next section we demonstrate how this can be done, in a relatively straightforward way, for certain models of the metal–vacuum interface.

We note in passing that Eq. (5.69) may, also, be useful in situations other than field emission. It has, for example, been applied (Modinos, Paranjape, and Aers, 1979) to the study of quantum-size-effects in metal–insulator–metal tunneling.

5.3. CALCULATION OF THE REFLECTION MATRIX R^c

Let us assume, for the sake of clarity, that any deviation from the bulk potential is limited to the top two layers at the surface of the semi-infinite crystal. This is usually satisfactory in practice, and in any case increasing this number from two to three or more does not present any serious difficulty. We assume, however, that the surface layers have the same 2D periodicity (the same surface unit cell) as the bulk layers. Reconstructed layers at the surface are most conveniently treated as overlayers (see Section 6.3). The present situation is described schematically in Fig. 5.2. The zeroth layer and the ones to the left of it are identical with those of the infinite crystal (bulk layers). The spherical potential within the muffin-tin spheres of the top layers (1 and 2) is different in each layer and different from that in the bulk layers. Also, the distance between the zeroth and the first layer, and that between the first and the second layer, may be different from each other and from the bulk interlayer distance. The metal–vacuum interface (the plane $z = 0$) is placed a distance $\frac{1}{2}d + D_s$ above the center of the surface layer (layer 2), where d is the bulk interlayer distance and D_s an adjustable parameter (the plane $z = 0$ is indicated by a broken line in Fig. 5.2). We assume

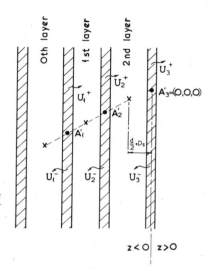

FIGURE 5.2

The three top layers of a semi-infinite crystal. The potential in the zeroth layer and the ones to the left of it is identical with that in the bulk of the crystal. The layers are separated by slots of constant potential (shaded regions) of infinitesimal thickness. The meaning of the various symbols is explained in the text.

that the potential outside the muffin-tin spheres right up to the metal–vacuum interface is constant and equal to the bulk interstitial potential V_0 (this, as we have already noted, may be the most limiting of our assumptions).† The potential for $z > 0$, which depends only on the z coordinate, does not enter into the calculation of the $\mathbf{R}^c$ matrix, and, therefore, there is no loss of generality if we assume that the constant potential extends to $z = +\infty$. The crosses in Fig. 5.2 denote the centers of the respective layers. $\mathbf{A}_1'(\mathbf{A}_2')$ denotes the midpoint of the vector from the center of the zeroth (first) layer to the center of the first (second) layer, and $\mathbf{A}_3'$ is the origin of coordinates, i.e., $\mathbf{A}_3' = (0,0,0)$.

Let us now assume that a wave described by

$$\Psi^{(i)} = \sum_{\mathbf{g}} U_{3\mathbf{g}}^{-} \exp(iK_{\mathbf{g}}^{-} \cdot (\mathbf{r} - \mathbf{A}_3'))$$ (5.70)

where

$$K_{\mathbf{g}}^{\pm} = \left(\mathbf{k}_{\parallel} + \mathbf{g}, \pm \left[\frac{2m}{\hbar^2} (E - V_0) - (\mathbf{k}_{\parallel} + \mathbf{g})^2 \right]^{1/2} \right)$$ (5.71)

is incident upon the crystal from the right. We assume without loss of generality [see comments following Eq. (5.65)] that $\mathbf{k}_{\parallel}$ lies in the SBZ of the surface under

†Usually [see Eq. (4.1)] we take the real part of V_0 as the zero of energy. We note, however, that V_0 may have a small (negative) imaginary part.

consideration. The corresponding reflected wave is described by

$$\Psi^{(r)} = \sum_{\mathbf{g}} U^+_{3\mathbf{g}} \exp[iK^+_{\mathbf{g}} \cdot (\mathbf{r} - \mathbf{A}'_3)] \tag{5.72}$$

By the definition of the $\mathbf{R}^c$ matrix, we have

$$U^+_{3\mathbf{g}'} = \sum_{\mathbf{g}} R^c_{\mathbf{g}'\mathbf{g}}(E, \mathbf{k}_{\parallel}) U^-_{3\mathbf{g}} \tag{5.73a}$$

or in matrix form

$$\mathbf{U}^+_3 = \mathbf{R}^c \mathbf{U}^-_3 \tag{5.73b}$$

We remember that in practice $\mathbf{U}^{\pm}_3$ are n-component vectors and therefore $\mathbf{R}^c$ is an $n \times n$ matrix where n is some finite number. The incident ($\mathbf{U}^-_3$) and reflected ($\mathbf{U}^+_3$) waves are indicated with appropriate arrows in Fig. 5.2. The wave field between the first and second layers is described by

$$\Psi^{(2)} = \sum_{\mathbf{g}} \{U^+_{2\mathbf{g}} \exp[i\mathbf{K}^+_{\mathbf{g}} \cdot (\mathbf{r} - \mathbf{A}'_2)] + U^-_{2\mathbf{g}} \exp[i\mathbf{K}^-_{\mathbf{g}} \cdot (\mathbf{r} - \mathbf{A}'_2)]\} \tag{5.74}$$

and that between the zeroth and the first layer by

$$\Psi^{(1)} = \sum_{\mathbf{g}} \{U^+_{1\mathbf{g}} \exp[i\mathbf{K}^+_{\mathbf{g}} \cdot (\mathbf{r} - \mathbf{A}'_1)] + U^-_{1\mathbf{g}} \exp[i\mathbf{K}^-_{\mathbf{g}} \cdot (\mathbf{r} - \mathbf{A}'_1)]\} \tag{5.75}$$

The $U^{\pm}_{2\mathbf{g}}$ coefficients are related to the $U^{\pm}_{1\mathbf{g}}$ coefficients through the scattering matrix elements of the first layer. We have, in matrix form,

$$\mathbf{U}^+_2 = \mathbf{Q}^{\mathrm{I}}(1)\mathbf{U}^+_1 + \mathbf{Q}^{\mathrm{II}}(1)\mathbf{U}^-_2 \tag{5.76a}$$
$$\mathbf{U}^-_1 = \mathbf{Q}^{\mathrm{III}}(1)\mathbf{U}^+_1 + \mathbf{Q}^{\mathrm{IV}}(1)\mathbf{U}^-_2 \tag{5.76b}$$

The above relations are demonstrated schematically in Fig. 5.3a. Similarly the $U^{\pm}_{3\mathbf{g}}$ coefficients are related to the $U^{\pm}_{2\mathbf{g}}$ coefficients through the scattering matrix elements of the second layer. We have

$$\mathbf{U}^+_3 = \mathbf{Q}^{\mathrm{I}}(2)\mathbf{U}^+_2 + \mathbf{Q}^{\mathrm{II}}(2)\mathbf{U}^-_3 \tag{5.77a}$$
$$\mathbf{U}^-_2 = \mathbf{Q}^{\mathrm{III}}(2)\mathbf{U}^+_2 + \mathbf{Q}^{\mathrm{IV}}(2)\mathbf{U}^-_3 \tag{5.77b}$$

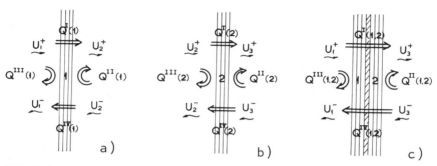

FIGURE 5.3

Schematic description of (a) the **Q**-matrix elements for layer 1; (b) the **Q**-matrix elements for layer 2; (c) the **Q**-matrix elements of the two layers (1 and 2) taken as a single unit.

The above relations are demonstrated schematically in Fig. 5.3b. We next define the scattering matrix elements (**Q** matrix elements) of the two layers (first and second), taken as a single unit, according to the following equations:

$$U_3^+ = Q^I(1\ 2)U_1^+ + Q^{II}(1\ 2)U_3^- \tag{5.78a}$$
$$U_1^- = Q^{III}(1\ 2)U_1^+ + Q^{IV}(1\ 2)U_3^- \tag{5.78b}$$

which are demonstrated schematically in Fig. 5.3c. Combining Eqs. (5.76) with Eqs. (5.77) we obtain

$$Q^I(1\ 2) = Q^I(2)[\mathbf{I} - Q^{II}(1)Q^{III}(2)]^{-1}Q^I(1) \tag{5.79a}$$
$$Q^{II}(1\ 2) = Q^{II}(2) + Q^I(2)Q^{II}(1)[\mathbf{I} - Q^{III}(2)Q^{II}(1)]^{-1}Q^{IV}(2) \tag{5.79b}$$
$$Q^{III}(1\ 2) = Q^{III}(1) + Q^{IV}(1)Q^{III}(2)[\mathbf{I} - Q^{II}(1)Q^{III}(2)]^{-1}Q^I(1) \tag{5.79c}$$
$$Q^{IV}(1\ 2) = Q^{IV}(1)[\mathbf{I} - Q^{III}(2)Q^{II}(1)]^{-1}Q^{IV}(2) \tag{5.79d}$$

Using Eqs. (5.79) one can obtain the Q-matrix elements for any two layers (considered as one unit) if the Q-matrices of the individual layers are already known. We have already shown (Section 3.5) how to calculate the Q-matrix for a layer of muffin-tin atoms. We note that whereas Eq. (3.98) for the $M_{\vec{g}'\vec{g}}^{\pm}$ elements is generally valid, Eqs. (3.102), which give the Q-matrix elements in terms of the M elements, must be modified because of the way we have chosen the origin of coordinates and, also, because of the variation in the interlayer distance (see Fig. 5.2). The required changes in the exponential factors of these equations are obvious and we shall not bother to write them down.

Returning to Eq. (5.75) we observe that the second term in this equation represents a wave incident on the "bulk crystal" (all layers from $z = -\infty$ up to and including the zeroth layer) from the right, and that the first term represents the reflected (diffracted) wave which arises in consequence. The two are related by the following equation:

$$U_{1g'}^+ = \sum_g R_{g'g}^B U_{1g}^-$$ (5.80)

which defines the reflection matrix $\mathbf{R}^B(E, \mathbf{k}_\parallel)$ of the "bulk crystal." The most general way of calculating $\mathbf{R}^B$ is the following: The wave function χ for $(z - A_{1z}) > 0$, corresponding to an incident beam given by $\exp[i\mathbf{K}_{1g}^- \cdot (\mathbf{r} - \mathbf{A}_1')]$, will consist of the incident plus a finite number (n) of diffracted beams. Hence

$$\chi = \exp[i\mathbf{K}_g^- \cdot (\mathbf{r} - \mathbf{A}_1')]$$
$$+ \sum_{g'} R_{g'g}^B \exp[i\mathbf{K}_{g'}^+ \cdot (\mathbf{r} - \mathbf{A}_1')], \qquad z - A_{1z} \geq 0 \quad (5.81)$$

The above expression must join smoothly to

$$\chi = \sum_j C_j \psi_j, \qquad z - A_{1z} \leq 0$$ (5.82)

where ψ_j are the Bloch waves (n in number) of the infinite crystal, for the given E and $\mathbf{k}_\parallel$, which propagate or decay in the negative z direction. In the region of constant potential around $z \simeq A_{1z}$ these Bloch waves are given by a plane wave expansion as in Eq. (3.52). We note that $\mathbf{A}_1$ is the midpoint of the vector from the center of the zeroth layer to the immediate next layer (first layer) when the interlayer distance between these two layers is the same as in the bulk of the metal. It follows that $\mathbf{A}_1$ may be different from $\mathbf{A}_1'$ as defined at the beginning of this section. The $2n$ coefficients C_j and $R_{g'g}$, for any one $\mathbf{g}$, are determined by the continuity of the wave function χ and its derivative at $z = A_{1z}$. The above method of calculating $\mathbf{R}^B$ is known as the Bloch wave method because it requires an explicit evaluation of the Bloch waves ψ_j of the infinite crystal. There is an alternative method of calculating $\mathbf{R}^B$, known as the doubling layer method (Pendry, 1974), which does not require any knowledge of the Bloch waves of the infinite crystal. It is a matter of straightforward algebra to obtain the scattering matrix elements of a stack of identical layers (all parallel to the surface but displaced relative to each other by a vector $\mathbf{a}_1$). We first find the scattering matrix of a pair of (consecutive) layers using Eqs. (5.79). Note that the waves on the right of the right layer are expressed with respect to a "right" origin at a point $\frac{1}{2}\mathbf{a}_1$ from the center of this layer, and the waves on the left of the left layer are expressed with respect to a "left" origin at a point $-\frac{1}{2}\mathbf{a}_1$ from the center of this layer. It is obvious that the scattering matrix elements of any pair of layers is the same. From the scattering matrix elements of the pair of layers we can find the scattering matrix elements of a pair of pairs of layers using Eq. (5.79). Note that the "left" origin of the right pair coincides with the "right" origin of the left pair for consecutive layers. Having found the scattering matrix elements for the four layers we can repeat the process to find the scattering matrix element of eight consecutive layers, then of 16 layers, and so on until the reflection matrix of the slab is practically identical with the reflection matrix R^B of the semi-infinite crystal. The incident and diffracted waves on the right of the slab refer to the "right" origin of the top layer (the zeroth layer in Fig. 5.2) at $\frac{1}{2}a_1$ from the center of this layer. They can be referred to another origin (A'_1 in Fig. 5.2) by a simple transformation. We emphasize that the doubling layer procedure converges only if the potential contains an imaginary component. Although for energies at the Fermi level the potential is real, adding a small imaginary component to this potential (-0.04 eV or so) does not change the results for the local density of states, or for the energy distribution of the field-emitted electrons, in any significant way.

Finally, knowing R^B and the $Q(1\ 2)$ matrices we can evaluate R^c as defined by Eq. (5.73) from the following formula:

$$R^c = Q^{II}(1\ 2) + Q^I(1\ 2)R^B[I - Q^{III}(1\ 2)R^B]^{-1}Q^{IV}(1\ 2) \qquad (5.83)$$

which is obtained by combining Eqs. (5.78) with Eq. (5.80)

5.4. MEASURED AND CALCULATED ENERGY DISTRIBUTIONS OF FIELD-EMITTED ELECTRONS

5.4.1. THE (100) PLANE OF TUNGSTEN

A typical experimental total energy distribution from the (100) plane of tungsten (Plummer and Gadzuk, 1970) is shown in Fig. 5.4. The broken curve in the same figure shows a calculated free-electron energy distribution. We shall, for the moment, disregard the wings at the high- and low-energy ends of the distribution. These are genuine (they are not a property of the analyzer) but the nature of their origin is not, at the moment, sufficiently understood (see Section 5.7). The experimental enhancement factor R, as defined by Eq. (1.93), is shown in Fig. 1.9. (Plummer and Bell, 1972). The same curve is shown by the broken line in Fig. 5.5. It is worth remembering that, for the reasons noted in section 1.7.1, this curve is known only within a constant scaling factor and therefore its vertical position in Fig. 5.5 is arbitrary. The solid line in Fig. 5.5 is the result of a theoretical calculation by the author (presented here for the first time) based on the theory

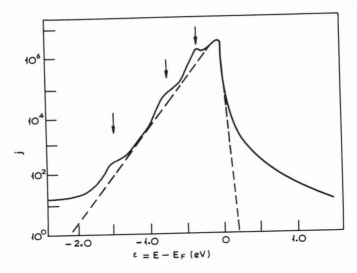

FIGURE 5.4

Energy distribution of field-emitted electrons from W(100) at $T = 78$ K. The dashed lines indicate a free-electron energy distribution. (From Plummer and Gadzuk, 1970.)

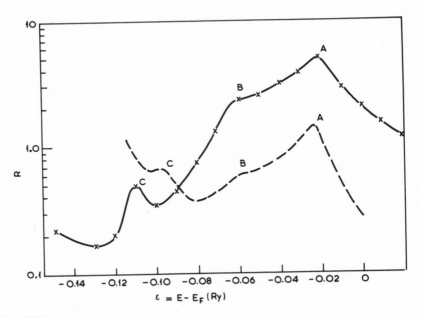

FIGURE 5.5

The experimental enhancement factor for field-emitted electrons from the (100) plane of tungsten is shown by the broken line. The solid line is the result of a calculation based on the theory described in Sections 5.2 and 5.3.

described in Sections 5.2 and 5.3. The self-consistent muffin-tin potential of the center layer of a seven-layer tungsten film, as calculated by Posternak *et al.* (1980), has been used throughout the semi-infinite crystal right up to the metal–vacuum interface. (We note that the potential at the center of a seven-layer film is practically identical with the bulk potential in the infinite crystal.) The (bulk) energy-band-structure obtained with this potential, in the energy region of interest to field emission, is shown in Fig. 5.6 along with the ΓH [normal to the (100) plane] and the ΓN [normal to the (110) plane] symmetry directions of the 3D Brillouin zone. In the field emission calculation (solid line in Fig. 5.5) the bulk potential of Posternak *et al.* (1980) was terminated abruptly at the metal–vacuum interface (the plane $z = 0$) taken at a distance $(d/2 + D_s)$ above the center of the top atomic layer (see Fig. 4.1). D_s was put equal to $D_s = 0.397$ Å. A small imaginary component $v_i = -0.004$ Ry (-0.054 eV) has been added to the potential on the metal side of the interface ($z < 0$), mainly for computational convenience (see comments at the end of Section 5.3). The potential for $z > 0$ was approximated by Eq. (5.1) with $V_s(z)$, the zero field potential, given by Eq. (4.2). [In the present case $E_F = 0.94$ Ry (12.789 eV), determined approximately from Fermi-surface data, $\phi = 4.5$ eV (known from experiment) and the adjustable parameter z_0 was put equal to $z_0 = 0.278$ Å.] After calculating the total energy distribution $j(E)$ (for an applied field $F = 0.37$ V/Å) we obtained the enhancement factor from the formula $R = j/j_0$. We remember that $j_0(E)$, the Fowler–Nordheim free-electron energy distribution as defined by Eq. (1.87), corresponds

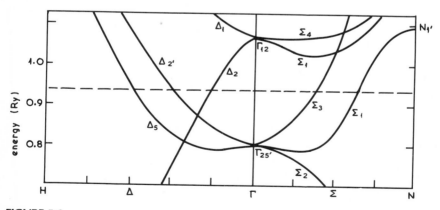

FIGURE 5.6

The energy band structure of tungsten in the energy region of interest to field emission; calculated with the self-consistent muffin-tin potential of Posternak *et al.* (1980). The constant potential between the muffin-tin spheres is taken as the zero of energy. The Fermi level is indicated by the broken line.

to an image barrier [Eq. (1.27)]. We emphasize that R is practically independent of the applied field for the W(100) plane (Vorburger *et al.*, 1975). We see from Fig. 5.5 that the calculation gives the Swanson–Crouser hump (peak A) and the smaller peak (B) very well. The calculation also gives the low-energy peak C, although the calculated position of the latter ($\epsilon \simeq -1.5$ eV) lies approximately 0.18 eV below its experimental value. We have already described the bands of surface states/resonances on the tungsten (100) plane (see Fig. 4.7). It is now clear that the Swanson–Crouser hump derives from the high-energy surface band, centered at Γ, and shown by the open circles in Fig. 4.7. We remember that this band is reproduced reasonably well by the model potential used in our calculation (broken line in Fig. 4.7). It is also clear that the structure at $\epsilon \simeq -0.78$ eV in the energy distribution (peak B) derives from the even component of the second high-energy surface band, shown by the squares in Fig. 4.7, which is also given reasonably well (broken line in Fig. 4.7) by our model potential. We note that the calculated surface bands (their position relative to E_F and relative to each other depends on the choice of surface barrier. The noted values of z_0 and D_s were chosen so that the calculated surface bands (broken lines in Fig. 4.7) approximate as closely as possible the experimentally determined band in the same figure. It is worth noting that the odd component of the second high-energy band (crosses in Fig. 4.7b), which is probably not reproduced by our model potential, does *not* contribute to the tunneling current. This becomes obvious when we look at Eqs. (5.25) and (5.26); the coefficients A_0^{B+} (similarly A_0^{S+}) are identically zero when the corresponding wave function has an odd symmetry.

The peak at $\epsilon \simeq -1.5$ eV (marked C in Fig. 5.5) derives from bulk density of states effects and is therefore much less sensitive to the details of the surface barrier. [It is worth noting (see Fig. 4.7) that no surface states (or resonances) exist on W(100) at this energy.] Plummer and Gadzuk (1970) note that

the wing of the energy distribution causes $R(\epsilon)$ to increase dramatically for $\epsilon \lesssim -1.17$ eV making the state at $\epsilon \simeq -1.5$ eV difficult to observe. At lower current density (low field) there is not enough current at $\epsilon \simeq -1.5$ eV and at higher current density the wing marks this structure.

This statement contains the explanation of the disagreement between the experimental and the theoretical enhancement factors (Fig. 5.5) for energies $\epsilon \lesssim -1.17$ eV. The upward turn in the experimental curve at about this energy is due to electron–electron interaction which is not taken into account by our (independent-electron) theory. There is no doubt, however, that a peak (C) does exist at this energy. It is possible that the calculated energy level of this peak relative to the Fermi level is a bit too low because of error in the estimated position of the latter.

The nature of peak C can be better understood by reference to Eq. (5.31). The contribution to the bulk density of states from the central region of the SBZ is given by

$$\left(\frac{dN}{dE}\right)_c = \frac{1}{Z\pi^3} \iint\limits_{\substack{\text{central} \\ \text{SBZ}}} d^2k_{\parallel} \sum_{\alpha} \left(\frac{\partial k_{\alpha z}}{\partial E}\right) \tag{5.84}$$

A peak in this quantity at a given energy can arise either because the number of states in the *central* SBZ becomes maximum at this energy, or because $\partial k_{\alpha z}/\partial E$ peaks at this energy (for a given α-band) near certain $k_{\parallel}$ values within the central SBZ. [We note that a peak in $(\partial k_{\alpha z}/\partial E)$ corresponds to a flat segment in the corresponding $E(k_z; k_{\parallel})$ dispersion curve.] A peak in $(dN/dE)_c$ does not, however, necessarily lead to a peak in $j(E)$, because of the modulation introduced by the value of the wave function at the surface, represented by $|A_0^{B+} + A_0^{B-}|^2$ in Eq. (5.31). The latter term, which depends to some degree on the surface potential even in the absence of resonances, may shift this peak, slightly in some cases, but in other cases it may eliminate it altogether. We note that in the case of W(100) the bulk states at the center of the SBZ ($k_{\parallel} = 0$), in the energy region of interest to field emission, belong (see Fig. 5.6) either to one of the two one-dimensional irreducible representations Δ_2, $\Delta_{2'}$ or one of the partner representations of the two-dimensional irreducible representation Δ_5 of the C_{4v} point group. The symmetry properties of these representations are well known (see, e.g., Slater, 1972), and it is easy to convince oneself that these properties lead to $|A_0^{B+} + A_0^{B-}| = 0$. Accordingly, the contribution to the tunneling current from bulk states with $k_{\parallel} = 0$ (and by continuity from states with $k_{\parallel}$ in the immediate vicinity of this point) is negligible. [An observation of this kind was first made by Politzer and Cutler (1972) in relation to the (100) plane of nickel.] We found that in the case of W(100) most of the contribution to $j_b(E)$ (at energies away from surface resonances) comes from states in the $0.1 \lesssim k_{\parallel} \lesssim 0.4$ Å^{-1} region of the SBZ. We have also established that peak C in the enhancement factor (Fig. 5.5) derives from a maximum of $(\partial k_{\alpha z}/\partial E)|A_0^{B+} + A_0^{B-}|^2$ at $\epsilon \simeq -1.5$ eV around certain points ($k_{\parallel} \simeq 0.2$ Å^{-1}) of the SBZ. It is clear that while the peak C is associated with a high bulk density of states, near the center of the 3D Brillouin zone, around the $\Gamma_{25'}$ energy level of the energy-band-structure (Fig. 5.6), its exact position depends (as we have already noted) on a number of other factors. It is interesting to note that a peak at about this energy occurs in the enhancement factor of the (111) and (112) planes as well (see Fig. 1.9), which suggests a common origin of these peaks (the high bulk density of states around the $\Gamma_{25'}$ level at $k \simeq 0$). We believe that this peak is more pronounced for the (111) and (112) planes, because the $g = 0$ component of the wave function does not vanish when $k_{\parallel} = 0$ (i.e., at the center of the SBZ of these planes) as it does in the case of the (100) plane.

It is worth noting that the earlier calculations of the energy distribution of the field-emitted electrons from W(100) (Nicolaou and Modinos, 1975; Modinos and Nicolaou, 1976; Kar and Soven 1976) based as they were on the same model potential [Eqs. (5.1), (4.1), and (4.2)] but employing the bulk potential of Mattheiss (1965), rather than the self-consistent potential of Posternak *et al.* (1980) used in the present calculation, failed to reproduce (this was not recognized at the time) the Swanson–Crouser hump. A typical R curve obtained with the Mattheiss (bulk) potential is shown in Fig. 5.1. The peak, seen in this figure at $\epsilon \simeq -0.65$ eV derives from the second high-energy surface band (denoted by crosses in Fig. 4.7) which is reproduced by that potential, and not from the first high-energy band (denoted by open circles in Fig. 4.7) which could *not* be obtained with the Mattheiss potential, in spite of the fact that the energy band structure obtained with the Posternak *et al.* potential is similar to that obtained by Mattheiss (1965). The band structure calculated by the latter for two different (non-self-consistent)

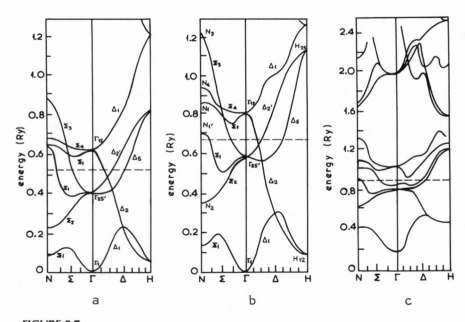

a   b   c

FIGURE 5.7

The energy band structure along symmetry directions in the 3D Brillouin zone of tungsten. (a) and (b) were calculated by Mattheiss (1965) using two different potentials. The exchange contribution to the potential in (b) is approximately ⅔ of that used in calculating the bands in (a). In both cases the zero of energy coincides with $E(\Gamma_1)$. (c) Energy band structure with the spin–orbit interaction taken into account (Christensen and Feuerbacher, 1974). In this case the energy is measured from the constant potential between the muffin-tin spheres. In all three figures the Fermi level is indicated by a broken line.

potentials is shown in Figs. 5.7a and 5.7b. [For the sake of comparison we show in Fig. 4.7c the energy band structure calculated by Christensen and Feuerbacher (1974) by means of the relativistic augmented-plane-wave method.] The first high-energy surface band could not be obtained with either of the two bulk potentials used by Mattheiss.

We note that the fact that the present calculation reproduces reasonably well the observed structure in $R(E)$ does not mean that the potential (especially in the surface region) used in this calculation is entirely satisfactory. It is not, for we have seen (Section 4.2.2) that it does not predict *all* the observed (in photo emission) surface states/resonances on the W(100) plane. We must also point out that, because the absolute magnitude (the vertical position of the R curve in Fig. 5.5) is not known experimentally, we cannot tell whether the calculated (absolute) value of R is indeed the correct one.

In the above-mentioned calculations of the electronic structure of W(100) and of the energy distribution of the field-emitted electrons spin–orbit interaction has been entirely neglected. This approximation is not obviously justified in the case of tungsten. It is well known (Mattheiss, 1965; Christensen and Fenerbacher, 1974) that spin–orbit interaction leads to a split of the d bands along the ΓH direction at an energy immediately below the Fermi level (see, e.g., Fig. 5.7c). It is also known (Pendry and Gurman, 1975) that a surface resonance is likely to appear in such a gap. As a matter of fact, the Swanson–Crouser hump in the energy distribution of the field-emitted electrons was originally (Plummer and Gadzuk, 1970) attributed to such a resonance. Feder and Sturm (1975) made an approximate calculation of spin–orbit effects in relation to the (100) plane of tungsten and found a surface resonance of this type at $\mathbf{k}_{\parallel} = 0$. However, the symmetry of this state is such that the $\mathbf{g} = 0$ component of the corresponding wave function (at $\mathbf{k}_{\parallel} = 0$) is zero and therefore such a state would not contribute to the field emission current (or to the photoemission current normal to the surface) and could not, therefore, explain the experimental data. Calculations by Kasowski (1975) suggest that spin–orbit interaction is not an important factor in the determination of the electronic properties of W(100). A similar conclusion has been reached by Desjonqueres and Cyrot Lockmann (1976). These authors calculated the local density of states on the (110) and (100) faces of Mo and W without and including the spin–orbit interaction. They concluded that the distribution of surface states/resonances on these planes is affected but only marginally by this interaction. Grise et al. (1979) found, however, that spin–orbit interaction may lead to significant changes in the dispersion of surface states/resonances and in the density of these states on W(100). [It is worth noting that the Green's function method, described in Sections 5.2 and 5.3, for calculating the $\mathbf{k}_{\parallel}$ resolved density of states at the surface, i.e., the matrix element defined by Eq. (5.43), and therefore $j(E)$, can be extended to take into account spin–orbit coupling and other relativistic effects. The problem reduces to that of calculating the reflection matrix $\mathbf{R}^c$ (Section

5.3) taking into account the above-mentioned effects. A method for doing this has been developed by Feder (1972, 1974; 1976).]

Finally, it is known that the W(100) surface reconstructs below 350 K (Debe and King, 1977; 1979). The fact that the calculated TED for the ideal (unreconstructed) surface is in very good agreement with the observed TED at $T = 78$ K suggests that either reconstruction does not occur on the field emission tip, or, if it does, it does not lead to a significant change in the surface density of states.

5.4.2. THE (110) PLANE OF TUNGSTEN

It is seen from Fig. 1.9 that the shape of the energy distribution of the emitted electrons from this plane agrees reasonably well with the prediction of the Fowler–Nordheim (FN) (free electron) theory. More recently, and for the first time in a field emission experiment of this kind, Ehrlich and Plummer (1978) were able to measure the *absolute* value of the tunneling current density from tungsten (110) (see Section 2.3). They found that for a given applied field this quantity is approximately five times less than the corresponding FN value. These experimental results are clearly demonstrated in Fig. 5.8 which is taken from the paper

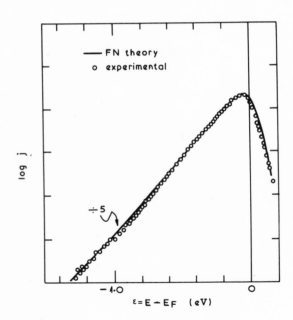

FIGURE 5.8

Total energy distribution of field-emitted electrons from W(110) at room temperature. $F = 0.327$ V/Å. (From Ehrlich and Plummer, 1978.)

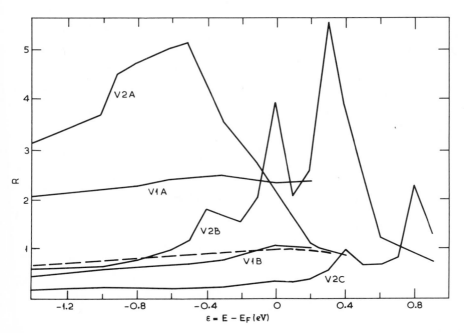

FIGURE 5.9

Calculated $R = j(E)/j_0(E)$ [$j_0(E)$ as defined by Eq. (1.87)] of field-emitted electrons from W(110). The solid lines were calculated using the $V1$ and $V2$ bulk potentials of Mattheiss and the broken line using the bulk potential of Posternak *et al.* A, B, and C denote different surface potential barriers (see text and Fig. 5.10).

of Ehrlich and Plummer. The same authors attributed the discrepancy between the measured current density (its magnitude) and the FN value to the inadequacy of the image barrier ($-e^2/4z$) near the metal–vacuum interface. Modinos and Oxinos (1981a) calculated the energy distribution of the emitted electrons $j(E)$ from W(110) corresponding to the two bulk potentials described by Mattheiss (1965), denoted (respectively) by $V1$ and $V2$. The $V1$ potential leads to the energy-band-structure shown in Fig. 5.7a and the $V2$ potential to that shown in Fig. 5.7b. Although the two band structures are qualitatively very similar, there are significant differences between the two particularly along the ΓN direction [normal to the (110) plane]. As a result of the above differences one finds (Modinos and Oxinos, 1981a) that, in the case of the $V1$ potential, no surface resonances exist in the central region of the SBZ for energies below the Fermi level but, in contrast, such resonances do exist in the case of the $V2$ potential (depending on the choice of surface barrier). These surface resonances are responsible for the significant structure in the calculated energy distributions shown in Fig. 5.9. It is worth noting that the energy-band-structure corresponding to the $V2$ potential is in better agreement (at least in the energy region around E_F) both with the relativistic calculation of Christensen and Feuerbaecher (1974) (Fig. 3.7c) and with the available experimental data on the shape of the Fermi surface.

In the field emission calculation the bulk potential ($V1$ or $V2$) was terminated abruptly at the interface (the plane $z = 0$) which is taken half an interlayer distance above the center of the top atomic layer. The surface potential barrier for $z > 0$ was approximated by Eq. (5.1) with $V_s(z)$, the zero field potential, given by Eq. (4.2). The results of the calculation are shown in Fig. 5.9. The curves denoted by $V2A$, $V2B$, and $V2C$ were obtained using the $V2$ potential inside the metal but different surface barriers corresponding to $z_0 = 0.17$ [$V_s(0+) = 0$], 0.34 [$V_s(0+) = (E_F + \phi)/2$] and 0.86 Å [$V_s(0+) = (4/3)(E_F + \phi)$]. These barriers denoted by A, B, and C, respectively, are shown in Fig. 5.10. The experimental results of Ehrlich and Plummer (Fig. 5.8) show that

$$R(E) \simeq 0.2 \qquad \text{for } E \lesssim E_F \qquad (5.85)$$

in agreement with $V2C$ curve, which suggests that the surface barrier rises sharply at the surface and that it tends to the image barrier at relatively large distances ($z > 3$ Å) from the interface (a barrier lying somewhere between the B and C barriers in Fig. 5.10). The peaks in the R curves $V2B$ and $V2C$ are due to surface resonances corresponding to values of $\mathbf{k}_\parallel$ at the periphery of the *central* SBZ. In contrast, the broad structure in the $V2A$ curve is associated with an increase in the surface density of states throughout the central region of the SBZ. We must emphasize that at the low temperatures used in a field emission experiment $j(E)$ decreases exponentially with the energy for $E - E_F > k_B T$ and its accurate measurement in the corresponding region of energy may not be possible. It may be possible, however, to observe structure such as the one in the $V2C$ curve, *if* it is in fact there, in a thermal-field emission experiment (see Section 1.6). We should point out, at this stage, that energy and angle-resolved measurements of the photo emission current and calculations of the surface density of states by Holmes, King, and Inglesfield (1979) reveal an abundance of surface states/resonances on W(110) in the energy region immediately below the Fermi level, but *not* within the central region of the SBZ (in agreement with the field emission results).

The curves $V1A$ and $V1B$ were obtained using the $V1$ potential inside the metal and two different surface barriers A and B. These correspond (respectively) to $z_0 = 0.21$ Å [$V(0+) = 0$] and $z_0 = 0.43$ Å [$V(0+) = (E_F + \phi)/2$]. For both barriers $R(E)$ is constant for $E \leq E_F$ but the $V1B$ curve is, obviously, nearer the experimental result as far as the magnitude of the emitted current is concerned, which again suggests that the surface barrier rises sharply at the surface.

We have seen in the previous section that the Mattheiss potentials fail to

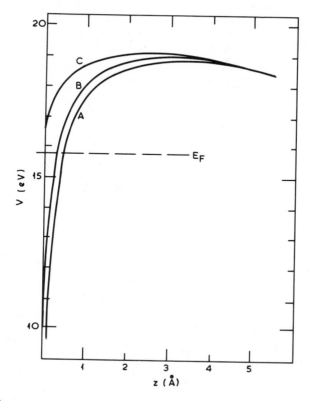

FIGURE 5.10

Surface potential barrier described by Eqs. (4.2) and (5.1). A, B, and C correspond to different values of z_0 and such that A: $V_s(0+) = 0$; B: $V_s(0+) = (E_F + \phi)/2$; C: $V_s(0+) = (4/3)(E_F + \phi)$.

produce the surface band responsible for the Swanson–Crouser hump in the energy distribution from the W(100) plane. We have, therefore, for the sake of completeness, calculated the energy distribution from the W(110) plane using the same *bulk* potential (Posternal *et al.,* 1980) which we used in the calculations of the energy distribution from the W(100) plane (Fig. 5.5). The surface potential barrier [Eqs. (4.2) and (5.1)] was parametrized as follows: $D_s = 0$, $E_F = 12.789$ eV, $\phi = 5.25$ eV, and $z_0 = 0.4$ Å, and $F = 0.327$ V/Å. The results of this calculation are shown by the broken line in Fig. 5.9. We see that $R(E)$ is practically constant for $E \lesssim E_F$, in agreement with the experimental results. Its value, (0.9) at E_F, is larger than the experimental value of ~ 0.2, but there is no doubt that a larger value of z_0, i.e., a steeper barrier, can lead to the experimental value without introducing any structure in the $R(E)$ curve. We note that the energy-

band structure along the ΓN direction obtained with the bulk potential of Posternak *et al.* (1980) (shown in Fig. 5.6) is very similar to that obtained with the $V1$
potential (Fig. 5.7a) which excludes the possibility of surface resonances in the
central SBZ in the energy region immediately below the Fermi level.

It is worth noting that our conclusion that the surface barrier rises sharply
at the surface does not depend on the choice of bulk potential. It is also worth
noting that a theoretical analysis of surface resonances observed in low-energy
electron diffraction from W(110) leads to the same conclusion (Baribeau and Carette, 1981).

5.4.3. THE (100) PLANE OF MOLYBDENUM

The energy-band structure of molybdenum (see, e.g., Moruzzi *et al.*, 1978)
is very similar to that of tungsten especially if spin–orbit coupling and other relativistic effects which are stronger for tungsten are disregarded. Moreover, the distribution of surface states/resonances on the Mo(100) plane (Kerker *et al.*, 1978;
Weng *et al.*, 1978) is very similar to that on W(100). It is, therefore, not surprising
that the energy distribution of the field-emitted electrons from Mo(100) (Swanson
and Crouser, 1967b; Weng, 1977; Richter and Gomer, 1979) is very similar to
that obtained from W(100). Figure 5.11 (Weng, 1977) shows the experimental
enhancement factor R. (Because the experimental TED is known only within a

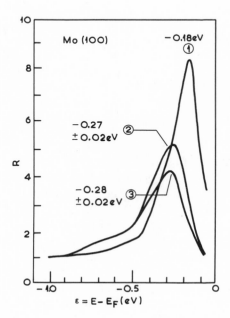

FIGURE 5.11

Experimental enhancement factor of field-emitted electrons from Mo(100). Curve 1 is obtained
at the center of the plane. Curves 2 and 3 are
obtained nearer the edge of the plane. The data
were taken at 78 K. (From Weng, 1977.)

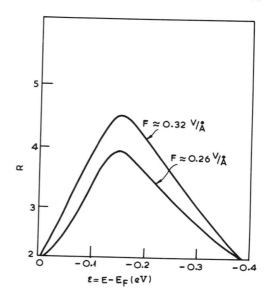

FIGURE 5.12

Experimental enhancement factor of field-emitted electrons from Mo(100) for two different values of the applied field. (From Richter and Gomer, 1979.)

constant scaling factor, so is R.) Curve 1 is typical of the Mo(100) plane in the sense that it is obtained when one probes electrons from the center of this plane. Its main features are a sharp peak at $\epsilon = -0.18$ eV and a shoulder at $\epsilon \simeq -0.58$ eV. The sharp peak derives [as in the case of peak A in the enhancement factor for W(100) (Fig. 5.5)] from a band of surfaces states/resonances centered at $\bar{\Gamma}$ ($\mathbf{k}_{\parallel} = 0$) and the shoulder at $\epsilon \simeq -0.58$ eV derives (like peak B in Fig. 5.5) from the even component of a double surface band which exists at finite values of $\mathbf{k}_{\parallel}$ (partly within the central region of the SBZ, but not at $\mathbf{k}_{\parallel} = 0$). Curves 2 and 3 are obtained when one probes electrons near the edge of the (100) plane (curve 3 is obtained from a region nearest to the edge). It is obvious that

i. the main resonance peak gets broadened, and
ii. the energy levels of the main peak and of the shoulder shift by approximately 0.1 eV as we go from curve 1 to curve 3.

These "edge-proximity" effects can be explained (Weng, 1977; Weng et al., 1978) as follows. The existence of the edge leads to scattering between the surface state (or resonance) at a given $\mathbf{k}_{\parallel}$ with bulk (nonresonant) states of *different* $\mathbf{k}_{\parallel}$ (the 2D periodicity is destroyed at the edge) leading to a broadening of the resonance. For some $\mathbf{k}_{\parallel}$ values (the surface band extends over a finite region of the SBZ) the surface resonance disappears altogether. Accordingly the peak or shoulder which derives from a given band is not only broadened by "edge" scattering but may be shifted in energy as well. In one respect the enhancement factor for the Mo(100) plane differs from that of the W(100) plane. The latter appears to be independent of the applied field (Vorburger *et al.*, 1975). This is not so for the Mo(100) plane. Figure 5.12 (Richter and Gomer, 1979) demonstrates this quite clearly; the sharp peak at $\epsilon \simeq -0.15$ eV increases with increasing field. It has been shown by Swanson and Crouser (1967b) that at sufficiently high fields the intensity of this peak exceeds that of the Fermi level. In general, a marked dependence of $R(E)$ on the applied field will arise when the *relative* contribution to $j(E)$ from the region of the SBZ which is responsible for a given structure in $R(E)$ changes with the applied field. If this region is centered around $\mathbf{k}_\| = 0$ the corresponding structure will be more pronounced at lower applied fields; the opposite will be true when the structure in $R(E)$ comes from a finite region of the SBZ which does not include the origin (Modinos and Nicolaou, 1976; Nicolaou and Modinos, 1976; see also next section). In this respect one would expect that the structure shown in Fig. 5.12 (which originates from a surface band centered at $\mathbf{k}_\| = 0$) to be more pronounced at the lower applied fields in contrast to the experimental results.

5.4.4. THE (111) PLANE OF COPPER

Figure 5.13 shows a theoretical enhancement factor R for this plane calculated by Kar and Soven (1976a) using a model potential for the semi-infinite crystal of the form described by Eqs. (5.1) and (4.1). The peak near the Fermi level

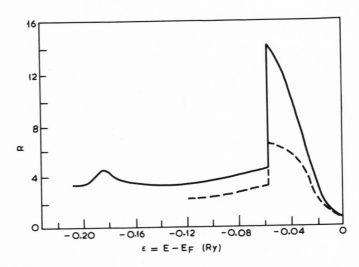

FIGURE 5.13

Theoretical enhancement factor of field-emitted electrons from Cu(111). Solid line: $F = 0.3$ V/Å; broken line: $F = 0.6$ V/Å. (From Kar and Soven, 1976a.)

is due to a band of (true) surface states which is known (from photoemission measurements) to exist at about this energy (Gartland and Slagsvold, 1975). The discontinuity in R at $\epsilon \simeq -0.055$ Ry corresponds to the bottom of the surface band (which occurs at $\mathbf{k}_{\parallel} = 0$) and derives from the fact that the density of states of a two-dimensional band does not go smoothly to zero at the edge of the band. This discontinuity may be difficult to observe experimentally due to the finite resolution of the analyzer and also because of the intrinsic broadening due to the noninfinite geometry of the emitting plane. Because the surface band responsible for the main peak in Fig. 5.12 is centered around $\mathbf{k}_{\parallel} = 0$, its relative contribution to the tunneling current is larger at the lower applied field and hence the peak is more pronounced at this field. Although the surface state peak in the energy distribution of the field-emitted electrons has been observed (Whitcutt and Blott, 1969) its physical origin has not been recognized at the time. The structure in the calculated $R(\epsilon)$ at the lower energy ($\epsilon \simeq -0.18$ Ry) is due to bulk-density-of-states effects. Unfortunately, we do not have an experimentally determined enhancement factor for this plane.

5.4.5. FIELD EMISSION ENERGY DISTRIBUTION FROM THE PLATINUM GROUP METALS

Dionne and Rhodin (1976) measured the energy distribution from different crystallographic planes of rhodium, palladium, iridium, and platinum. The (normalized) enhancement factors obtained by these authors are shown in Fig. 5.14. Dionne and Rhodin argue that these results can be understood in terms of the bulk-energy-band-structure if some band edge contraction at the surface is assumed.

5.5. FIELD EMISSION FROM FERROMAGNETIC METALS

In the last few years reliable measurements of the spin polarization of the field emission current from a number of ferromagnetic metal surfaces have been obtained (Landolt and Campagna, 1977; 1978; Landolt, Campagna, Chazalviel, Yafet, and Wilkens, 1977; Landolt and Yafet, 1978). The apparatus used by the above authors is basically a modified field emission microscope. The new element in the experimental setup consists of a magnetic field at the cathode site. This field determines the quantization axis, along the measured (hkl) direction. It also acts as an image steering element controlling the effective acceptance angle of the probe hole on the field emission screen (anode), which selects the electrons emitted from (the central region of) the measured plane from the field emission pattern. [The field emission pattern on the screen (which lies outside the region of the magnetic

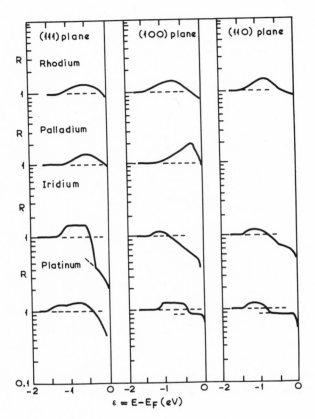

FIGURE 5.14

Experimental enhancement factors of field-emitted electrons from low index planes of the platinum group metals. (From Dionne and Rhodin, 1976.)

field) is distorted by the magnetic field. Because of this distortion meaningful measurements can be carried out only for field values below ⌐10 kOe.] The spin polarization of the electrons which pass through the probe hole is detected by Mott scattering. So far, energy-resolved spin-polarization measurements have not been reported so only the spin polarization of the (total) current density emitted from certain surfaces (crystallographic planes) are known with reasonable accuracy. The results for iron and nickel are summarized in Table 5.1 (Campagna, Alvavado, and Kisker, 1979). The spin polarization P is defined by

$$P = \frac{J^\uparrow - J^\downarrow}{J^\uparrow + J^\downarrow} \tag{5.86}$$

TABLE 5.1

Spin Polarization of the Field Emission Current[a]

Ni (hkl)	P (%)	Fe (hkl)	P (%)
100	-3 ± 1	110	-5 ± 10
110	$+6 \pm 1$	100	$+26 \pm 5$
111	-6 ± 1	111	$+20 \pm 5$
321	$+7.3 \pm 0.5$		

[a]Campagna, Alvarado, and Kisker (1979).

where $J^{\uparrow(\downarrow)}$ is the field-emitted current density of spin-up (-down) electrons. We have

$$J^{\uparrow(\downarrow)} = e \int j^{\uparrow(\downarrow)}(E) \, dE \tag{5.87}$$

One can calculate the energy distribution $j^{\uparrow(\downarrow)}(E)$ of the field-emitted electrons of spin up (down) from a ferromagnetic metal using the formulas in Section 5.2, provided ferromagnetism is treated on the basis of the band theory of ferromagnetism (Slater, 1936; Stoner, 1938; 1939). According to this theory the majority electrons (spin up) see a different one-electron potential from the minority electrons (spin down). [In the $X\alpha$ method, described in Section 3.2, the dependence of the potential on the spin is described by Eqs. (3.1) and (3.4).] The difference between the number of electrons with spin up and that of the electrons with spin down determines the magnetic moment (per atom) for the given metal. Self-consistent calculations of the spin-up and -down one-electron potentials (in the muffin-tin approximation) and of the corresponding one-electron states (of spin up and down) have been performed for nickel, iron, and cobalt by a number of authors (see, e.g., Wakoh and Yamashita, 1966; 1970; Moruzzi et al., 1978).

An approximate one-electron potential for the semi-infinite crystal (with or without an externally applied electric field) can be constructed as in the case of nonmagnetic metals (see Sections 4.2 and 5.2). The bulk potential for spin up (down) is terminated abruptly at the metal–vacuum interface (the plane $z = 0$), which is taken at a distance $(d/2 + D_s)$ above the center of the top atomic layer (d and D_s have the same meaning as in Fig. 4.1). The potential between the muffin-tin spheres of the top layer and the plane $z = 0$ is constant (equal to the interstitial bulk potential) and the surface barrier (the potential for $z > 0$) depends only on z. [The shape of the spin-up surface barrier may be different from that of the spin-down barrier. The difference, however, between the two barriers is likely to be small and very difficult to estimate in the absence of self-consistent calculations of the surface potential (see also, Nagy, 1979).] Because the spin does not enter explicitly into the formulas which determine the one-electron states or their contribution to the tunneling current, the formulas given in Chapter 4 and those in Sections 5.2 and 5.3 apply equally well for spin up or spin down. We note, however, that when a formula has been obtained by summing over states of both spin, it must now be multiplied by (½). For example, the contribution to the field-emitted current from bulk states is

$$j_b(E) = j_b^\uparrow(E) + j_b^\downarrow(E) \tag{5.88}$$

where $j_b^{\uparrow(\downarrow)}(E)$ is given by Eq. (5.18) multiplied by (½). It is understood, of course, that in calculating $j_b^\uparrow(E)$ and $j_b^\downarrow(E)$ all quantities entering Eq. (5.18) must be evaluated using the spin-up and the spin-down potential respectively. The above implies that the magnetic field at the cathode site does not affect in any significant way the local density of states at the surface of the emitting plane; this has been shown to be the case by Caroli (1975). It is also assumed that the spin state of the electron does not change as it crosses the surface barrier. Schmit and Good (1977) have shown that this is indeed the case for the magnetic fields used in the field emission experiments. A general discussion of these and other factors that might affect the spin polarization of the field-emitted electrons can be found in the articles of Nagy, Cutler, and Feuchtwang (1979), Feuchtwant, Cutler, and Schmit (1978), and Feuchtwang, Cutler, and Nagy (1978). Numerical calculations of the spin polarization of the field emission current have so far been performed only for the (100) and (111) planes of nickel.

5.6. SPIN-POLARIZED FIELD EMISSION FROM THE (100) AND (111) PLANES OF NICKEL

The calculation of the spin polarization of the field emission current from Ni(100) by Politzer and Cutler (1972) is well known. In their model, electron emission is treated as originating from the unhybridized $3d$ bands and independently from the $4s$-p bands. The $4s$-p bands are approximated by free-electron bands (the exchange splitting between the spin-up and spin-down bands is neglected). The $3d$ bands are divided into identical spin-up and spin-down band separated by an exchange splitting of 0.408 eV. The $3d$ bands and wave functions are calculated in the tight-binding approximation but the energy bands are modified to agree with the minority bands as calculated by the $X\alpha$ method (Connolly, 1967). Tunneling from the d bands is treated by matching the wave function on the vacuum side of the interface to a superposition of Bloch waves inside the metal, but the latter are evaluated in an approximate manner starting from a tight-binding description of the wave functions. In view of these approximations their result ($P = -4\%$), in very good agreement with the experimental result (see Table 5.1), must be attributed, at least partly, to a fortuitous cancellation of errors. Chazalviel and Yafet (1977) have shown that s-d hybridization plays an important role in the determination of the spin polarization of the field emission current, and concluded that in a quantitative analysis of the experimental data the hybridized nature of the electronic wave functions must be taken properly into account. Such a calculation has been performed for Ni(100) by Modinos and Oxinos (1981b) and for Ni(111) by Zavadil and Modinos (1982). These calculations were done by means of the Green's function method, described in Sections 5.2 and 5.3, using a one-electron potential for spin up (down) of the form described in Section 5.5. The bulk potential used in these calculations was determined by the following considerations. The self-consistent calculations of the energy-band structure of ferromagnetic nickel (Wakoh, 1965; Wang and Callaway, 1977; Moruzzi *et al.*, 1978; Anderson *et al.*, 1979) give an exchange splitting between the spin-up and spin-down bands (measured at the symmetry point X_5) $\Delta > 0.6$ eV. These calculations (in reasonable agreement with each other) give a good account of the ground state properties of nickel (magnetic moment per atom, Fermi surface) but they fail to reproduce the spin polarization of the photoemission current. The latter quantity has been measured as a function of photon energy (the photons are incident normally on the surface) by Eib and Alvarado (1976) for the Ni(100) plane and by Kisker *et al.* (1979) for the Ni(111) plane. The results for the (100) plane are shown in Fig. 5.15. Moore and Pendry (1978) and Kisker *et al.* (1978) were able to reproduce (respectively) the measured spin polarization of the photoelectrons emitted from the (100) and (111) nickel planes, using a spin-dependent potential which differs from that evaluated self-consistently in that the exchange splitting between the spin-up and the spin-down d-bands (specified by its value Δ at the X_5 symmetry point) is treated as an adjustable parameter. The parame-

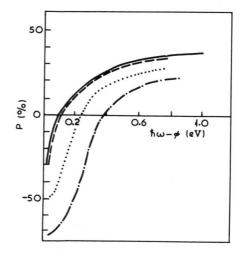

FIGURE 5.15

Experimental and calculated spin polarization of the photoemission current from Ni(100) as a function of photon energy $\hbar\omega$. The work function $\phi = 5.1$ eV. Solid curve: experimental results. Dashed curve: calculated with $\Delta = 0.33$ eV; dotted curve: calculated with $\Delta = 0.4$ eV; dot–dashed curve: calculated with $\Delta = 0.5$ eV. (From Moore and Pendry, 1978.)

trized potential of MP (Moore and Pendry, 1978), used in the calculations of the theoretical curves in Fig. 5.15, is a weighted average of Wakoh's (Wakoh, 1965) spin-up and spin-down (muffin-tin) potentials giving the required value of Δ. The energy band structure (for $\Delta = 0.33$ eV) along the ΓX and ΓL directions, normal (respectively) to the (100) and (111) planes, is shown in Fig. 5.16. [The Fermi level for a given potential (specified by the value of Δ in the corresponding band structure) is determined from Fermi surface data.] The theoretical curves in Fig. 5.15 show that the self-consistent potentials ($\Delta > 0.6$ eV) cannot explain the spin polarization of the photoemission current from the (100) plane, but when the value of Δ is reduced to 0.33 eV one obtains a very good fit to the experimental data. Very similar results were obtained for the (111) plane.

In the field emission calculations the potential on the vacuum side of the interface ($z > 0$) (for spin up and down) was approximated by [see Eqs. (4.2) and (5.1)]

$$V(z) = E_F + \phi - \frac{e^2}{4(z + z_0)} - eFz \qquad (5.89)$$

where the zero of energy coincides with the constant potential between the muffin-tin spheres inside the metal. [In general this potential (the muffin-tin zero) may

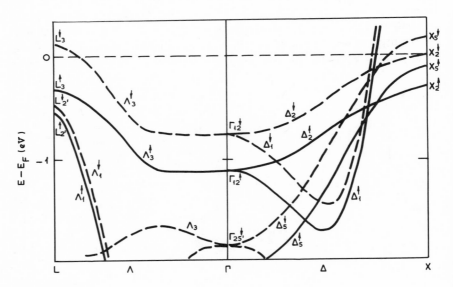

FIGURE 5.16

Energy band structure of nickel along the ΓX and ΓL symmetry direction ($\Delta = 0.33$ eV). Solid curves: majority bands (spin up); dashed curves: minority bands (spin down).

depend on the spin, so the numerical value of E_F (the Fermi level) in Eq. (5.89) may be different for the two spins. In the case of the MP potential the muffin-tin zero is the same for spin up and spin down]. The work function taken from independent experiments is denoted by ϕ. In contrast to the photoemission results shown in Fig. 5.15 (which are determined mostly by the bulk density of states and are, therefore, not very sensitive to the shape of the surface barrier), the spin polarization is very sensitive to the shape of this barrier. We examined the effect of different barriers on the (calculated) value of P, by treating z_0 in Eq. (5.89) as an adjustable parameter. It and D_s determine the shape of the surface barrier in the vicinity of the metal–vacuum interface (see Fig. 4.1).

We found (Modinos and Oxinos, 1981b) that the (calculated) spin polarization P of the field emission current from Ni(100) for the MP potential ($\Delta = 0.33$ eV) depends on the surface barrier as follows. A barrier given by Eq. (5.89) with $D_s = 0.0$ and $z_0 = (e^2/2)/(E_F + \phi)$ (barrier A) gives $P = -0.993\%$. A barrier given by Eq. (5.89) with $D_s = 0.0$ and $z_0 = (e^2/4)/(E_F + \phi)$ (barrier B) gives $P = -4.04\%$. It is obvious that with a surface barrier between barriers A and B one can obtain the measured value of $P = -3\%$. A similar calculation using the self-consistent bulk potential ($\Delta = 0.68$ eV) of MJK (Moruzzi et al., 1978) gave for the same two barriers A and B, $P = -10.004\%$ and $P = -11.198\%$, respectively. In the case of the MJK potential the majority d-bands lie well below E_F and hence contribute relatively little to the tunneling current. Therefore, as far as the Ni(100) plane is concerned, one may conclude that (a) the self-consistent potential of MJK (the same applies to any other potential leading $\Delta \gtrsim 0.6$ eV) is inconsistent with the observed value of P; (b) the MP potential is consistent with the experimental result but, in this case, the value of P depends critically on the shape of the surface barrier. It follows that if a self-consistent calculation of the surface potential (possibly spin dependent) becomes available, the value of P that it leads to can provide a good test of the accuracy of this potential. We note that the calculated $j^{\uparrow(\downarrow)}(E)$ from Ni(100) is free-electron-like for the MP and the MJK potentials (Modinos and Oxinos, 1981b). The absence of structure in $j^{\uparrow(\downarrow)}(E)$ implies that no surface states/resonances exist on the Ni(100) surface in the region of E and $\mathbf{k}_\parallel$ which contributes to the field emission current ($\mathbf{k}_\parallel$ in the central region of the SBZ and $-0.5 < E - E_F < 0$ eV). This is in agreement with the calculations of Moore and Pendry (1978), but in disagreement with the calculations of Dempsey and Kleinmann (1977) and Dempsey, Grise, and Kleinman (1978). If a band of surface states/resonances exists on Ni(100), near the Fermi level and in the central region of the SBZ, as suggested by these authors, this will greatly influence the spin polarization of the field-emitted electrons. Experimental angle resolved photoelectron spectra from Ni(100) show a surface sensitive peak in the normal direction ($\mathbf{k}_\parallel = 0$), but the physical origin of this peak is not clear (Plummer and Eberhardt, 1979).

We now consider the spin polarization of the field emission current from the nickel (111) surface (Zavadil and Modinos, 1982). Using the MP potential (Δ = 0.33 eV) and the surface barrier described by Eq. (5.89) and Fig. 4.1 with D_s = 0.0 and z_0 = $(e^2/2)/(E_F + \phi)$ (barrier A), we found a spin polarization of P = -23.1% which is much larger (in magnitude) than the measured value of P = $(-6 \pm 1)\%$. No surface states/resonances are found for this barrier (in the E and $\mathbf{k}_\parallel$ region that matters in field emission) and the corresponding $j^{\uparrow(\downarrow)}(E)$ are structureless (free-electron-like). The large negative value of P is consistent with the energy band structure along the ΓL direction (shown in Fig. 5.16). It is clear that the minority d-bands at and immediately below E_F contribute most of the tunneling current. We note, in particular, that the s-p bands (which contribute approximately equally to the spin-up and spin-down current) lie well below E_F and therefore their contribution to the tunneling current, which tends to reduce the spin polarization of this current, is relatively small. This is in contrast to the situation of the (100) plane where the s-p bands (these cross E_F for $\mathbf{k}_\parallel \simeq 0$) contribute most of the tunneling current. A second calculation using the same bulk potential (Δ = 0.33 eV) but a different surface barrier (A'), described by Eq. (5.89) with D_s = 0.2 a.u. and z_0 = $(e^2/2)/(E_F + \phi)$, gave very different results. The calculated $j^{\uparrow(\downarrow)}(E)$ for this barrier exhibit considerable structure. The corresponding enhancement factors, defined by

$$R^{\uparrow(\downarrow)}(E) = j^{\uparrow(\downarrow)}(E)/\tfrac{1}{2}j_0(E) \tag{5.90}$$

$$R(E) = [j^{\uparrow}(E) + j^{\downarrow}(E)]/j_0(E) \tag{5.91}$$

where $j_0(E)$ is the Fowler–Nordheim energy distribution [defined by Eq. (1.87)], are shown in Fig. 5.17. The structure in the energy distribution of the emitted electrons arises from a surface band (a band of surface states/resonances) which exists in the energy region immediately below E_F. [It is a true surface state at $\mathbf{k}_\parallel$ = 0 (an s-p state in the Λ_1 gap) and becomes a surface resonance for $k_\parallel > 0$. It ceases to exist for $k_\parallel \gtrsim 0.1$ Å^{-1}]. We note that this surface band is actually there because it has been observed in angle-resolved photoemission experiments (Himpsel and Eastman, 1978). We note, however, that according to the photoemission results the surface state at $\mathbf{k}_\parallel$ = 0 has energy E = E_F − 0.25 eV and (as a resonance) disperses downwards in energy in every $\mathbf{k}_\parallel$ direction, whereas we find that this (spin-up) surface state at $\mathbf{k}_\parallel$ = 0 has energy E = E_F − 0.2 eV, dispersing upwards in energy for $k_\parallel > 0$ [along the (110) direction it persists as a weak resonance up to $k_\parallel \simeq 0.07$ Å^{-1} where its energy coincides with E_F]. We have already noted that the energy distribution of the emitted electrons could not be measured in the spin-polarization experiments of Campagna et al. (see references at the beginning of this section). Measurements of the energy distribution $[j^{\uparrow}(E) + j^{\downarrow}(E)]$ from Ni(100) and Ni(111) have been reported by Rihon (1978b). The energy distribution from the (100) plane appears to be free-electron-like in agreement with our calculation. In the case of the (111) plane, although Rihon

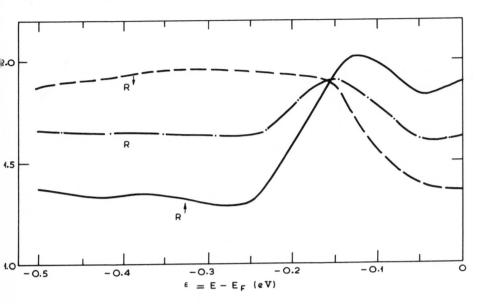

FIGURE 5.17

Calculated enhancement factors, as defined by Eqs. (5.90) and (5.91), of field-emitted electrons from Ni(111).

observed some deviation from free-electron behavior for $(E - E_F) \leq -1$ eV, he did not see any structure in the region covered by our calculation (Fig. 5.17).

A numerical integration of the $j^{\uparrow(\downarrow)}(E)$ obtained with barrier A' [corresponding to the $R^{\uparrow(\downarrow)}(E)$ curves of Fig. 5.17] gives a positive spin polarization of the field emission current $P \simeq 6.55\%$. In other words, the presence of the surface band has changed the spin polarization from -23.1% to 6.55%. We have already noted that the calculated surface band does not accurately represent the actual band as determined by angle-resolved photoemission. It is reasonable to assume that a modified surface potential could lead to a better description of the surface band and to a value of P in agreement with the experimental result. It is worth remembering that the (self-consistent) potential in the surface layer (the top one or two atomic layers of the semi-infinite crystal) may be different from the bulk potential (Nagy, 1979; Weling, 1980), although our calculations suggest that for the nickel (100) and (111) planes our assumption, namely, that the bulk potential remains the same up to the metal–vacuum interface (the plane $z = 0$), is a reasonably good first approximation. It is also clear that the top (surface) layer of these planes is not magnetically dead, because if that were the case the spin polarization of the field emission current would be zero.

In conclusion, we may say that the one-electron (bulk) potential ($\Delta = 0.33$

eV) which reproduces the measured spin polarization of the photoelectrons from Ni(100) and Ni(111) (Fig. 5.15) is also consistent with the observed spin polarization of the field-emitted electrons from these two planes, although in the latter case the exact value of the spin polarization depends critically on the surface potential barrier. The problem remains, of course, of how to justify theoretically the MP potential. Why is the exchange splitting ($\Delta = 0.33$ eV), seen in photo-emission and field emission, much smaller than that ($\Delta > 0.6$ eV) obtained in self-consistent calculations of the energy-band structure (which describe well the ground-state properties of the metal)? It is possible that an improved treatment of correlation within an (effectively) one-electron theory will lead to a band structure with a reduced Δ (Kleinmann, 1979). It may be, however, that many-body effects (e.g., electron–magnon interactions) play a more important role in determining the quasiparticle excitation spectrum seen in the photoemission and field emission experiments than they do in determining the ground-state properties of the metal, so that the exchange splitting in the excitation spectrum is different from the band splitting (Edwards and Hertz, 1973; see also, panel discussion in Lee, Perz, and Fawcett, 1978). Recent papers relevant to this, as yet unresolved, problem include Penn (1979); Liebsh (1979; 1981); Treglia, Ducastelle, and Spanjard (1980).

5.7. MANY-BODY EFFECTS

We have already discussed the effect of residual electron–electron interaction (correlation) on the one-electron states of a normal (nonmagnetic) metal in Sections 3.2 and 4.5. We have seen that because of this interaction a (sharp) energy level acquires a finite width $2\Gamma = \hbar/\tau$ [τ is the lifetime of the electron (or hole) in the corresponding state]. We have also seen how a finite Γ can be taken into account in an (effectively) one-electron theory by adding an energy-dependent imaginary part to the one-electron potential. The effect of a finite Γ on the energy distribution of the field-emitted electrons comes through $\langle \chi_{q\parallel} | G(E) | \chi_{q\parallel} \rangle$ [defined by Eq. (5.43)] whose evaluation is facilitated rather than hindered by the addition of a small imaginary component to the potential (see, e.g., Section 5.3). We must remember, however, that for energies in the region of the Fermi level Γ is very small [see Eq. (3.20) and Fig. 3.1]. If there are sharp resonances in the surface density of states (leading to corresponding peaks in the energy distribution of the emitted electrons) these are softened by a finite Γ, but otherwise this has no significant effect on the energy distribution. [It would seem that a comparison between the experimental enhancement $R(E)$ (at a surface resonance) and theoretical curves corresponding to different values of Γ could provide information on the magnitude of this quantity. This, however, is not possible in practice because Γ (for field-emitted electrons) is of the same order of magnitude (or smaller) as the energy resolution of present day energy analyzers.]

A more subtle manifestation of electron–electron correlation in the energy distribution of field-emitted electrons has been invoked by Lea and Gomer (1970) in order to explain a high-energy tail observed by these authors in the energy distribution from the (111), (112), and (120) planes of tungsten. Their experimental results have been confirmed by Gadzuk and Plummer (1971b; 1973), who measured the energy distribution over a wider energy region (up to 3 eV above E_F). Their results for the W(111) plane are shown in Fig. 5.18. The dotted line in this figure shows the energy distribution expected from the Fowler–Nordheim theory (the shape of this distribution for $E - E_F > 0$ is determined by the Fermi–Dirac distribution function). It is clear that adding an imaginary component to the potential inside the metal which is equivalent to replacing the surface density of states by a modified one in the manner described by Eq. (4.45) cannot lead to the (observed) high-energy tail. In principle, Coulomb scattering on the vacuum side of the barrier might lead to deviations from the Fowler–Nordheim distribution and as we have already mentioned in Section 1.7.1, this interaction is held responsible for the very wide distributions observed at high emitted current den-

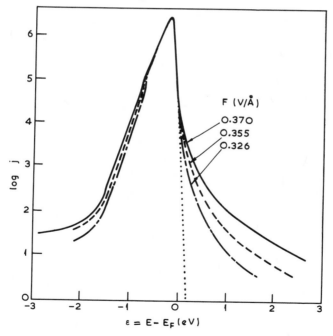

FIGURE 5.18

Total energy distributions (normalized) of field-emitted electrons from W(111) at 78 K. (From Gadzuk and Plummer, 1971b.)

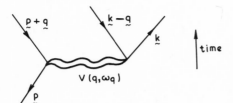

FIGURE 5.19

Diagram illustrating the process leading to an electron with energy $E_k > E_F$ and an electron with energy $E_p < E_F$ tunneling out of the metal.

sities. Lea and Gomer (1970) argue, however, that in their experiments, the current emitted from the entire tip is sufficiently small leading to an average interaction of an electron with the space charge on the vacuum side of the barrier which is orders of magnitude smaller than the width of the observed tail. Therefore, to obtain the latter, one would have to postulate the existence of extremely large fluctuations in the space charge without obvious justification. According to Lea and Gomer, the high-energy tail in the energy distribution of the emitted electrons results from electron–electron scattering in the metal. Their arguments can be described best by reference to the diagram of Fig. 5.19 (Gadzuk and Plummer, 1971). An electron with energy $E_{p+q} > E_F$ and momentum $\mathbf{p} + \mathbf{q}$ "jumps" into a state of energy $E_p < E_{p+q}$ and momentum $\mathbf{p}$, the latter state emptied by electron tunneling; this process is coupled via a momentum and frequency dependent screened Coulomb interaction (denoted by $V(q,\omega_q)$ and represented by the wavy line in the diagram) to the creation of a secondary hole–electron pair: an electron with energy $E_{k-q} < E_F$ takes up the energy $(E_{p+q} - E_p)$ "jumping" into a state with energy $E_k > E_F$. The latter electron may tunnel out of the metal (before recombination occurs) and in this way a high-energy tail (such as the one shown in Fig. 5.18) is obtained. We note that the "event" described by Fig. 5.19 can take place only when the electron scattered downward in energy can tunnel out of the metal. Therefore, the overall probability of observing an electron of energy $E_F + \Delta$ in the high-energy tail is given by a dynamical factor $f(\Delta)$, determined by the scattering process in the metal, multiplied by the probability P_{tunnel} of *both* electrons (represented by $\mathbf{p}$ and $\mathbf{k}$ in Fig. 5.19) tunneling out of the metal. Assuming (Lea and Gomer, 1970) that these two electrons have energy $E_F - \Delta$ and $E_F + \Delta$, respectively, one obtains (see Section 1.4)

$$P_{\text{tunnel}} \simeq \exp[-C(\phi + \Delta)^{3/2}/F] \exp[-C(\phi - \Delta)^{3/2}/F] \qquad (5.92)$$

where ϕ and F denote, as usual, the work function and the applied field, respectively, and $C = 0.683$ in eV Å units. To first order in Δ, the energy distribution $j(\Delta)$ in the high-energy tail is given by

$$j(\Delta) = P_{\text{tunnel}} \simeq \exp(-2C\phi^{3/2}/F) \qquad (5.93)$$
$$\simeq \text{const} \cdot I^2$$

where I is the total emitted current. [The latter step in Eq. (5.93) follows from Eq. (1.60); the image correction factor v has been put equal to unity.] Therefore, the current in the high-energy tail is given by

$$i^* = \int j(\Delta)\, d\Delta \propto I^2 \qquad (5.94)$$

Lea and Gomer (1970) and independently Gadzuk and Plummer (1971; 1973) have shown that i^* is indeed proportional to the square of the measured (probe-hole) current. Gadzuk and Plummer (1971b) have also derived a semiempirical formula for the high-energy tail of the TED curve based on the above ideas, by treating approximately the screened Coulomb interaction responsible for the scattering process shown in Fig. 5.19. A summary of their results (in reasonable agreement with the experimental data) and references to relevant work by other authors is to be found in their review article (Gadzuk and Plummer, 1973). Although the wing at the low-energy end of the energy distribution (seen in Figs. 5.4 and 5.18) has not been carefully analyzed, it is likely that it also derives from processes similar to those described in the present section. It appears that because of this tail, it will not be possible (even with a perfect analyzer) to probe the surface density of states by means of TED measurements at energies below a certain level (2 eV or so from E_F). On the other hand, it appears that accurate measurements of the high-energy tail of the TED of the field-emitted electrons could be very useful in the study of many-body effects.

5.8. PHOTOFIELD EMISSION

The process of photofield emission is shown schematically in Fig. 5.20. An electron is first excited by a photon and subsequently tunnels out of the metal. A theory of photofield emission based on a simple model of the metal has been described by Bagchi (1974). Measurements of the photofield emission current have been reported by Radon and Kleint (1976, 1977), Lee and Reifenberger (1978), and Radon (1980). The observation of a periodic electric-field-dependent component in the photocurrent (at a given incident photon energy) is very inter-

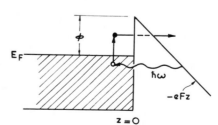

FIGURE 5.20

Schematic description of photofield emission.

esting because of the unusually small period ($\sim$0.02 V/Å) of this oscillation, which cannot be explained by barrier-induced resonances in the transmission coefficient. It has been suggested (Reifenberger, Haavig, and Egert, 1981) that dynamical effects (related to the time involved in the formation of the image charge on the metal surface) may be responsible for the observed oscillations.

A review of photofield emission studies lies beyond the scope of the present book. There are a number of review articles on photoemission (see Section 4.2.2 for references), and photofield emission is in reality a special case of photoemission.

CHAPTER SIX

FIELD EMISSION FROM ADSORBATE-COVERED SURFACES

6.1. INTRODUCTION

The current I emitted from a given area of an adsorbate-covered metal surface, when an external voltage V is applied between it and the anode in a field emission diode, obeys the Fowler–Nordheim (FN) law (as in emission from clean metals), i.e.,

$$I = b^{\text{ad}} V^2 \exp(m_f^{\text{ad}}/V) \tag{6.1}$$

m_f^{ad} and $\ln(b_{\text{ad}})$ are given by the slope and intercept (respectively) of the experimental FN plot (a straight line) of $\ln(I/V^2)$ versus V^{-1}. A comparison between m_f^{ad} and b_{ad} and the corresponding quantities, m^{cl} and b_{cl}, obtained from measurements on the clean metal surface can provide us, when properly analyzed, with valuable information on the electronic structure of the adlayer–substrate complex. We assume that, by experimental design, the current is drawn from the same area before and after adsorption and that this area is part of a single (substrate) crystal plane. (The development of suitable current probes made such measurements perfectly possible.) By analogy to Eq. (2.14), we have

$$m_f^{\text{ad}} = -0.683\phi_{\text{ad}}^{3/2}s'/\beta' \tag{6.2}$$

where ϕ_{ad} denotes the work function of the adsorbate-covered surface, and s' and β' have the same meaning as s and β in Eq. (2.14). The adsorption of a uniform layer of atomic thickness does not change in any significant way the value of the local field factor [which is determined by the geometry of the field emission tip (whose radius is much larger than the adlayer thickness)], and, therefore, we may assume that $\beta' = \beta$. If we further assume that $s' \simeq s$ (we justify this assumption later in our discussion) we obtain

$$\phi_{\text{ad}}/\phi_{\text{cl}} = (m_f^{\text{ad}}/m_f^{\text{cl}})^{2/3} \tag{6.3}$$

169

Hence, if the work function ϕ_{cl} of the clean surface is known, one obtains the work function ϕ_{ad} of the adsorbate-covered surface directly from the experimentally determined slopes m_f^{cl} and m_f^{ad}. In adsorption studies one is particularly interested in the change of the work function following adsorption, defined by

$$\Delta\phi = \phi_{ad} - \phi_{cl} \qquad (6.4)$$

The intercept of the FN plot depends on the shape of the surface barrier in the immediate vicinity of the emitter–vacuum interface and on the density of states at this interface at the Fermi level. Therefore, the change in the intercept of the FN plot, given by

$$B = \ln(b_{cl}/b_{ad}) \qquad (6.5)$$

reflects the change in the above quantities following adsorption. Additional information on the electronic structure of the adlayer–substrate complex is obtained by comparing the total energy distribution of the electrons emitted from the adsorbate-covered surface with that obtained from the clean surface.

Unfortunately, a quantitative analysis of field emission data from adsorbate-covered surfaces, comparable to that from clean surfaces (Chapter 5) is very difficult and much remains to be done in this direction. In the present chapter we provide some rules for a semiquantitative interpretation of such data, we indicate those cases for which a reasonable theory has been, or can be formulated, hoping that some explicit calculations will be undertaken along these lines in future work, and we point out some of the existing problems and difficulties. We begin, however, with a summary of some typical experimental data. An extensive list of relevant papers can be found in review articles by Gadzuk and Plummer (1973) and by Jones (1980).

6.2. FIELD EMISSION FROM ADSORBATE-COVERED SURFACES. SOME EXPERIMENTAL RESULTS

6.2.1. INERT ATOMS ON TUNGSTEN

Our first example, that of krypton on tungsten, is typical of inert atom adsorption which has been thoroughly studied by Gomer and his co-workers. Figure 6.1 taken from a paper by Lea and Gomer (1971) shows $\Delta\phi$ and B for Kr on W(112) as a function of fractional coverage θ, from zero to a full monolayer. [We note that the direct determination of the absolute coverage in the first layer, mea-

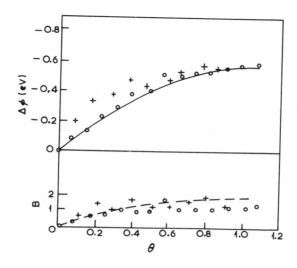

FIGURE 6.1

$\Delta\phi$ and B, as defined by Eqs. (6.4) and (6.5) for Kr on W(112) as a function of fractional coverage, from zero to a full monolayer. The circles denote experimental results taken by Lea and Gomer, and the crosses denote experimental results taken by Engel and Gomer (1970). The solid line for $\Delta\phi$ is a best fit to the data (circles) by the Topping formula [Eq. (6.6)]. The dashed line for $B(\theta)$ was calculated from Eq. (6.24) using the HCP value of n_0 and the value of α determined from fitting $\Delta\phi(\theta)$ by the Topping formula. (From Lea and Gomer, 1971.)

sured in adatoms per unit area, on a plane of a field emission tip is not possible at the present moment. Usually one obtains, in the first instance, $\Delta\phi$ versus the number of doses impinged on a given area of the emitter. While these doses may be well defined and reproducible they cannot be translated unambiguously into absolute coverage for the following reasons. The sticking coefficient is not always unity or even constant but, in general, is a function of the temperature and of the coverage itself. Even in such cases, e.g., immobile adsorption at very low temperatures when the sticking coefficient is unity, the coverage in the first layer is not necessarily equal to the total coverage (equal to the number of atoms impinged on unit area) since it is possible for incipient second (and higher) layer formation to occur before the first layer is completed. This is demonstrated quite clearly in Fig. 6.2, taken from a paper by Wang and Gomer (1980), which shows the coverage in the first layer versus total coverage for xenon on *macroscopic* tungsten (110) and (100) surfaces at 20 K (by "macroscopic" we mean extended surfaces of single crystals in contrast to the small area planes of a field emission tip). Nevertheless, it is possible in some instances to translate impinged doses into a fractional coverage θ in the first layer within reasonably good limits of approximation. If, for

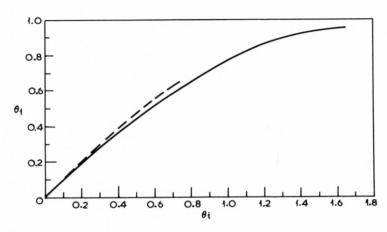

FIGURE 6.2

Coverage in the first layer θ_1 versus total coverage θ_i. Both quantities are normalized: $\theta_1 = n_1/n_0$ and $\theta_i = n_i/n_0$, where n_1 and n_i are the Xe atoms per cm^2 in the first layer and total impinged Xe atoms per cm^2, respectively, and $n_0 = 5.5 \times 10^{14}$ atoms per cm^2 is the maximum coverage in the first layer. Solid curve: Xe on W(110). Dashed curve: Xe on W(100). (From Wang and Gomer, 1980.)

example, $\Delta\phi$ levels off beyond monolayer coverage, as appears to be the case in inert atom adsorption, $\theta = 1$ can be taken as the point where the curve $\Delta\phi$ versus impinged doses levels off. If it is further assumed that the ratio of the coverage in the first layer to the total coverage is constant over the range $0 < \theta < 1$ (this is only approximately true as can be seen from Fig. 6.2), one can translate the number of impinged doses into fractional coverage in the first layer. This is how the curve in Fig. 6.1 was obtained. Obviously, if the density of adsorbed atoms at full monolayer coverage is known, one can translate the fractional coverage θ into absolute coverage. We note that methods have been devised (see, e.g., Wang and Gomer, 1978) which allow one to measure the absolute coverage on macroscopic surfaces, and it is, therefore, possible to determine, for such surfaces, the change of the work function on adsorption as a function of absolute coverage. When such data are available for a given substrate–adatom system, they can be used in a field emission experiment, or for that matter in any other similar situation, to obtain the absolute coverage corresponding to a measured value of $\Delta\phi$, if one assumes a one-to-one correspondence between $\Delta\phi$ and absolute coverage.] The solid curve in Fig. 6.1 represents a fit to the experimental data by the Topping formula (Topping, 1927)

$$\Delta\phi = eC_1\theta/(1 + C_2\theta^{3/2}) \qquad (6.6a)$$

with C_1 and C_2 treated as adjustable parameters. It is common to express C_1 and C_2 in terms of two other parameters α and μ as follows:

$$C_1 = \frac{1}{e}\left(\frac{d\Delta\phi}{d\theta}\right)_{\theta=0} = 4\pi n_0 \mu \qquad (6.6b)$$

$$C_2 = 9\alpha n_0^{3/2} \qquad (6.6c)$$

where n_0 is the number of adatoms per unit area at full monolayer coverage. μ has the dimensions of dipole moment and α of polarizability. We note that in the limit of low coverages ($\theta \ll 1$), when no appreciable interaction between the adatoms exists, Eqs. (6.6) give

$$\Delta\phi/e \simeq C_1\theta = 4\pi\mu n_0\theta \qquad (6.7)$$

where $n_0\theta$ represents the absolute coverage for the given θ. The expression on the right-hand side of Eq. (6.7) gives the potential difference between the two sides of an infinitely thin surface dipole layer, with a continuous distribution of ($\mu n_0\theta$) dipole moment per unit area. If we disregard the fact that in the present case we have a two-dimensional array of discrete dipoles of finite length instead of a continuous distribution of point dipoles, we can use Eq. (6.7) and the experimentally determined value of C_1 to estimate the dipole moment μ associated with one adatom, if we know the adatom density n_0. Unfortunately, n_0 has so far been measured directly for only one of the inert atoms, xenon, on W(100) and W(110). It has been established (Wang and Gomer, 1979) that for these two planes at $T <$ 66 K, we have $n_0 = 5.5 \times 10^{14}$ Xe atoms per cm^2, corresponding effectively to hexagonal closepacking (HCP). In the case of the W(100) plane the above number is reduced at 66 K to 4.5×10^{14} Xe atoms per cm^2, almost the same with the density (5×10^{14} atoms per cm^2) corresponding to a commensurate ($\sqrt{2} \times \sqrt{2}$) 45° structure. Assuming an HCP value for the density of Kr atoms (at full monolayer coverage) on W(112) at the bath temperature (15 K), one obtains, given the atomic radius of krypton (2.01 Å), a density $n_0 = 7.2 \times 10^{14}$ atoms/cm^2, which in turn leads to a dipole moment (normal to the surface) $\mu = -0.43$ debye units.† The denominator in Eq. (6.6a) takes into account in a semiempirical way depolarization due to dipole–dipole interaction between the adatoms, and α in Eq. (6.6c) may be thought of as the polarizability of the "adatom complex." (The "adatom complex" may be loosely defined as that region of space inside and around the adatom where a charge redistribution occurs following adsorption.) It is obvious that α cannot be identified with the polarizability of the free atom, although it might be of comparable magnitude, and that in general it depends on the properties of the metal substrate as well as those of the adsorbed atom. Sub-

---

†The negative sign defines the direction of the dipole moment: An array of such dipoles leads to a reduction in the value of the work function.

stitution of the experimentally determined value of C_2 and the HCP value of n_0 into Eq. (6.6c) gives, for the Kr on W(112) complex, $\alpha = 5.76$ Å³ which is larger than the atomic polarizability by a factor of approximately 2. This is not surprising in view of the above-mentioned differences between the two systems and the semiempirical nature of Eq. (6.6).

Lea and Gomer (1971) obtained results similar to those shown in Fig. 6.1 for Kr adsorption on other planes of tungsten, namely, the (110), (100), and (111) planes. Similar results have also been obtained by Engel and Gomer (1970) for Ar, Xe, and Kr adsorption on the (110), (120), (100), (211), and (111) planes of tungsten. It is worth noting that in the case of Kr the results obtained by Lea and Gomer are in good agreement with those of Engel and Gomer, although the apparatus used was different in each case. In Table 6.1 we give the FN work function change [obtained on the basis of Eqs. (6.2) and (6.3)] at monolayer coverage at the bath temperature (4.2 K for Ar, 15 K for Kr, and 20 K for Xe). The dipole moments in the third column were obtained by Lea and Gomer from a best fit to the Topping formula, in the manner we have described in relation to Fig. 6.1, assuming an HCP value for the density of adatoms (at $\theta = 1$) in each case. The numbers in the fourth column are the results of a theoretical calculation by Oxinos and Modinos (1979); the basic assumptions of this calculation will be described in a later paragraph. In the last column of Table 6.1 we give approximate values of $B(\theta = 1)$, as defined by Eq. (6.5), extracted from experimental curves, like Fig. 6.1, obtained by Engel and Gomer and Lea and Gomer for krypton. We must emphasize at this point that the large values of $\Delta\phi$ (and hence of μ) obtained on the basis of Eq. (6.2) for Kr and Xe on the (110) plane are erroneous. Subsequent measurements (not based on field emission) by Wang and Gomer (1980) of the work function change associated with the adsorption of Xe (at $\theta = 1$) on macroscopic W(110) and W(100) surfaces gave $\Delta\phi(100) = -1.1$ eV in good agreement with the FN value of this quantity, but a value for the (110) plane, $\Delta\phi = -0.45$ eV, which is much smaller than the FN value of this quantity (-2.4eV) shown in Table 6.1. [It is worth noting that the same large FN value has been obtained in more recent experiments by Chen and Gomer (1980)]. By implication, the large value of $B(\theta = 1)$ for the W(110) plane shown in Table 6.1 is also erroneous. Wang and Gomer concluded that "in the case of the W(110) plane the FN equation breaks down for Xe adsorption." Obviously the same applies to Kr and probably, though to a lesser degree, to Ar adsorption on the same plane. There are also other known instances of adsorption on W(110), e.g., virgin CO adsorption (Engel and Gomer, 1969), where again the FN equation breaks down for no apparent (understood) reasons.

Before discussing the possible reasons for the reduction in the preexponential term of the FN formula [Eq. (6.1)] following inert atom adsorption on tungsten (Table 6.1), let us consider rather more generally how the surface potential barrier may change when a layer of foreign atoms is adsorbed on a metal surface. We need to do that not only because of the effect of this change on B, but also in order

TABLE 6.1

Adsorption of Inert Atoms on Tungsten. Experimental Data[a]

Plane	$\Delta\phi(\theta = 1)$ (eV)	μ (D)	μ theory (D)	$B(\theta = 1)$
Argon				
(110)	−0.85	−0.40	−0.183	
(100)	−0.46	−0.41	−0.215	
(112)	−0.34	−0.23	−0.228	
(111)	−0.40	−0.24	−0.244	
Krypton				
(110)	−1.97	−1.58	−0.298	9.0–10.0
(100)	−0.83	−0.65	−0.325	1.0–1.5
(112)	−0.58	−0.43	−0.335	1.0–2.0
(111)	−0.83	−0.56	−0.349	1.0–3.0
Xenon				
(110)	−2.4	−2.94		
(100)	−1.35	−1.16		
(112)	−0.92	−0.90		
(111)	−1.13	−0.60		

[a] Engel and Gomer (1970); Lea and Gomer (1971).
[b] Oxinos and Modinos (1979).

to clarify the assumption, made in our derivation of Eq. (6.2), that the correction factor s' in Eq. (6.2) is practically the same as s (the corresponding factor before adsorption). We assume, for the sake of simplicity, that we can treat a single layer of atoms as if it were a continuous film of thickness l and dielectric constant K. Imagine an electron sitting at a distance z from the emitter–vacuum interface (the latter is taken as the xy plane and the z axis points outwards into the vacuum). The force pulling the electron towards the interface can be described in terms of an "image potential energy" field given by

$$U(z) = E_F + \phi - \frac{(K-1)}{(K+1)}\frac{e^2}{4z} - \frac{2}{(1+K)}\frac{e^2}{4(z+l)} \quad (6.8)$$

for $z \gg 0$. The derivation of the above formula is based on a straightforward extension of the method of images as applied to the case of one electron sitting in front of a dielectric surface (see, e.g., Jeans, 1941). Note that in the limit $l \to 0$, corresponding to an electron sitting in front of a clean metal surface, Eq. (6.8) reduces to the expression given by Eq. (1.26), and in the limit $l \to \infty$, corresponding to an electron sitting in front of a dielectric surface, it reduces to the equally well-known formula

$$U(z) = E_F + \phi - \frac{(K-1)}{(K+1)}\frac{e^2}{4z} \quad (6.9)$$

In Fig. 6.3 the potential energy barrier described by Eq. (6.8), is compared with the potential barrier described by Eq. (1.26). When an external applied field is applied to the adsorbate-covered surface the following term must be added to the potential energy field:

$$V_F(z) = -(eF/K)(l + z), \quad -l < z < 0 \quad\quad (6.10)$$
$$= -(eF/K)l - eFz, \quad z > 0$$

Hence, the total barrier on the vacuum side of an adsorbate covered surface is given by

$$V(z) = E_F + \phi_{ad} - \left(\frac{eF}{K}\right)l - eFz - \frac{(K-1)}{(K+1)}\frac{e^2}{4z} - \frac{2}{(1+K)}\frac{e^2}{4(z+l)}$$

$$(6.11)$$

We assume, as we have done in the case of the clean metal surface [Eq. (1.27)], and for the purposes of the present argument, that the above formula is valid for $z > z_c$ where z_c is such that $V(z_c) = 0$.

We recall (see Section 1.4) that the field-emitted current is proportional to the tunneling probability T_F of an electron with energy $E = E_F$ incident normally

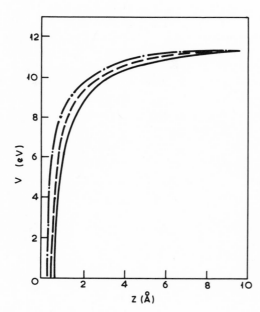

FIGURE 6.3

The potential barrier described by Eq. (6.8) with $E_F + \phi = 12$ eV, $l = 2$ Å, $K = 2$ (dash–dot curve) and $K = 4$ (dashed curve) is compared with the simple image barrier ($K = \infty$) represented by the solid curve.

on the surface barrier, and that

$$T_F = \exp\left[-\frac{4}{3e}\left(\frac{2m}{\hbar^2}\right)^{1/2} \phi^{3/2}\frac{v'}{F}\right] \tag{6.12}$$

where v' is a correction factor depending on the shape of the barrier, $v' = 1$ for a triangular barrier [Eq. (1.63)], and $v' = v$ (listed in Table 1.1) for the image barrier [Eq. (1.27)]. Noting that the potential barrier described by Eq. (6.11) lies somewhere between the triangular and the image barrier (one can see that this is so from Fig. 6.3), we conclude that the tunneling probability of an electron with $E = E_F$ incident normally on this barrier will be given by

$$T_F' = \exp\left\{-\frac{4}{3e}\left(\frac{2m}{\hbar^2}\right)^{1/2}\left[\phi_{ad} - \left(\frac{eF}{K}\right)l\right]^{3/2}\frac{v'}{F}\right\} \tag{6.13}$$

where v' has a value between unity and v as listed in Table 1.1. Putting $v' = 1$, and using the fact that $(eF/K)l \ll \phi_{ad}$, we obtain

$$T_F' \simeq M' \exp\left[-\frac{4}{3e}\left(\frac{2m}{\hbar^2}\right)^{1/2}\frac{\phi_{ad}^{3/2}}{F}\right] \tag{6.14}$$

$$M' \simeq \exp\left[2\left(\frac{2m}{\hbar^2}\right)^{1/2}\left(\frac{l}{K}\right)\phi_{ad}^{1/2}\right] \tag{6.15}$$

The current emitted from the adsorbate covered surface is proportional to $T_F'F^2$, where $F = \beta V$, which leads to a value of m_f^{ad} given by Eq. (6.2) with $s' \simeq 1$. Since the corresponding factor $s(y)$ (Table 1.1) in m_f^{cl} is almost unity, our assumption that $s' \simeq s$ [see comments following Eq. (6.2)] may lead to an error in the estimation of $\Delta\phi$, as defined by Eq. (6.4), not greater than 5% or so.

Let us now consider B as defined by Eq. (6.5). According to Eq. (6.14) penetration of the applied field through the adlayer to the surface of the metal substrate at $z = -l$ [see Eq. (6.10)], tends to increase the preexponential term in Eq. (6.1) by a factor M'. However, there are other factors to be taken into account in evaluating B. From our discussion in Sections 1.7.1 and 5.2 (see, in particular, Fig. 5.1) we have

$$\frac{b_{cl}}{b_{FN}} \simeq R_{cl}(E_F) \tag{6.16}$$

$$R_{cl}(E) = \gamma_{cl}(E)\langle\rho_{cl}(E, z = 0)\rangle_{c;0} \tag{6.17}$$

The same equation with "cl" replaced by "ad" is valid for the adsorbate covered surface. $R_{cl}(R_{ad})$ is the enhancement factor [Eq. (1.93)] for the clean (adsorbate covered) surface. To obtain R_{ad} we divide the measured TED by the FN distribution [Eq. (1.87)] for the same work function ϕ_{ad}. b_{FN} is the *theoretical* value of $b_{cl}(b_{ad})$ obtained on the basis of Eq. (1.57) for the given work function and the given emitting area. [We note that b_{FN} is not identical with A' in Eq. (1.57) because of the dependence of B', in the exponent of that equation, on the applied field.] $\langle \rho(E, z = 0) \rangle_{c;0}$ for the clean surface is given by Eq. (5.30). A formal definition of this quantity valid for clean and adsorbate-covered surfaces is given by Eq. (6.33). γ_{cl} (similarly γ_{ad}) is a *slowly* varying function of the energy which depends on the shape of the barrier at the interface but not (directly) on the surface density of states. From Eq. (6.17) and the corresponding equation for the adsorbate-covered surface we obtain

$$\frac{b_{cl}}{b_{ad}} \simeq \gamma_F \frac{\langle \rho_{cl}(E_F, z = 0) \rangle_{c;0}}{\langle \rho_{ad}(E_F, z = 0) \rangle_{c;0}} \tag{6.18}$$

which tells us that the difference between b_{cl} and b_{ad} is partly due to a difference in the density of states, at the Fermi level, at the emitter–vacuum interface before and after adsorption (we note that in the latter case the interface, i.e., the plane $z = 0$, is taken above the adsorbed layer of atoms and not between it and the metal substrate), and partly due to the fact that the potential barrier for $z > 0$ is not the same before and after adsorption. The latter determines the value of γ_F in Eq. (6.18).

In the case of inert atom adsorption we can obtain a rough estimate of (b_{cl}/b_{ad}) by assuming that $\gamma_F = 1/M'$, where M' is given by Eq. (6.15), which implies that field penetration is the dominant factor in the determination of this quantity, and by treating the second factor in Eq. (6.18), $\langle \rho_{cl}(\ldots) \rangle_{c;0} / \langle \rho_{ad}(\ldots) \rangle_{c;0}$, approximately as follows. We assume (we shall seek a justification for this assumption later in our discussion) that the adsorbed layer of inert atoms may be replaced for the purpose of calculating $\langle \rho_{ad}(E_F, z = 0) \rangle_{c;0}$ by an effective square barrier of thickness l and a height (above E_F) equal to ϕ_{ad}. In that case we may write

$$\langle \rho_{ad}(E_F, z = 0) \rangle_{c;0} \simeq \exp \left[-2 \left(\frac{2m\phi_{ad}}{\hbar^2} \right)^{1/2} l \right] \langle \rho_{cl}(E_F, z = 0) \rangle_{c;0} \quad (6.19)$$

and, therefore,

$$B \simeq \ln \left(\frac{b_{cl}}{b_{ad}} \right) \simeq 2 \left(\frac{2m\phi_{ad}}{\hbar^2} \right)^{1/2} l \frac{(K - 1)}{K} \tag{6.20}$$

According to Eq. (6.10) the electric field in the adlayer region ($-l < z < 0$) has the value F/K, where F is the externally applied field. By definition (see, e.g., any book on electrostatics) we have

$$4\pi P = (K - 1)F/K \tag{6.21}$$

where P is the polarization (field-induced dipole moment per unit volume) in the adlayer region. On the other hand, using Topping's formula we obtain, by analogy to Eqs. (6.6),

$$P = \frac{\mu_i n_0^\theta/l}{1 + 9\alpha n^{3/2}\theta_0^{3/2}} \simeq \frac{\alpha F n_0 \theta/l}{1 + 9\alpha n^{3/2}\theta_0^{3/2}} \tag{6.22}$$

We note that $\mu_i = \alpha F$ is a field-induced dipole moment in contrast to μ in Eq. (6.6b) which is for practical purposes field independent. It is also worth noting that μ in the case of inert atom adsorption (Table 6.1) has the opposite direction of μ_i. By comparing Eq. (6.21) with Eq. (6.22) we find

$$l(K - 1)/K \simeq \frac{4\pi\alpha n_0\theta}{1 + 9\alpha n^{3/2}\theta_0^{3/2}} \tag{6.23}$$

which we substitute into Eq. (6.20) to obtain

$$B(\theta) = \frac{8\pi(2m\phi_{\rm ad}/\hbar^2)^{1/2}\alpha n_0\theta}{1 + 9\alpha n_0^{3/2}\theta^{3/2}} \tag{6.24}$$

The above formula was first derived by Schmidt and Gomer (1965) by somewhat different arguments. Lea and Gomer (1971) obtained a good fit to the experimental data for $B(\theta)$ for Kr adsorption on the (100), (111), and (112) planes of tungsten using this formula with the HCP monolayer density ($n_0 = 7.2 \times 10^{14}$ atoms/cm^2) and the value of α obtained by fitting Topping's formula [Eqs. (6.6)] to the experimental data for $\Delta\phi(\theta)$ [see the discussion following Eq. (6.7)]. This is demonstrated in Fig. 6.1 for the case of Kr on the (112) plane. The broken curve in this figure shows $B(\theta)$ calculated in the above manner. Although the good fit between Eq. (6.24) and the experimental data may be to some degree the result of a fortuitous cancellation of errors, our simple model [whereby the layer of inert atoms is replaced by an effective square barrier which, in turn, leads to Eq. (6.19)] appears to be reasonable. It is interesting to consider briefly what this simple model implies for the total energy distribution of the emitted electrons. Let $j_{\rm ad}(E)$ denote this quantity for the adsorbate-covered surface and $j_{\rm cl}(E)$ that of the clean surface. The ratio of these two distributions is, by analogy to Eqs. (6.17) and (6.18), give by

$$r(E) = \frac{j_{\rm cl}(E)}{j_{\rm ad}(E)} \simeq \gamma(E) \frac{\langle \rho_{\rm cl}(E, z = 0)\rangle_{c;0}}{\langle \rho_{\rm ad}(E, z = 0)\rangle_{c;0}} \tag{6.25}$$

where $\gamma(E)^{-1}$ is given by Eq. (6.15) with ϕ_{ad} replaced by $\phi_{ad} - E + E_F$ and $\langle \rho_{ad}(E,z = 0)\rangle_{c;0}/\langle \rho_{cl}(E,z = 0)\rangle_{c;0}$ is given by Eq. (6.19) with E_F replaced by E and ϕ_{ad} replaced by $\phi_{ad} - E + E_F$. Repeating the steps leading from Eq. (6.18) to Eq. (6.24), we obtain

$$\ln r(E) \simeq \left(\frac{\phi_{ad} - \epsilon}{\phi_{ad}}\right)^{1/2} B(\theta), \qquad \epsilon \equiv E - E_F \qquad (6.26)$$

with $B(\theta)$ given by Eq. (6.24). The above equation suggests that over a narrow region of energy ($|\epsilon| \ll \phi_{ad}$), r is practically constant and equal approximately to b_{cl}/b_{ad}. Lea and Gomer measured j_{ad} over the range -0.6 eV $< \epsilon < 0$ from Kr-covered W(112), W(111), W(100), and W(110) planes. Their results for W(110) must be disregarded (see comments in relation to Table 6.1). In the case of W(112) and W(111) they found $\ln r(E) \simeq B(\theta)$ in agreement with Eq. (6.26) within the limits of experimental error.

An approximate estimate of the dipole moment μ associated with an adatom (see Table 6.1) can be made, and a "reason" as to why an adlayer of inert atoms behaves like an effective square barrier can be given as follows. We calculated (Oxinos and Modinos, 1979) the local density of states in the neighborhood of a singly adsorbed inert atom assuming (i) that inside the metal the potential is constant (equal to the minimum of the conduction band which is taken as the zero of energy); (ii) that the surface potential barrier is a step barrier (the step equals $E_F + \phi$); and (iii) that the adsorbed atom is adequately represented by a spherical potential of the muffin-tin type superimposed on the constant potential on the vacuum side of the interface. The variation of this potential along the z axis (normal to the surface and passing through the adatom nucleus) is shown schematically in Fig. 6.4. Contact with reality was made by replacing the potential inside the adatom sphere by that of the free atom as calculated self-consistently by

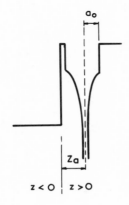

FIGURE 6.4

Variation (schematic) along the z direction of the one-electron potential used in the calculation of the local density of states in the neighborhood of an inert atom adsorbed on a jellium (free-electron–metal) substrate.

Schwarz (1972). We note that the results obtained with this model potential are in good qualitative agreement with the result of a self-consistent calculation by Lang (1981). We found that the valence E_{4p} level of Kr gives rise to a sharp adatom resonance (see Section 7.2) at $E \simeq 4.39$ eV which lies well below the region accessible by field emission ($E_F = 8.5$ eV; $\phi = 4.5$ eV). For Ar the corresponding resonance occurs at $E_{3p} \simeq 3.21$ eV. (The calculation was not done for Xe.) The dipole moment associated with the adatom is given by

$$\mu = \mu(E_F) \tag{6.27}$$

$$\mu(E) = \int_{-\infty}^{E} p(E)\, dE \tag{6.28}$$

$$p(E) = \int e(z - Z_a)[\rho(E,\mathbf{r}) - \rho_M(E,\mathbf{r})]\, d^3r \tag{6.29}$$

where Z_a is the distance of the atomic nucleus from the interface assumed equal to the atomic radius, and $\rho_M(E,\mathbf{r})$ is the local density of states for the clean metal (represented by the potential shown in Fig. 6.4 *without* the adatom well). $p(E)\, dE$ represents the contribution to the dipole moment from occupied states with energy between E and $E + dE$. The evaluation of the integral in Eq. (6.28) is made easier by the fact that $\rho(E,\mathbf{r}) - \rho_M(E,\mathbf{r}) \simeq 0$ except in the neighborhood of the adsorbed atom. The solid line in Fig. 6.5 shows $p(E)$ and $\mu(E)$ calculated in the above manner, for krypton. The corresponding dipole moment calculated on the basis of Eq. (6.27) is $\mu = -0.25$ D. The broken curve is the result of a semiempirical correction (Oxinos and Modinos, 1979) which takes into account the lattice structure of the substrate. The calculated value of μ corrected in this way is the one listed in the fourth column of Table 6.1 for Kr and Ar adsorption on four different tungsten planes. The agreement between these theoretical values and the experimental values in the third column of the same table, though not entirely satisfactory, is certainly not bad in view of the uncertainty in the experimental values [see the discussion following Eq. (6.7)] and the approximate character of our model calculation.

We note from Fig. 6.5 that $p(E)$ is quite large near the resonance at $E \simeq 4.39$ eV but much smaller at higher energies, nearer the Fermi level. This implies that for these energies $\rho(E,\mathbf{r}) \simeq \rho_M(E,\mathbf{r})$, i.e., that to a first approximation the atomic potential well does not affect the corresponding electron wave functions. If we assume that when we go from a single adatom to an adlayer ($0 < \theta < 1$) we may still disregard, in a first approximation, the scattering of the electron by the atomic cores, we come to the conclusion that the adlayer may, in effect, be represented by a rectangular barrier, as postulated in our earlier discussion. (We emphasize that this is valid only for electron energies in the neighborhood of the Fermi level.) While the above approximation [which leads to Eq. (6.26)] appears to work reasonably well for adsorption on the W(111) and W(112) planes it fails in the case of adsorption on the W(100) plane. This is demonstrated quite clearly

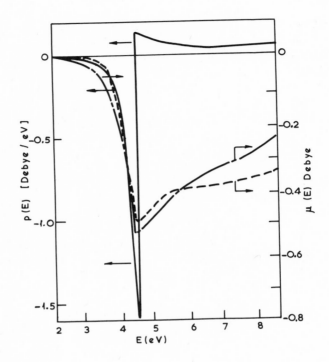

FIGURE 6.5

Solid line: $\mu(E)$ and $p(E)$ as defined by Eqs. (6.28) and (6.29), respectively, for Kr on a jellium substrate. Broken line: Corrected for lattice effects $\mu(E)$ for Kr on tungsten. The zero of the energy coincides with the minimum of the conduction band in the metal.

in Fig. 6.6 (Lea and Gomer, 1971) which shows how the total energy distribution from W(100) changes with increasing Kr coverage. At $\theta = 0$ one sees the familiar Swanson–Grouser hump at $E - E_F \simeq -0.37$ eV (see Fig. 5.4). As the coverage increases the hump diminishes and disappears completely at monolayer coverage, which suggests that scattering of the electron by the adlayer, although weak, destroys the surface resonance which is responsible for this hump. Obviously, the approximate expression for $r(E)$ given by Eq. (6.26) is not good enough in this case (see, also, Section 6.2.2).

6.2.2. METALLIC ADSORBATES ON TUNGSTEN: $\Delta\phi(\theta)$ AND $B(\theta)$

Figures 6.7a and 6.7b show experimental results for $\Delta\phi$ and B for gold on tungsten (110) and (112) planes, respectively, taken by Richter and Gomer

(1976b). We note that by experimental design, 7.5 doses correspond to a mono-layer on the (110) plane and 6 doses correspond to the same coverage for the (112) plane. Results similar to those of Au on W(112) have been obtained for Au on W(111) by Young and Gomer (1974) and for Au on W(100) by Richter and Gomer (1976a; 1979). We note that the sign of $\Delta\phi$ and B is different for the W(110) and W(112) planes, which shows that lattice effects play a significant role in the determination of these quantities. The results for the W(110) plane are also remarkable for another reason. Richter and Gomer point out that when metal atoms are adsorbed on a metal surface, a *decrease* in the work function ($\Delta\phi < 0$) is usually accompanied by a negative value of B, i.e., by an increase in the preex-ponential of the FN formula [Eq. (6.1)] but, as can be seen from Fig. 6.7a, this rule is not obeyed in the case of Au adsorption on W(110). (It is worth noting that a similar rule does not apply for $\Delta\phi > 0$.) One may attribute the considerable

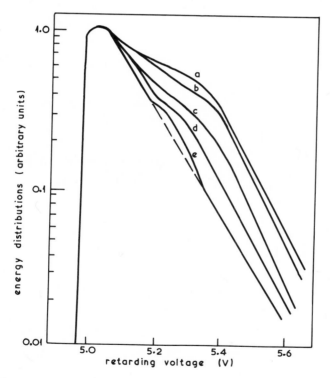

FIGURE 6.6

Variation of the energy distribution of field emitted electrons from W(100) with increasing Kr cov-erage. (a) $\theta = 0$; (b) $\theta \simeq 0.1$; (c) $\theta \simeq 0.4$; (d) $\theta \simeq 0.9$, and (e) $\theta \simeq 1.2$. (From Lea and Gomer, 1971.)

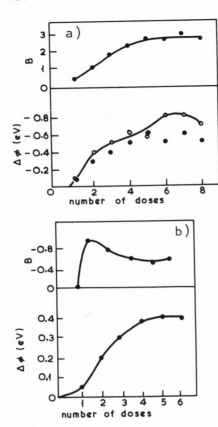

FIGURE 6.7

Experimental results for $\Delta\phi(\theta)$ and $B(\theta)$ for (a) gold on tungsten (110) and (b) gold on tungsten (112). Closed circles are values obtained from FN plots. Open circles are values obtained from TED data using Eq. (1.88). (From Richter and Gomer, 1976b.)

decrease of the work function when a monolayer of Au is adsorbed on W(110) to the large value of the work function of the clean W(110) plane, but it is much more difficult to understand the large reduction in the preexponential of Eq. (6.1). According to Eq. (6.18), which remains valid whatever the nature of the adsorbate (metallic or nonmetallic) a value of $b_{cl}/b_{ad} \gg 0$ implies that either $\gamma_F \gg 1$ or that $\langle \rho_{ad}(E_F, z = 0) \rangle_{c;0} \ll \langle \rho_{cl}(E_F; z = 0) \rangle_{c;0}$. We note that at monolayer coverage the number of Au adatoms per cm^2 is roughly the same with the number of W atoms per cm^2 in the top layer of the clean tungsten surface and, therefore, the surface potential barrier on the vacuum side of the interface for the adsorbate-covered surface cannot be much different from that of the corresponding clean surface [in view of the metallic nature of the adlayer $K \simeq \infty$ in Eq. (6.11)]. So, although γ_F may not be exactly unity, it will be of that order, which suggests that the partial density of states $\langle \rho_{ad}(E_F, z = 0) \rangle_{c;0}$ at the interface between the Au adlayer and vacuum, is considerably less (by ten times or so) than $\langle \rho_{cl}(E_F, z = 0) \rangle_{c;0}$ for the clean W(110) surface. A self-consistent calculation of the potential and of the density of states at the surface [corresponding, say, to a (1 × 1) overlayer] should explain this fact.

It is obvious that in general, measurements of $\Delta\phi$ and B provide a very good test for any calculation of the potential and density of states at the surface of the metal–adlayer complex.

6.2.3. METALLIC ADSORBATES ON TUNGSTEN: ENERGY DISTRIBUTIONS

We have already pointed out (see Fig. 6.6) that adsorption of a fraction of a monolayer of Kr on W(100) destroys the surface resonance peak at $\epsilon = E - E_F \simeq -0.37$ eV in the TED from this plane. The question arises as to whether there are other adsorbates which preserve this surface resonance, and if a criterion can be established as to the conditions which lead to the quenching or otherwise of this resonance. Figure 6.8 shows R, the enhancement factor as defined by Eq. (1.93), obtained by Richter and Gomer (1979) from Au-covered W(100); the Au

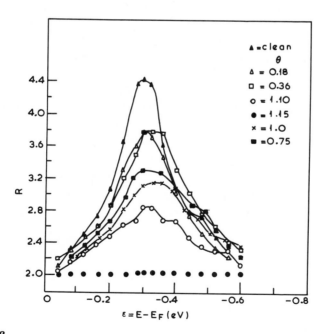

FIGURE 6.8

Experimental enhancement factor of field-emitted electrons from Au adsorbed on W(100) at 20 K and annealed at 300 K. (From Richter and Gomer, 1979.)

was deposited at 20 K and annealed to 300 K (in the case of unannealed deposits very similar results were obtained). It is seen that the resonance peak is only slightly shifted in energy, that its intensity is reduced as the coverage is increased, and that it disappears altogether when the coverage exceeds that of a monolayer (i.e., at approximately $\theta \simeq 1.15$). TED measurements from Au-covered W(100) have also been obtained by Billington and Rhodin (1978), for coverages extending from zero to *two* monolayer at $T = 78$ K. Their results are summarized in Fig. 6.9a. In this case, the resonance peak disappears at partial coverages and reemerges when a monolayer coverage is reached. On further adsorption it goes away to reappear at a coverage corresponding to two monolayers. At present, the reason for the discrepancy between the results shown in Figs. 6.8 and 6.9a for partial coverages ($0 < \theta < 1$) is not fully understood, but the following tentative explanation has been proposed by Billington and Rhodin (1978) and by Richter (1978). For reasons which will become apparent later in our discussion (see con-

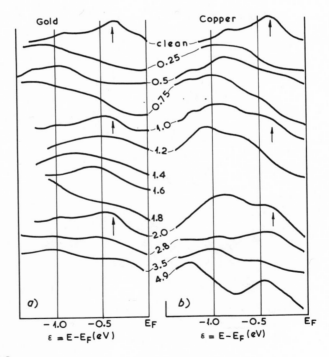

FIGURE 6.9

Experimental enhancement factor of electrons field-emitted from (a) Au covered W(100); (b) Cu covered W(100). The number associated with each of the curves indicates fractional coverage ($\theta = 1$ corresponding to monolayer coverage, as usual). $T = 78$ K. (From Billington and Rhodin, 1978.)

cluding paragraphs of Section 6.3.1), it is believed that the preservation of the surface resonance is associated with a (1 × 1) overlayer structure. It is then argued that in the Richter–Gomer experiments the conditions were favorable to the formation of islands with a (1 × 1) structure at partial coverages, whereas in the Billington–Rhodin experiments a more dispersed adsorption occurs so that a (1 × 1) overlayer is possible only at monolayer coverage or, for the same reasons, at two monolayers coverage but not for coverages in between. The above difference in ordering at partial coverages may be due to a different distribution of defects [these are more likely to occur on an annealed substrate (Richter–Gomer) than on a field evaporated one (Billington–Rhodin)] which act as nucleation centers for island formation, and to somewhat different thermal conditions, i.e., it is suggested that an absence of defects coupled with greater kinetic energy leads to a more dispersed adsorption in the Billington–Rhodin experiments, in contrast to (1 × 1) island formation in the Richter–Gomer experiments.

Richter and Gomer (1979) found that when copper is adsorbed on W(100) the resonance peak at $\epsilon \simeq -0.37$ eV disappears at $\theta \simeq 0.3$ and that it does not reemerge at monolayer coverage. In contrast, Billington and Rhodin (1978) found that the resonance peak does reemerge (though with reduced intensity) at mono-layer coverage, that it goes away on further adsorption in order to reappear at *two* monolayers coverage as was the case (in their experiments) for Au adsorption. Their results are shown in Fig. 6.9b. We note that whereas in the case of Au adsorption, the enhancement factor $R(E;\theta = 1,2)$, shown in Fig. 6.9a, appears to be very much the same as $R(E;\theta = 0)$ over the whole of the energy region considered, this is not so in the case of Cu adsorption. The peak between $\epsilon = -0.5$ eV and $\epsilon = -1.0$ eV from the Cu-covered surface at $\theta = 1$ and $\theta = 2$, seen in Fig. 6.9, is probabily due to a copper-induced surface resonance as suggested by Jones and Roberts (1977). They also measured the TED of field-emitted electrons from copper-covered W(100). Their results show a small peak in R at $\epsilon \simeq -0.6$ eV at monolayer coverage, but otherwise agree with those of Richter and Gomer. At the moment we do not know what causes the discrepancy between the results shown in Fig. 6.9b and those reported by the other authors.

Richter and Gomer (1979) measured also the TED distribution from W(100) covered with Ir, Ta, Nb, and Mo. They found, as in the case of Cu adsorption on this plane, that the surface resonance is quenched for $\theta > (0.3$ or so). In the case of Mo, however, they found that when layers in the coverage range $0.8 < \theta < 1.2$ are heated to 750 K a resonance peak appears at $\epsilon \simeq -0.18$ eV, the intensity of which increases with increasing applied field. In this respect this peak is very similar to the resonance peak at $\epsilon \simeq -0.15$ eV in the TED distribution of a clean Mo(100) surface whose intensity also increases with the applied field (see Fig. 5.12). [As we have already pointed out in Section 5.4.3, the intensity of the resonance peak, at $\epsilon \simeq -0.3$ eV, in the TED of the clean W(100) plane does not depend on the applied field.] It is worth noting that the adsorption of a disordered partial second layer of molybdenum destroys the above resonance at about $\theta = 1.4$, but adsorption of a complete second layer and heating to 750 K leads to a TED which is practically identical with that of a pure Mo(100) emitter. The above results suggest that the band of surface states/resonances responsible for the peak at $\epsilon = -0.18$ eV (-0.15 eV) in the TED of the Mo covered W(100) at $\theta \simeq 1$ ($\theta \simeq 2$) are very much localized in the Mo adlayer itself (for $\theta = 2$, in the top layer).

6.2.4. HYDROGEN ON TUNGSTEN

Plummer and Bell (1972) measured the field emission current and energy distribution from the (110) and (100) planes of tungsten as a function of coverage by hydrogen and deuterium. Their results for $\Delta\phi$ and B, as defined by Eqs. (6.4) and (6.5), respectively, are shown in Fig. 6.10 as a function of exposure (number of molecules per cm^2 impinged on the surface) of hydrogen (deuterium) on W(100) at 300 K. The solid line in the figure is a calculated curve which has been

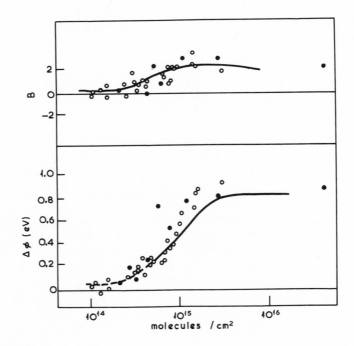

FIGURE 6.10

$\Delta\phi$ and B, as defined by Eqs. (6.4) and (6.5), as a function of exposure of hydrogen (closed circles) and deuterium (open circles) on W(100) at 300 K. The solid line for $\Delta\phi$ is a calculated curve for a macroscopic surface. (From Plummer and Bell, 1972.)

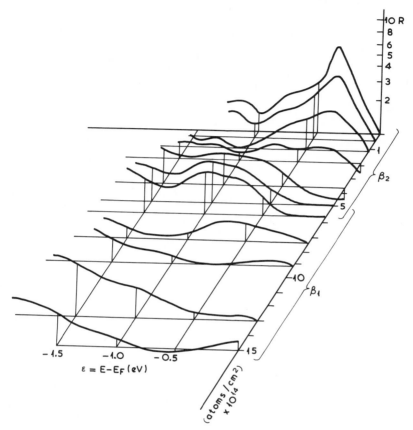

FIGURE 6.11

Experimental enhancement factor for hydrogen on W(100) at $T = 300$ K as a function of absolute coverage. (From Plummer and Bell, 1972.)

obtained (for a macroscopic surface) using a linear relationship between $\Delta\phi$ and (absolute) coverage (Madey and Yates, 1970), the experimental data of Tamm and Schmidt (1971) which give the relative sticking coefficient as a function of coverage and an initial sticking coefficient of 0.38. It is seen from this figure that when the exposure exceeds about 5×10^{14} molecules per cm^2 the sticking coeff-icient on the (100) plane of a field emission tip is larger than on the macroscopic (100) surface. This effect, which is attributed to diffusion from adjacent regions of the tip where weakly bound states exist, demonstrates that the condensation kinetics for a macroscopic single-crystal plane and for a small single plane on a field emission tip may differ significantly.

Figure 6.11 shows the enhancement factor R for hydrogen on W(100), at T

= 300 K, as a function of absolute coverage. The latter were determined from the measured value of $\Delta\phi$ using the established relation between the work function change and absolute coverage. [R was obtained, as usual, by dividing the measured TED by the corresponding FN energy distribution j_0, evaluated on the basis of Eq. (1.87). The value of the work function, needed in the evaluation of $j_0(E)$ was obtained from $\phi = \phi_{cl} + \Delta\phi$, using the measured value of $\Delta\phi$ and putting $\phi_{cl} = 4.64$ eV.] We see that as the coverage increases the structure characteristic of the clean surface disappears. At a coverage of about 5×10^{14} H atoms per cm^2 a new well-defined structure develops which consists of a peak positioned at $\epsilon = -1.0$ eV. The peak which has a full width at half-maximum of ~ 0.6 eV is superimposed on a background curve which increases gradually with decreasing energy. This peak disappears with further hydrogen adsorption. At full monolayer coverage (estimated to be in the region 1.5×10^{15} to 2.0×10^{15} atoms/cm^2) there is no structure in R, only a gradual increase in the curve as the energy decreases below $\epsilon = -1.0$ eV.

The evidence from LEED experiments (Estrup and Anderson, 1966) from flash desorption studies (Madey and Yates, 1970; Tamm and Schmidt, 1969, 1970) and from electron-energy-loss spectroscopy measurements (Froitzheim, Ibach, and Lehwald, 1976; Ho, Willis, and Plummer, 1978) and theoretical estimates of the binding energy for different overlayer geometries by Bullett and Cohen (1977), indicate the existence of two binding states β_1 and β_2 for hydrogen (deuterium) on W(100). The β_2 state is populated first and it saturates at about 5×10^{14} atom/cm^2 [We note that there are 10^{15} tungsten atoms per cm^2 on the clean W(100) surface.] It has been suggested (see, e.g., Bullett and Cohen, 1977) that in the β_2 state hydrogen (deuterium) atoms sit on top of tungsten atoms, and that the overlayer structure of the β_2 phase (at saturation) is that shown in Fig. 6.12a. Figure 6.13 shows the local density of states on a hydrogen atom and on the neighboring tungsten atom corresponding to this structure, as calculated by Bullet and Cohen (1977). It is seen that the theoretical local density of states on the hydrogen atom exhibits a peak, 0.7 eV wide, at about 1.7 eV below E_F. Because of the essentially s-type character of the electron orbitals about the hydrogen nucleus, we expect a peak at the same energy and of approximately the same width in $\langle \rho_{ad}(E, z = 0)\rangle_{c;0}$, the average density of states at the surface defined by Eq. (6.33), and in the corresponding (field emission) enhancement factor [see Eq. (6.17)]. Bullet and Cohen (1977) point out that in their calculation the Fermi level lies too high and that therefore the peak found at $\epsilon \simeq -1.7$ eV lies (for the overlayer structure shown in Fig. 6.12a) nearer the Fermi level, and they suggest that this peak in the local density of states is responsible for the peak at $\epsilon = -1.0$ eV in the experimental enhancement factor shown in Fig. 6.11. We should also point out that the overlayer structure shown in Fig. 6.12a is at best a first approximation. It is believed (Willis, 1979; Griffiths, King, and Thomas, 1981) that at low coverage (β_2 phase) hydrogen induces a reconstruction of the under-

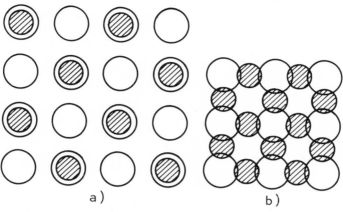

FIGURE 6.12

Most probable structure of hydrogen overlayer on W(100) (a) β_2 phase at saturation; (b) β_1 phase at saturation.

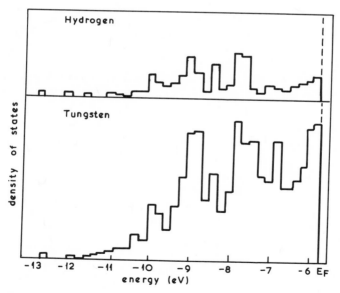

FIGURE 6.13

Calculated local density of states on a hydrogen atom and on the neighboring tungsten atom corresponding to the β_2 phase at saturation (overlayer structure shown in Fig. 6.12a). (From Bullett and Cohen, 1977.)

lying W(100) surface similar to that which occurs on the clean surface below 350 K (Debe and King, 1977; 1979). This is, apparently, a weak reconstruction and it is possible that the local density of states on the hydrogen atoms for the reconstructed surface is not significantly different from that calculated by Bullet and Cohen (1977). It is also possible that whereas a macroscopic W(100) surface reconstructs, a small single crystal plane on the field emission tip does not (see comments at the end of Section 5.4.1).

All available evidence (see, e.g., Willis, 1979; Bullet and Cohen, 1977) suggests that the location of the H atoms in the β_1 phase, which begins to develop for coverages above 5×10^{14} atoms/cm^2, is different from that in the β_2 phase. Although it is not clear how the transition from the β_2 to the β_1 phase occurs, it is almost certain that the structure of the β_1 phase at saturation is the one shown in Fig. 6.12b (we note that there are two H atoms per surface W atom in this $p(1 \times 1)$ structure). The local density of states on a H atom as calculated by Bullet and Cohen (1977) for this structure does not show any peak like the one found in the β_2 phase (Fig. 6.13), which could lead to structure in the R curve, in agreement with the experimental spectrum shown in Fig. 6.11. It is also worth noting that a self-consistent calculation by Kerker, Yin, and Cohen (1979) of the electronic structure of H on Mo(100) for the same geometry (Fig. 6.12b), shows no hydrogen-induced structure in the surface density of states near the Fermi level.

6.3. CALCULATION OF THE ENERGY DISTRIBUTION OF THE EMITTED ELECTRONS FROM ADSORBATE-COVERED METAL SURFACES

In this section we shall be concerned with emission from metal surfaces (single-crystal planes) covered with a monolayer of adsorbed atoms. (The case of field emission from a site containing a *single* adsorbed atom is considered in the following chapter.) We assume that the adlayer and the metal substrate are described by 2D layers of muffin-tin atoms parallel to the xy plane. The spherical potential within the muffin-tin spheres of the adlayer and the top one or two substrate layers must, at least in principle, be determined self-consistently. That in the layers lying further into the metal is practically identical with the potential in the infinite metal. The plane $z = 0$ is chosen in such a way that for $z > 0$ the potential depends only on z. The potential between the muffin-tin spheres of the adlayer (with its center at $z = -d'$) and the emitter–vacuum interface (at $z = 0$) is in general a function of both z and $\mathbf{r}_{\parallel}$ (d' is of the order of 2 Å or so). We may assume that in the region $z < -d'$ the potential between the muffin-tin spheres is constant throughout the crystal. When an electric field is applied to the surface a field term must be added to the above (zero field) potential. When the adlayer is metallic, we may approximate this term by

$$V_F = -eF(z - z_i), \qquad z > z_i$$
$$= 0, \qquad z < z_i \tag{6.30}$$

where $-d' < z_i < 0$. The case of nonmetallic adlayers is more difficult because of the penetration of the electric field through the adlayer to the metal substrate [see Eq. (6.10)]. Perhaps the easiest way to take field penetration into account is to add a constant term to the potential in the adlayer region, equal to the average value of the field term in this region.

If we assume (there is no loss of generality in doing so) that at $z = 0$ there exists a slot of infinitesimal thickness where the potential is constant (zero), we can use the same formulas to describe the total energy distribution of the emitted electrons from an adsorbate-covered surface as from a clean metal surface. One can easily convince himself that Eq. (5.42) gives the total energy distribution of the emitted electrons for an adsorbate-covered surface as well as for a clean surface, and that this is so whether the adlayer has a periodic structure parallel to the interface or not. The Green's function in Eqs. (5.42) and (5.43) must, of course, be replaced by a Green's function, denoted by $G_{ad}(E,\mathbf{r},\mathbf{r}')$ appropriate to the adsorbate-covered surface. We have

$$j_{ad}(E) = -\left(\frac{\hbar}{2m\pi^3}\right) f(E) \iint d^2q_{\parallel} \, \mathrm{Im}\langle\chi_{\mathbf{q}_{\parallel}} | G_{ad}(E | \chi_{\mathbf{q}_{\parallel}}\rangle \tag{6.31}$$
$$\times \frac{|T_{s;ad}(E,\mathbf{q}_{\parallel})|^2}{|1 + R_{s;ad}(E,\mathbf{q}_{\parallel})|^2}$$

$$\langle\chi_{\mathbf{q}_{\parallel}} | G_{ad}(E) | \chi_{\mathbf{q}_{\parallel}}\rangle = \iint \chi_{\mathbf{q}_{\parallel}}^*(\mathbf{r}) G_{ad}(E;\mathbf{r} \cdot \mathbf{r}')\chi_{\mathbf{q}_{\parallel}}(\mathbf{r}') \, d^3r \, d^3r' \tag{6.32}$$

where $\chi_{\mathbf{q}_{\parallel}}(r)$ is defined by Eq. (5.37). As in the case of clean metals [see comments following Eq. (5.43)] the integrand in Eq. (6.31) is practically zero for $q_{\parallel} > q_{\parallel}^0$ ($q_{\parallel}^0 \simeq 0.5$ Å^{-1} or so). The surface barrier transmission matrix element $T_{s;ad}$ and the reflection matrix element $R_{s;ad}$ are defined in exactly the same way as the corresponding elements T_s and R_s for the clean metal surface. The subindex "ad" indicates that the surface barrier may be different for different adsorbed species. It is also worth noting that in some instances, i.e., when the emitter–vacuum interface (the plane $z = 0$) lies well above the center of the adlayer the $R_{s;ad}$ element may depend to some degree on the applied field.

The partial density of states $\langle\rho_{ad}(E,z = 0)\rangle_{c;0}$ introduced earlier in our discussion (Section 6.2.1) can now be formally defined as follows:

$$\langle\rho_{ad}(E;z = 0)\rangle_{c;0} = -\frac{1}{\pi} \iint\limits_{q_{\parallel} < q_{\parallel}^0} d^2q_{\parallel} \, \mathrm{Im}\langle\chi_{\mathbf{q}_{\parallel}} | G_{ad}(E) | \chi_{\mathbf{q}_{\parallel}}\rangle \tag{6.33}$$

Although Eqs. (6.31)–(6.33) are valid for either ordered or disordered overlayers, the actual evaluation of the matrix element defined by Eq. (6.32) is possible, at the present moment, only for ordered overlayers.

CALCULATION OF $\langle \chi_{q_\parallel} | G_{ad}(E) | \chi_{q_\parallel} \rangle$ FOR ORDERED OVERLAYERS

Let $a_2^{(b)}$, $a_3^{(b)}$ be the primitive vectors of the 2D periodic lattice (b) associated with the clean (unreconstructed) metal surface (see Section 3.3), and let $a_2^{(a)}$, $a_3^{(a)}$ be the primitive vectors of the overlayer lattice (a). We note that the overlayer may consist of the same atoms as the substrate as in the case of a reconstructured clean surface, or it may consist of different atoms as in adsorption studies. We shall assume that the two lattices are *rationally related*, i.e., that there exists a lattice (c), with primitive vectors $a_2^{(c)}$, $a_3^{(c)}$, such that every point in (c) is a point in both the (a) and the (b) lattice. From the above definition it follows that a primitive cell of (c) contains an integer number n_{cb} of primitive cells of (b) and, similarly, an integer number n_{ca} of primitive cells of (a). It is also evident that the (b) lattice can be described as follows:

$$R_j^{(b)}(s) = R_j^{(c)} + R_s^{(0)} \tag{6.34}$$

where $R_j^{(c)}$ ($j = 1,2,3, \ldots$) is the jth lattice point of (c) and $R_s^{(0)}$ is the position vector of the sth lattice point of (b) within the jth unit cell of (c) relative to the "origin" of that cell (the point $R_j^{(c)}$). Since there are n_{cb} such points, Eq. (6.34) tells us that the (b) lattice can be broken into n_{cb} sublattices corresponding to the different $R_s^{(0)}$ ($s = 1,2, \ldots, n_{cb}$) vectors. The reciprocal of lattice (c) is constructed in the usual manner (see Section 3.3). One can easily show that the primitive cell of the reciprocal of (b) contains n_{cb} primitive cells of the reciprocal of (c) and that the vectors of the reciprocal of (c) are

$$g_j^{(c)}(s) = g_j^{(b)} + g_s^{(0)} \tag{6.35}$$

where $g_j^{(b)}$ ($j = 1,2,3, \ldots$) is the jth vector in the reciprocal of (b) and $g_s^{(0)}$ denotes the sth lattice vector of the reciprocal of (c) within the primitive cell of the reciprocal of (b) relative to the origin of that cell (the point $g_j^{(b)}$). Since there are n_{cb} such points, Eq. (6.35) tells us that the reciprocal lattice of (c) can be broken into n_{cb} sublattices corresponding to the different $g_s^{(0)}$ ($s = 1,2, \ldots, n_{cb}$) vectors. Obviously, a relation analogous to Eq. (6.34) exists between lattice (a) and lattice (c) and a relation analogous to Eq. (6.35) between the reciprocal of (c) and the reciprocal of (a). In Fig. 6.14a we show a simple example of an ordered overlayer. The

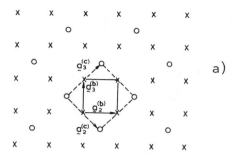

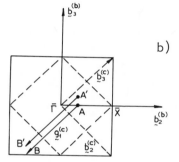

a)

b)

FIGURE 6.14

(a) $(\sqrt{2} \times \sqrt{2})\ 45°$ overlayer on a square (substrate) lattice. The crosses indicate the lattice points of the substrate and the circles those of the overlayer (b) primitive vectors and cells of the reciprocal lattices of the substrate and overlayer structures shown in (a).

crosses indicate the points of lattice (b) (a square lattice corresponding to the substrate) and the circles those of lattice (a) (a square lattice corresponding to the overlayer) and of lattice (c) which coincides with lattice (a) (when this is so the structures of substrate and overlayer are said to be simply related). Usually the relation of (c) to (b) is described as follows:

$$\left(\frac{|\mathbf{a}_2^{(c)}|}{|\mathbf{a}_2^{(b)}|} \times \frac{|\mathbf{a}_3^{(c)}|}{|\mathbf{a}_3^{(b)}|} \right) \phi' \tag{6.36}$$

where ϕ' is the angle of rotation of the primitive cell of (c) relative to that of (b). [It is assumed that the primitive cells of (c) and (b) have the same included angle. When $\phi' = 0$ it is omitted.] The metal–adlayer structure shown in Fig. 6.14a is, according to the above rule, described as a $(\sqrt{2} \times \sqrt{2})\ 45°$ structure. In Fig. 6.14b we show the primitive vectors and cells of the corresponding reciprocal lattices [$\mathbf{b}_2^{(b)}$, $\mathbf{b}_3^{(b)}$ are the primitive vectors of the reciprocal of (b) and $\mathbf{b}_2^{(c)}$, $\mathbf{b}_3^{(c)}$ those of the reciprocal of (c)]. It is obvious from Fig. 6.14a that the primitive cell of (b) (solid line) goes two times into the primitive cell of (c) (broken line) and, from Fig. 6.14b, that the primitive cell (SBZ) of the reciprocal of (c) (broken line) goes two times into the primitive cell (SBZ) of the reciprocal of (b) (solid line), as expected from the preceding general statements. A description of other overlayer structures can be found in a variety of books dealing with surface crystallography (see, e.g., Pendry, 1974; Van Hove and Tong, 1979).

In the case of an ordered overlayer we find, by repeating the arguments of Section 5.2.2 leading to Eq. (5.69),

$$\langle \chi_{\mathbf{q}_\parallel} | G_{ad}(E) | \chi_{\mathbf{q}_\parallel} \rangle = \alpha_{q_\parallel} \{ 1 + (1 + 2R_{s;ad}) \sum_{\mathbf{g}^{(c)}} [\mathbf{I} - \mathbf{R}^{c;ad}\mathbf{R}^{s;ad}]^{-1}_{\mathbf{0}\mathbf{g}^{(c)}} R^{c;ad}_{\mathbf{g}^{(c)}\mathbf{0}}$$
$$+ R_{s;ad}(E,\mathbf{q}_\parallel)[\mathbf{I} - \mathbf{R}^{c;ad}\mathbf{R}^{s;ad}]^{-1}_{\mathbf{0}\mathbf{0}} \} \qquad (6.37)$$

$\alpha_{q_\parallel}$ is defined by Eq. (5.57) $R^{s;ad}$ is the diagonal matrix which describes electron reflection by the surface barrier and is given by

$$R^{s;ad}_{\mathbf{g}^{(c)}\mathbf{g}^{(c)\prime}} = R_{s;ad}(E,\mathbf{q}_\parallel + \mathbf{g}^{(c)}) \, \delta_{\mathbf{g}^{(c)}\mathbf{g}^{(c)\prime}} \qquad (6.38)$$

$\mathbf{R}^{c;ad}(E,\mathbf{q}_\parallel)$ is defined by [note the analogy with Eq. (5.65)]

$$\int G_0(E,\mathbf{r},\mathbf{r}'') T_{c;ad}(\mathbf{r}'',\mathbf{r}') \exp(i\mathbf{K}^-_{\mathbf{g}^{(c)\prime}} \cdot \mathbf{r}') \, d^3r'' \, d^3r'$$
$$= \sum_{\mathbf{g}^{(c)}} R^{c;ad}_{\mathbf{g}^{(c)}\mathbf{g}^{(c)\prime}} \exp(i\mathbf{K}^+_{\mathbf{g}^{(c)}} \cdot \mathbf{r}), \qquad z \geq 0 \quad (6.39)$$

where

$$\mathbf{K}^\pm_{\mathbf{g}^{(c)}} = \{ \mathbf{q}_\parallel + \mathbf{g}^{(c)}, \pm[2mE/\hbar^2 - (\mathbf{q}_\parallel + \mathbf{g}^{(c)})^2]^{1/2} \} \qquad (6.40)$$

$T_{c;ad}(\mathbf{r}'',\mathbf{r}')$ describes to all orders the scattering of the electrons by the metal–adlayer potential field to the left of the interface (the plane $z = 0$). Because the periodicity of this field is described by the lattice (c) the diffracted beams [(exp $i\mathbf{K}^+_{\mathbf{g}^{(c)}} \cdot \mathbf{r}$) in Eq. (6.39)] have a parallel wave vector which differs from that of the incident beam [exp($i\mathbf{K}_{\mathbf{g}^{(c)\prime}} \cdot \mathbf{r}$) in Eq. (6.39)] by a vector of the reciprocal of (c). For the purpose of calculating $\mathbf{R}^{c;ad}(E,\mathbf{q}_\parallel)$ we may think of the region between the muffin-tin spheres of the adlayer and the plane $z = 0$ as another "scattering layer." Suppose that by some means or other we have obtained the scattering matrix elements for this "layer." We may then combine these scattering matrix elements with those of the top layer of muffin-tin atoms (adatoms) to obtain the scattering matrix elements of these two layers considered as one unit. We then combine these matrix elements with those of the top layer of the substrate to obtain the scattering matrix elements for a slab consisting of the top substrate layer, the adatom layer and the interface region. In the same manner we may obtain the scattering matrix elements for a slab which contains the surface region, the adatom layer, and those (two or three) layers of the substrate where the potential is different from the bulk. [The way one obtains the scattering matrix of two consecutive layers (slabs) when those of the individual layers (slabs) are known has been described in Section 5.3.] Finally we may combine the scattering matrix elements of this slab with the reflection matrix $\mathbf{R}^B$, defined by Eq. (5.80) of the bulk crystal to obtain $\mathbf{R}^{c;ad}$ in the same manner that $\mathbf{R}^c$ is obtained for the clean metal surface (see Section 5.3). The calculation of $R^B_{\mathbf{g}^{(c)}\mathbf{g}^{(c)'}}$ is greatly simplified by using Eq. (6.35). It is obvious from this equation that scattering by the bulk potential, whose periodicity is that of lattice (b), will not mix $\mathbf{g}^{(c)}$ vectors corresponding to different $\mathbf{g}_s^{(0)}$; therefore, assuming that one deals with n $\mathbf{g}^{(b)}$ vectors, the calculation of a matrix like $\mathbf{R}^B$ with dimensions $(n \times n_{cb}) \times (n \times n_{cb})$ is reduced to that of calculating n_{cb} matrices each having dimensions $(n \times n)$, which saves considerable computing time.

Kar and Soven (1976b) and Kar (1978) calculated the total energy distribution from a (1×1) monolayer of Au on W(100) using a method which, though not identical with it, is equivalent with the one we have just described. In the Kar–Soven model the Au adatoms take the place of the tungsten atoms in the top layer of a clean unreconstructed W(100) surface. The Mattheiss potential (Mattheiss, 1965) for bulk tungsten was used in all tungsten layers. The muffin-tin potential in the Au layer was obtained from a superposition of Au and W atomic potentials as prescribed by Mattheiss (1964). The metal–vacuum interfaces (the plane $z = 0$) was taken at the edge of the Au muffin-tin spheres, and the potential between the muffin-tin spheres for $z < 0$ was put equal to the interstitial potential (zero) in the bulk of the crystal. It is noted by Kar and Soven that although the muffin-tin potential in the Au adlayer was not calculated self-consistently, it gave reasonable results for the scattering phase shifts with the d-resonance below the Fermi level (as determined in the bulk of the substrate), where it must be if the Au atom is to have its proper share of electrons. The main limitation of the calculations of Kar and Soven arises from their choice of the substrate potential. This is the same potential used by Modinos and Nicolaou in the calculation of the total energy distribution of the emitted electrons from the clean W(100) surface shown in Fig. 5.1. We now know that the pronounced peak in the R-curve obtained with this potential is not to be associated with the experimentally observed peak at $\epsilon \simeq -0.37$ eV (the Swanson–Crouser hump), but with the shoulder observed at $\epsilon = -0.78$ eV (peak B in Fig. 5.5). It is this latter peak which is reproduced in Kar's model calculation (broken line in Fig. 6.15) and it is the effect on this peak of a (1×1) overlayer of Au that the solid curve in the same figure shows. For this reason, the above result cannot be compared with the experimental results (Richter and Gomer, 1979; Billington and Rhodin, 1978) on the effect of Au adsorption on the Crouser hump, as intended by its author at that time, but it remains useful because it demonstrates quite clearly that a band of surface resonances, and the corresponding structure in the total energy distribution of the emitted electrons, may survive the adsorption of a monolayer of foreign atoms

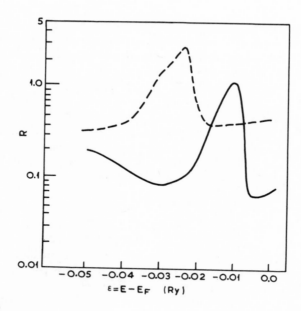

FIGURE 6.15

Calculated enhancement factor of field-emitted electrons for a model tungsten (100) surface (broken line) and for the same surface covered with a (1 × 1) gold overlayer (solid line). (From Kar, 1978.)

although it may be shifted in energy. A hint as to what might be the single most important factor determining whether a given adlayer will quench a given surface state or resonance is provided by an approximate calculation of Kar and Soven (1976b) of the effect of Kr adsorption on the resonance peak, shown by the broken line in Fig. 6.15. The ideas which underlie their calculation can be summarized as follows. The Kr atoms have a large radius, consequently a (1 × 1) overlayer is not possible. If one assumes that they form a ($\sqrt{2} \times \sqrt{2}$) 45° overlayer the SBZ for the adlayer will be half the size of that of tungsten (100). The two zones are shown in Fig. 6.14b. A true surface state (of the substrate) at the point denoted by A in Fig. 6.14b will mix with an extended (bulk) state, of the same energy, at B which differs from A by the reciprocal vector $\mathbf{g}_1^{(c)}$, because of electron scattering by the overlayer. Similarly a virtual surface state (a surface resonance) at A′ will mix with a bulk state, of the same energy, at B′. (We note that the surface states/ resonances which give rise to the peak in the R curve, shown in Fig. 6.15, have values of $k_\parallel \simeq 0.4$ Å^{-1} in the region approximately of points A′ and A (Fig. 6.14b), and that bulk states of the same energy do exist in the region of points B and B′). The interaction between a surface state (true or virtual) with a bulk state of the same energy provides an *escape* channel for the electron in the surface state.

As a result the true surface state will become a virtual surface state (a surface resonance) with an energy width which can be small or large depending on the strength of the interaction. If the latter is strong enough, the surface state will disappear altogether. The same argument applies for a virtual surface state. Its width will increase by the interaction with bulk states via the overlayer potential, and again it might possibly disappear altogether. Kar and Soven have actually calculated, using perturbation theory, the width of the resulting resonances and have shown that this can be so large as to lead to the disappearance of the corresponding peak in the energy distribution of the emitted electrons. We should point out that a ($\sqrt{2} \times \sqrt{2}$) 45° overlayer is not the only structure that leads to "mixing" of a surface state/resonance with a bulk state. A disordered layer, for example, is perfectly capable of effecting the same result, and, by the same arguments, it also can lead to the quenching of a surface state or resonance. The preservation of the surface resonance by a (1 × 1) Au overlayer (Fig. 6.15) and the quenching of the same resonance by a ($\sqrt{2} \times \sqrt{2}$) 45° Kr overlayer (in spite of the fact that a single Au atom scatters the tunneling electron more strongly than a Kr atom) is believed by some authors (see, e.g., Billington and Rhodin, 1978) to be a manifestation of a rather general rule which can be stated, more or less, as follows. A disordered overlayer, or an overlayer with a unit cell larger than that of the substrate, leads through scattering by the overlayer potential field to a mixing of the surface state with bulk states of the same energy and *most probably* to the disappearance of the surface state (or resonance). According to the same rule, a (1 × 1) overlayer with a potential not very different from that of the substrate will preserve a surface state (or resonance). We have already explained in Section 6.2.3 how this rule has been employed by Billington and Rhodin in order to explain their experimental data on Au and Cu covered W(100) surfaces (Fig. 6.9).

6.4. INELASTIC ELECTRON TUNNELING

It is well known (see, e.g., Lambe and Jaklevic, 1968; Kirtley and Hall, 1980) that when a molecule is placed at one of the interfaces of a M–I–M (metal–insular–metal) junction, it produces significant structure in the tunneling current versus voltage characteristic. When the second derivative of the current (d^2I/dV^2) is plotted against the applied voltage V one sees peaks at certain values of V corresponding to vibrational excitation energies of the metal–adsorbed-molecule complex. Flood (1970) was the first to point out that a similar situation could arise in a field emission experiment.

A vibrational excitation associated with a molecule adsorbed on a metal surface may originate from an adsorbate–substrate mode or from an adsorbate internal mode. In either case we can approximate the vibrating metal–adsorbed-molecule complex by a localized vibrator. Let ω_p be the frequency and $\hbar\omega_p$ the excitation energy of the pth vibrational mode. In a field emission experiment, an electron incident from within the metal with energy E may scatter inelastically off the vibrator (a metal–adatom or a metal–admolecule complex) leaving the metal with energy $E - \hbar\omega_p$. If the interaction between the tunneling electron and the vibrator is weak (so that it can be treated by first-order approximation theory) and if it is assumed that the vibrations associated with different atoms (molecules) do not interact with each other, the energy distribution of the field-emitted electrons is, to a first approximation, given by (Flood, 1970; Gadzuk and Plummer, 1973)

$$j_T(E) \simeq j(E) + n \sum_p \sigma_p j(E + \hbar\omega_p) \qquad (6.41)$$

where $j(E)$ is the elastic TED, σ_p is the cross section for excitation of the pth mode, and n is the number of adsorbed atoms (molecules) per unit area. Equation (6.41) tells us that the inelastic contribution to the TED associated with the pth oscillator is equal to the elastic TED multiplied by a constant $(n\sigma_p)$ and displaced in energy by $\hbar\omega_p$. The energy distribution described by Eq. (6.41) is shown, schematically, in Fig. 6.16. The corresponding enhancement factor is given by

$$R_T(E) = j(E)/j_0(E) + n \sum_p \sigma_p j(E + \hbar\omega_p)/j_0(E) \qquad (6.42)$$

where $j_0(E)$ denotes as usual the FN distribution for the given parameters (the experimentally determined work function, applied field and temperature).

The magnitude of the step in $R_T(E)$ associated with the excitation of the pth vibrational mode depends of course on the value σ_p of the cross section for excitation. In the Flood treatment of inelastic field emission, which constitutes an extention of the Scalapino and Marcus (1967) theory of inelastic tunneling in M–I–M junctions, the interaction potential between the tunneling electron and the vibrator is approximated by a dipole potential,

$$V_i^{(p)} = -2e\mu_z z/(z^2 + r_\perp^2)^{3/2} \qquad (6.43)$$

where μ_z is the component of the dipole moment of the vibrator normal to the surface, and $r_\perp$ is the distance between the electron and the oscillator in the plane of the surface. It is assumed that the component of the dipole field parallel to the surface is negligible. It follows from Eq. (6.43) that

$$\sigma_p \propto |\langle\phi_p|\mu_z|\phi_0\rangle|^2 \qquad (6.44)$$

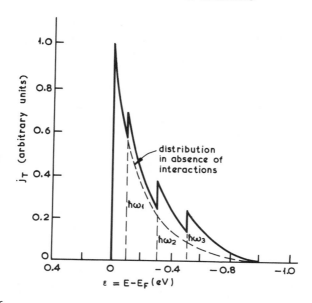

FIGURE 6.16

The inelastic tunneling contribution [second term in Eq. (6.41)] is superimposed on the elastic TED, leading to a steplike structure in the energy distribution of the field-emitted electrons.

where ϕ_0 denotes the ground state of the vibrator (zero energy) and ϕ_p the excited state of the vibrator (pth mode, energy $\hbar\omega_p$). It turns out that for light atoms and molecules, such as H, H_2, CO, OH, σ_p is of the order of 10^{-17} cm^2, leading to a ratio of inelastic to elastic tunneling current of the order of 10^{-2} (for an adsorbate density $n \simeq 10^{15}$ cm^{-2}). It is obvious that the observation of vibrational energy loss structure of such a small magnitude requires very accurate measurements (a signal-to-noise ratio of 10^3). The only such measurements reported so far are those of Plummer and Bell (1972) for hydrogen (and deuterium) on tungsten (111) and (100) planes. Their results for hydrogen (deuterium) adsorption in the β_2 state (see Section 6.2.4) on the W(100) plane are shown in Fig. 6.17. The isotopic shift of the vibrational excitation energies (in the present case this means that the vibrational excitation energies of hydrogen are larger than those of deuterium by a factor of $\sqrt{2}$) provides a means for distinguishing vibrational energy loss structure from similar structure, that might exist, due to electronic excitation of the metal–adsorbed-molecule complex. The signal from deuterium is stronger than that of hydrogen in Fig. 6.17, because experimentally it proved possible to saturate the β_2 state of deuterium (on a field emitter) but not that of a hydrogen.

It is worth comparing the spectrum shown in Fig. 6.17 with the vibrational-energy loss spectra shown in Fig. 6.18 obtained by scattering electrons off a hydrogen covered W(100) surface (Willis, 1979). The step at $\epsilon \simeq -0.14$ eV (corresponding to $\hbar\omega = 0.14$ eV) in the field emission TED curve for hydrogen and the peak in Fig. 6.18a (specular scattering) correspond to the same vibrational mode of the adatom, normal to the surface. We note that the excitation energy depends on the coverage and that for $\beta \simeq 0.15$ (β_2 state) the energy loss peak in Fig. 6.18a occurs at about the same energy as in the field emission curve. The electron energy loss spectra off-specular directions show excitation losses corresponding to lateral vibrations as well as perpendicular vibrations. In the spectra shown in Fig. 6.18b the lateral vibrations are responsible for the peaks denoted by ν_{as} (asymmetric stretch) and ν_w (wag), and the normal mode for the ν_s peak. The much weaker peak around $2\nu_{as}$ is associated with a broad envelope of overtone and combination bands of these three fundamental modes (Willis, 1979). It appears, from these spectra, that the step at about $\epsilon \simeq -0.07$ eV in the field emission TED curve shown in Fig. 6.17, is due to a lateral vibration (with $\hbar\omega \simeq 0.07$ eV. [We note that an incident (from within the metal) electron which scatters off the specular direction by the vibrator does tunnel out of the metal provided its reduced wave vector $\mathbf{k}_\parallel$ (after the collision) lies in the central region

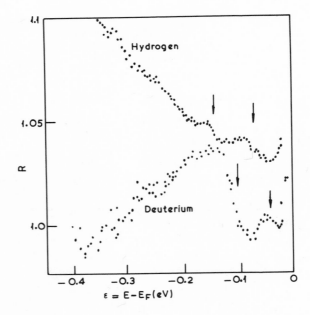

FIGURE 6.17

Experimental enhancement factor of field-emitted electrons for hydrogen and deuterium adsorption in the β_2 state on W(100) at 300 K. The total energy distribution was obtained at 78 K. (From Plummer and Bell, 1972.)

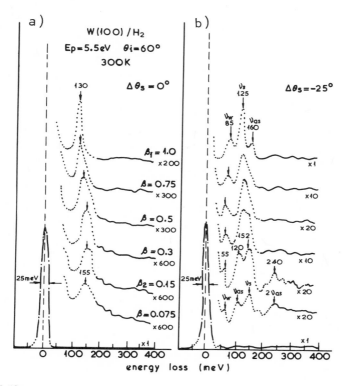

FIGURE 6.18

Vibrational-energy loss spectra obtained by scattering electrons off a hydrogen-covered W(100) surface. The fractional coverage is denoted by β. Saturation coverage (β_1) corresponds to 17×10^{14} H cm^{-2}. (a) specular scattering, and (b) 25° off the specular beam direction towards the surface normal direction, $\Delta\theta_s \simeq -25°$. The spectra are normalized to constant elastic beam intensity, the gain shown for each displaced curve. Impact energy $E_p = 5.5$ eV, angle of incidence $\theta_i = 60°$ and energy resolution $\Delta E_p = 25$ meV. (From Willis, 1979.)

of the SBZ.] This suggests that Flood's hypothesis [Eq. (6.43)], according to which only normal vibrations contribute significantly to the inelastic tunneling current, may not be valid.

6.5. MEASUREMENTS OF DESORPTION ENERGIES AND DIFFUSION COEFFICIENTS

Many applications of field emission to adsorption and desorption studies are based on the one-to-one correspondence between adsorbate coverage and $\Delta\phi$. For example, the change in the coverage, obtained from the measured $\Delta\phi$ (see Sections 6.1 and 6.2), induced by variation in the external parameters (temperature, pressure) can be used in conjunction with standard equations of thermodynamics to estimate adsorption (desorption) energies (see, e.g., Gomer, 1961; Domke, Jahnig, and Drechsler, 1974).

A rather interesting application of field emission to diffusion is based on the fact, pointed out by Kleint (1971) and Gomer (1973), that measurable fluctuations in the probe current are related to adsorbate density fluctuations over the region (part of a single plane) which is observed. Gomer has shown that the relaxation time of these fluctuations is related to the surface diffusion coefficient in a manner which allows, under certain conditions, a determination of the latter in a relatively simple manner.

A discussion of these applications of field emission is beyond the scope of the present book. For a brief review and some references see Gomer (1978).

FIELD EMISSION SPECTROSCOPY OF SINGLY ADSORBED ATOMS

7.1. INTRODUCTION

Duke and Alferieff (1967) were the first to point out that a resonance in the local density of states, corresponding to a broadened atomic energy level, may lead to specific structure, e.g., a peak, in the energy distribution of the electrons which are field-emitted from the neighborhood of a singly adsorbed (on a metal surface) atom. In their calculation the adatom was represented by a one-dimensional potential well, superimposed on the surface potential barrier of a jellium (free-electron) metal substrate, as shown schematically in Fig. 7.1. In the same figure we show schematically the energy distribution of the emitted electrons from the neighborhood of the adatom, when a resonance, a broadened atomic level centered at E_s lies in the energy region (within 2 eV or so from the Fermi level) which is acccessible through field emission. On the experimental side a very promising beginning has been made by Clark and Young (1968) and by Plummer and Young (1970). The latter measured the relative change in the total energy distribution of the emitted electrons upon adsorption of single Ba, Sr, and Ca atoms on various crystal planes of tungsten (we describe some of their results in Section 7.3). Unfortunately, no further work has been done along these lines during the last ten years. On the theory side a number of papers have been published (Gadzuk, 1970; Penn, Gomer, and Cohen, 1972; Modinos and Nicolaou, 1971) following the pioneering work of Duke and Alferieff. In all these papers the adatom is represented by a more or less realistic three-dimensional potential, but in none of these theories is the electronic structure of the metal substrate properly taken into account. [For a summary of these theories the reader is referred to the review article of Gadzuk and Plummer (1973).]

The calculation of the local density of states at and around a singly adsorbed atom is a problem of central importance in the theory of adsorption and has

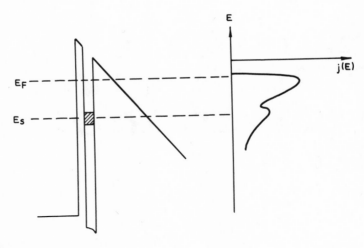

FIGURE 7.1

Schematic description of the energy distribution $j(E)$ of electrons field-emitted from the neighborhood of an adatom when a resonance, a broadened atomic level E_s lies near the Fermi level E_F. The adatom potential is represented by a one-dimensional square well (Duke and Alferieff model).

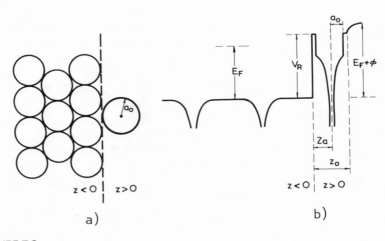

FIGURE 7.2

One-electron potential (schematic) for a metal–adatom complex. (a) The circles represent muffin-tin spheres. The broken line defines the metal–vacuum interface (the plane $z = 0$). (b) Variation of the potential along a direction normal to the surface passing through the nucleus of the adatom.

attracted considerable attention in recent years, but a lot remains to be done in this direction. It is worth noting in this respect, that, so far, self-consistent calculations of the potential and local density of states have been performed only for atoms adsorbed on a jellium (free-electron) metal substrate (Lang and Williams, 1978). A review of the relevant literature is beyond the scope of the present book. For that we refer the reader to existing articles; see, e.g., Lang (1973), Gomer (1975), Lang and Williams (1978); Rhodin and Ertl (1979). Here, our principal aim is to show that field electron emission from the neighborhood of a singly adsorbed atom can provide us with valuable information about the electronic structure of the metal–adatom complex provided we have a good theory for the analysis of the relevant experimental data. We shall base our discussion on a model one-electron potential (Modinos and Oxinos, 1978) which is a direct extension of the one defined by Eq. (4.1) for the clean surface. Because of the postulated form [Eq. (7.1)] of this potential, it cannot be made fully self-consistent. It incorporates, however, most of the essential features of the potential which the electron sees in a metal–adatom complex, and has the advantage of being relatively simple [for a critical analysis of the model assumptions see, Oxinos and Modinos (1980)].

7.2. ADATOM RESONANCES

We assume that, in the absence of an applied field, the one-electron potential of the metal–adatom system has the form shown schematically in Fig. 7.2. In the region $z < 0$ it is identical with that of the semi-infinite crystal as defined by Eq. (4.1a). (The plane $z = 0$ is taken half an interlayer distance above the center of the top atomic layer in the metal.) In the region $0 < z < z_0$ the potential is approximated by a constant V_R, with a spherical potential of the muffin-tin type U_a, representing the adatom, superimposed on it. In the region $z > z_0$ the potential rises *slowly* to its asymptotic value given by $(E_F + \phi)$

$$V = V_R + U_a(r_a), \qquad 0 < z < z_0 \tag{7.1}$$
$$U_a(r_a) = 0 \qquad \text{for } r_a > a_0 \tag{7.1a}$$
$$V = V_s(z), \qquad z > z_0 \tag{7.2}$$
$$V_s(z_0) = V_R \tag{7.2a}$$
$$V_s(z) \rightarrow E_F + \phi \qquad \text{as } z \rightarrow \infty \tag{7.2b}$$

We define $\mathbf{r}_a$, $\mathbf{R}_a$, and z_0 as follows:

$$\mathbf{r}_a \equiv \mathbf{r} - \mathbf{R}_a \tag{7.3}$$
$$\mathbf{R}_a \equiv (X_a, Y_a, Z_a) = (X_a, Y_a, a_0) \tag{7.4}$$
$$\mathbf{z}_0 \equiv (0,0,z_0) = (0,0,2a_0) \tag{7.5}$$

where $\mathbf{R}_a$ is the position of the atomic nucleus, and a_0 the radius of the atomic sphere. We assume that the interstitial potential inside the metal equals zero, and that

$$E_F < V_R < E_F + \phi \tag{7.6}$$

as shown in Fig. 7.2.

Let us further assume, for the sake of simplicity, that the wave funtions of the one-electron states, with energy $E < E_F$, in the region $z < z_0$ are not significantly affected by the shape of the slowly varying potential in the region $z > z_0$. In that case, for the purpose of calculating the local density of states for $E < E_F$ and $z < z_0$, we can approximate the potential on the vacuum side of the interface $(z > 0)$ by

$$V \simeq V_R + U_a(r_a), \qquad z > 0 \tag{7.7}$$

Let us consider, for a moment, the bound states of the adatom potential U_a, when it is superimposed on a constant potential V_R extending over *all* space, i.e., when the metal potential in the region $z < 0$ is switched off. The corresponding energy levels are given by the poles of $S_l^a(E)$, as defined by Eq. (3.62), in the energy region $E < V_R$. We note that q, which appears in Eq. (3.62), is given by

$$q = i[(2m/\hbar^2)(V_R - E)]^{1/2} \tag{7.8}$$

in the present case, and that $[v(r) - E)]$ in Eq. (3.61), which determines the radial part of the wave function inside the atomic sphere, must be replaced by $[U_a(r) + V_R - E]$. We note, also, that outside the atomic sphere $(r_a > a_0)$ the wave function which describes any one of the above bound states consists of a spherical wave given (apart from a multiplying constant) by $h_l^{(1)}(qr_a)Y_{lm}(\Omega_a)$, where $h_l^{(1)}(qr_a)$ decays exponentially as $r_a \to \infty$. Let us now consider what happens when the metal potential in the region $z < 0$ is switched on. If no electron states exist in the metal in the given energy region (this could be an energy gap below the conduction band), the electron cannot "escape" into the metal and, therefore, an electron state may exist, in this energy region, which is truly localized on the adsorbed atom. The corresponding wave function in the region between the metal–vacuum interface and the adatom sphere $(z > 0, r_a > a_0)$ will consist of an outgoing (from the atom) wave given by

$$\psi_{\text{out}} = \sum_{lm} A_{lm} h_l^{(1)}(qr_a) Y_{lm}(\Omega_a) \tag{7.9}$$

and an incident (on the atom) wave ψ_{inc} which is generated when ψ_{out} is scattered by the metal. We can obtain an explicit expression for ψ_{inc} as follows. Using Eq.

(3.93) we can expand the spherical waves in Eq. (7.9) into plane waves. We have

$$h_{l'}^{(1)}(qr_a) Y_{l'm'}(\Omega_a) = \frac{(i)^{-l'}}{2\pi q} \iint_{SBZ} d^2 k_{||} \sum_{g} \frac{\exp[-i\mathbf{K}^{\pm}(\mathbf{k}_{||} + \mathbf{g}) \cdot \mathbf{R}_a]}{K_z^+(\mathbf{k}|| + \mathbf{g})}$$
$$\times Y_{l'm'}(\Omega_{\mathbf{K}\pm(\mathbf{k}_{||}+\mathbf{g})}) \exp[i\mathbf{K}\pm(\mathbf{k}_{||} + \mathbf{g}) \cdot \mathbf{r}] \quad (7.10)$$

where the $+(-)$ sign applies to $z_a > 0$ (< 0). The sum includes a finite number of terms corresponding to different reciprocal vectors $\mathbf{g}$ of the substrate surface, and the integral is over the SBZ of the same surface:

$$\mathbf{K}^{\pm}(\mathbf{q}_{||}) \equiv \{\mathbf{q}_{||}), \pm i[(2m/\hbar^2)(V_R - E) + q_{||}^2]^{1/2}\} \quad (7.11)$$
$$K_z^+(\mathbf{q}_{||}) \equiv +i[(2m/\hbar^2)(V_R - E) + q_{||}^2]^{1/2} \quad (7.12)$$

Each term in Eq. (7.10) proportional to $\exp[i\mathbf{K}^-(\mathbf{k}_{||} + \mathbf{g}) \cdot \mathbf{r}]$ represents a plane wave incident on the metal. This wave is scattered by the metal giving rise to a number of diffracted beams. The calculation of the diffracted amplitudes, denoted by $R_{\mathbf{g}'\mathbf{g}}$, can be calculated, as in LEED theory, in a number of ways, e.g., by matching at the surface ($z = 0$),

$$\chi = \exp[i\mathbf{K}^-(\mathbf{k}_{||} + \mathbf{g}) \cdot \mathbf{r}] + \sum_{g'} R_{\mathbf{g}'\mathbf{g}}(E, \mathbf{k}_{||})$$
$$\times \exp[i\mathbf{K}^+(\mathbf{k}_{||} + \mathbf{g}') \cdot \mathbf{r}], \quad z \geq 0 \quad (7.13)$$

to a wave function inside the metal ($z < 0$) which consists of a superposition of Bloch waves, of the same E and $\mathbf{k}_{||}$, all of which decay exponentially inside the metal since there are no propagating waves in an energy gap (see Section 4.2.2). The wave scattered from the metal when $h_{l'}^{(1)}(qr_a)Y_{l'm'}(\Omega_a)$ is incident upon it is given by

$$\psi_{l'm'} = \frac{i^{-l'}}{2\pi q} \iint_{SBZ} d^2 k_{||} \sum_{g',g} Y_{l'm'}(\Omega_{\mathbf{K}-(\mathbf{k}_{||}+\mathbf{g})}) \frac{\exp[-i\mathbf{K}^-(\mathbf{k}_{||} + \mathbf{g}) \cdot \mathbf{R}_a]}{K_z^+(\mathbf{k}_{||} + \mathbf{g})} \quad (7.14)$$
$$\times R_{\mathbf{g}'\mathbf{g}}(E, \mathbf{k}_{||}) \exp[i\mathbf{K}^+(\mathbf{k}_{||} + \mathbf{g}') \cdot \mathbf{r}]$$
$$= \sum_{lm} G_{lm;l'm'}(E) j_l(qr_a) Y_{lm}(\Omega_a) \quad (7.15)$$

where

$$G_{lm;l'm'}(E) \equiv \frac{2(i)^{l-l'}}{q} \iint_{SBZ} d^2 k_{||} \sum_{g',g} Y_{l'm'}(\Omega_{\mathbf{K}-(\mathbf{k}_{||}+\mathbf{g})}) \frac{\exp[-i\mathbf{K}^-(\mathbf{k}_{||} + \mathbf{g}) \cdot \mathbf{R}_a]}{K_z^+(\mathbf{k}_{||} + \mathbf{g})}$$
$$\times R_{\mathbf{g}'\mathbf{g}}(E, \mathbf{k}_{||}) \exp[i\mathbf{K}^+(\mathbf{k}_{||} + \mathbf{g}') \cdot \mathbf{R}_a](-1)^m Y_{l-m}(\Omega_{\mathbf{K}+(\mathbf{k}_{||}+\mathbf{g}')}) \quad (7.16)$$

Equation (7.15) is obtained by expanding the plane waves, $\exp[i\mathbf{K}^+(\mathbf{k}_{\parallel} + \mathbf{g}) \cdot \mathbf{r}]$, in Eq. (7.14) into spherical waves around $\mathbf{R}_a$, using Eq. (3.58). Finally, ψ_{inc}, the wave scattered from the metal when ψ_{out} is incident upon it, is given by

$$\psi_{\text{inc}} = \sum_{\substack{lm \\ l'm'}} j_l(qr_a)Y_{lm}(\Omega_a)G_{lm;l'm'}(E)A_{l'm'} \tag{7.17}$$

Now, ψ_{out} is the wave scattered from the atom when ψ_{inc} is incident upon the latter and, therefore

$$\psi_{\text{out}} = \sum_{\substack{lm \\ l'm'}} S_l^a h_l^{(1)}(qr_a)Y_{lm}(\Omega_a)G_{lm;l'm'}(E)A_{l'm'} \tag{7.18}$$

The above equation follows from the definition of S_l^a [see, e.g., Eq. (3.60a)]. Comparing Eq. (7.18) with Eq. (7.9) we obtain

$$\sum_{l'm'} \{\delta_{ll'}\,\delta_{mm'} - S_l^a G_{lm;l'm'}(E)\}A_{l'm'} = 0 \tag{7.19}$$

which tells that a nontrivial solution of the Schrödinger equation, which describes an electron state localized on and around the adatom, exists only for (real) values of the energy such that the determinant of the coefficients in Eq. (7.19) vanishes. This condition can be written in the following form:

$$\det\{[\mathbf{I} - \mathbf{G}\mathbf{S}^a]\mathbf{S}^{a-1}\} = 0 \tag{7.20}$$

where $\mathbf{I}$ is the unit matrix, $\mathbf{G}$ is the matrix defined by Eq. (7.16), and $\mathbf{S}^a$ is the diagonal matrix $S_{lm;\,l'm'}^a = S_l^a\,\delta_{ll'}\delta_{mm'}$. Equation (7.20) shows that the energy eigenvalues of localized (on the adatom) states correspond to poles on the real energy axis of the matrix

$$\mathbf{\Pi}(E) \equiv \mathbf{S}^a\,(\mathbf{I} - \mathbf{G}\mathbf{S}^a)^{-1} \tag{7.21}$$

which, as we shall see in the next section, plays an important role in our analysis of the energy distribution of the field-emitted electrons from the neighborhood of the adatom. It is also worth noting that the existence of *real* roots of Eq. (7.20) [poles of $\mathbf{\Pi}(E)$ on the real energy axis] implies that these roots lie within an energy gap of the metal, because only then does the corresponding eigenfunction decay into the metal [see comments following Eq. (7.13)]. When the condition (7.20) is satisfied for a (real) energy we can solve the system of Eqs. (7.19) to obtain A_{lm}, and, therefore, we obtain the wave-function (apart from a normalization constant),

$$\Psi = \psi_{inc} + \psi_{out} \tag{7.22}$$

for $z > 0$ and $r_a > a_0$. $\Psi(\mathbf{r})$ inside the adatom spheres can be written as a superposition of spherical waves,

$$\Psi(\mathbf{r}_a) = \sum_{lm} A^0_{lm} R_l(r_a) Y_{lm}(\Omega_a) \tag{7.23}$$

where $R_l(r)$ is obtained by numerical integration of Eq. (3.61) [with $v(r) - E$ replaced by $U_a(r) + V_R - E$] at the time when one evaluates $S^a_l(E)$. For $r_a = a_0$ formula (7.23) must give the same value as formula (7.22) and, therefore, we determine A^0_{lm}. The wave function inside the metal is given by

$$\Psi(\mathbf{r}) = \iint_{SBZ} d^2 k_{\parallel} \sum_j C(\mathbf{k}_{\parallel}, j) \psi_{\mathbf{k}_{\parallel}; j}(\mathbf{r}), \qquad z \leq 0 \tag{7.24}$$

where $\psi_{\mathbf{k}_{\parallel}; j}$ are the Bloch waves (which decay "exponentially" as $z \to -\infty$) of the infinite crystal for the given energy. (Note that we have denoted explicitly the dependence of these waves on $\mathbf{k}_{\parallel}$ in contrast to our convention in Chapter 4 where the same waves are denoted by ψ_j.) The coefficients $C(\mathbf{k}_{\parallel}, j)$ can be determined from the continuity of the wave function at the metal–vacuum interface (the plane $z = 0$), without much difficulty, if an explicit knowledge of these coefficients is required. Finally, if need be, the wave function can be normalized so that $\int |\Psi(\mathbf{r})|^2 d^3r = 1$.

The physical significance of the $\Pi(E)$ matrix becomes clearer in the simple case when U_a, the adatom potential, has a single bound state, an s state, in the energy region under consideration. In this case $S^a(E)$ is approximately given by

$$S^a_l(E) \simeq \frac{\alpha_s}{E - E^0_s} \delta_{l0} \tag{7.25}$$

where E^0_s is the energy level of the above-mentioned s state, and α_s is a constant (a real number) with dimensions of energy. One can easily show that in this case

$$\Pi(E) \simeq \frac{\alpha_s}{E - E^0_s - G_{00;00}(E)\alpha_s} \tag{7.26}$$

If we further assume that $G_{00;00}(E)$ is practically constant over an energy region around E^0_s of the order of $2G_{00;00}(E^0_s)\alpha_s$ (we note that this may not be so for a real metal), we can approximate $G_{00;00}(E)$, in this region, by $G_{00;00}(E_s^0)$, in which case $\Pi(E)$ has a pole at

$$E_s = E_s^0 + G_{00;00}(E_s^0)\alpha_s \tag{7.27}$$

If $G_{00;00}(E_s^0)\alpha_s$ is real (this implies that E_s lies within an energy gap of the metal) the metal–adatom complex has a state localized on the adatom with energy E_s. In other words, the interaction with the metal has shifted the "atomic" energy level by an amount $G_{00;00}(E_s^0)\alpha_s$.

The question naturally arises as to what is the physical meaning of a pole of $\Pi(E)$, in the complex energy plane. It could be, for example, that E_s, as defined by Eq. (7.27), has a *small* (negative) imaginary part. In this case no true localized (on the adatom) state exists, and by implication $\text{Re}(E_s)$ lies within an allowed energy region of the metal energy-band structure. We obtain instead (the situation is analogous to the transition from *true* surface states to surface resonances which we discussed in Section 4.2.2) an "adatom resonance." What happens is that for real energies near a complex pole of $\Pi(E)$, the one-electron eigenfunctions of the metal–adatom complex, which we shall denote by Ψ^M (these are delocalized wave functions extending over the entire system) peak near the adsorbed atom, giving rise to a concentration of probability density which in many ways is physically indistinguishable from a true localized (on the adatom) state. We note, however, that these resonances are not independent one-electron states. The situation is best described in terms of the local density of states defined, as usual, by

$$\rho(E,\mathbf{r}) \equiv 2 \sum_M |\Psi^M(\mathbf{r})|^2 \, \delta(E - E^M) \tag{7.28}$$

where M denotes a set of quantum numbers, which does not include the spin, which specify the electron state, and the factor of 2 takes into account spin degeneracy. We now define an adatom density of states as follows:

$$\rho_{\text{ADS}}(E) \equiv \int_{|\mathbf{r}-\mathbf{R}_a|<a0} \rho(E,\mathbf{r}) \, d^3r \tag{7.29}$$

$\rho_{\text{ADS}}(E)$ can be broken into "partial densities" corresponding to different angular moment (l,m) as follows. Inside the adatom sphere the wave function $\Psi^M(\mathbf{r})$ can be written as

$$\Psi^M(\mathbf{r}_a) = \sum_{lm} A_{lm}^M R_l(r_a) Y_{lm}(\Omega_a) \tag{7.30}$$

where A_{lm}^M are constant coefficients. Using the above formula and the orthogonality property of the spherical harmonics one can easily show that

$$\rho_{\text{ADS}}(E) = \sum_{lm} \rho_{\text{ADS};lm}(E) \tag{7.31}$$

$$\rho_{\text{ADS};lm}(E) = \sum_{M} |A_{lm}^M|^2 I_l(E)\, \delta(E - E^M) \tag{7.32}$$

$$I_l(E) = \int_0^{a_0} [R_l(r)]^2 r^2 \, dr \tag{7.33}$$

We (Modinos and Oxinos, 1978; Oxinos and Modinos, 1980) calculated $\rho_{\text{ADS}}(E)$ and the partial densities $\rho_{\text{ADS};lm}(E)$ for a simple metal–adatom potential. Using the notation of the present chapter we can describe this potential as follows. The potential inside the metal equals zero (i.e., the electrons are free inside the metal). The potential on the vacuum side of the metal–vacuum interface equals $V_R = E_F + \phi = 16.05$ eV ($E_F = 11.55$ eV; $\phi = 4.5$ eV). The adatom potential U_a, which is superimposed on the constant potential on the vacuum side of the interface and is centered on $\mathbf{R}_a = (0,0,Z_a)$ with $Z_a \geq a_0$, is given by

$$U_a = \sum_l v_{a;l} P_l, \qquad r_a < a_0$$
$$= 0, \qquad r_a > a_0 \tag{7.34}$$

where $v_{a;l}$ are constants and P_l is a projection operator which picks out the lth spherical harmonic component of the wave function. In other words, it is assumed that the lth spherical harmonic component of the wavefield around the atom sees a spherical well potential whose depth $(v_{a;l})$ depends on l. By varying $v_{a;l}$ one can generate an atomic s state with an energy E_s^0 (when $Z_a = \infty$) in the immediate vicinity of the Fermi level with no other atomic states in the same region of energy. We refer to this atomic state as an isolated s state. For a narrow well, say $a_0 \simeq 1.35$ Å or so, this can be achieved by choosing the depth of the well independent of l and such that $E_s^0 \simeq E_F$. States of higher angular momentum either do not exist or if they exist they lie sufficiently away from E_s^0. In the same manner one can generate an isolated p state, with E_p^0 (when $Z_a = \infty$) $\simeq E_F$ and no other state in the same energy region, or an isolated d state, etc. On the other hand by choosing different values of $v_{a;l}$ for different l, one can generate a pair of atomic states, say an s and a p state or an s and a d state in the vicinity of the Fermi level. Figure 7.3 shows the resonance which results from an isolated (in the sense described above) s state of the adsorbed atom. The parameter values (given in the figure caption) were chosen to approximate as far as possible those of Na adsorbed on Al. The numerical results for the full width of the resonance at half-maximum, denoted by Γ_s, are in very good agreement with those of Muscat and Newns

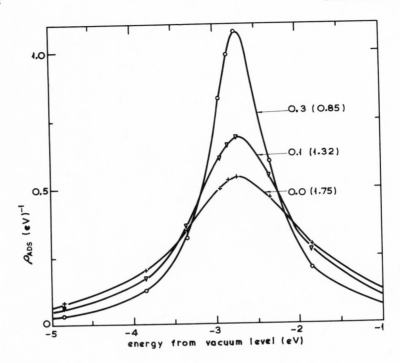

FIGURE 7.3

Adatom density of states for an isolated s-resonance for three different metal–adatom separations. Curves are labeled by the value of $(Z_a - a_0)$ in Å. The full width (in eV) at half-maximum is given in parentheses. The parameter values used in the calculation are $v_{a;l} = -11.91$ eV, $a_0 = 1.35$ Å; $E_s^0 = -2.85$ eV (relative to the vacuum level).

(1978) who have also shown that some important features (e.g., the width of the resonance) of alkali chemisorption on simple metals, as calculated by more elaborate methods (Lang and Williams, 1978), are reproduced quite well by the simple model potential used here. We note that the $\rho_{ADS}(E)$, shown in Fig. 7.3 is described well by

$$\rho_{ADS}(E) \simeq \text{const} \, \frac{1}{(E - E_s)^2 + (\Gamma_s/2)^2} \tag{7.35}$$

where E_s and $\Gamma_s/2$ are the real and imaginary parts of the pole of $\Pi(E)$, given by Eq. (7.27); i.e.,

$$E_s \simeq E_s^0 + \text{Re}[G_{00;00}(E_s^0)\alpha_s] \tag{7.36}$$

$$\Gamma_s \simeq -2 \, \text{Im}[G_{00;00}(E_s^0)\alpha_s] \tag{7.37}$$

We note that our assumption leading to Eq. (7.27), namely, that $G_{00;00}(E)$ is practically constant over an energy region of order Γ_s, is true for a free-electron metal. The decrease in Γ_s with increasing metal–adatom separation, shown in Fig. 7.3, is a standard result (see, e.g., Gadzuk, 1970) and applies equally well to adatom resonances of higher angular momentum.

Figure 7.4 shows the resonance deriving from an isolated atomic p state. Some of the parameter values are given in the figure caption; the values of the other parameters are the same as in Fig. 7.3. The different components of the adatom density of states, given by Eq. (7.32), are also shown in this figure. The difference in the width at half maximum, denoted by Γ_{lm}, of these components is

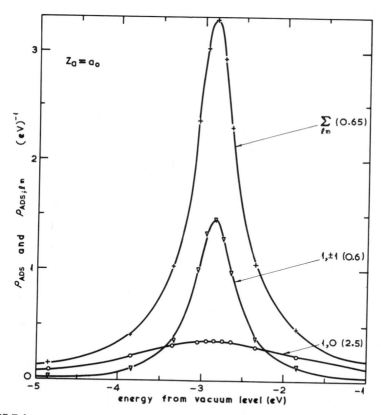

FIGURE 7.4

Adatom density of states ρ_{ADS} and partial components $\rho_{ADS;lm}$ for an isolated p resonance. The curves for $\rho_{ADS;lm}$ are labeled by the quantum numbers l,m. The full width (in eV) at half-maximum is given in parentheses. $v_{a;l} = -25.88$ eV; $E_p^0 = -2.85$ eV (relative to the vacuum level).

obvious. We see that $\Gamma_{1\pm1} \simeq (1/4)\Gamma_{10}$. Similar results have been obtained for an isolated d-type resonance, i.e., $\Gamma_{2\pm2} < \Gamma_{21} < \Gamma_{20}$. It has also been established by comparing the results for the isolated s-, p-, and d-type resonances for the given model, that $\Gamma_d \simeq 0.1\Gamma_s$ and $\Gamma_p \simeq 0.4\Gamma_s$ in agreement with earlier estimates of these quantities by Gadzuk (1970). We have also examined the case of an adatom with adjacent-in-energy s and p (or d) states, and found that the interaction of the

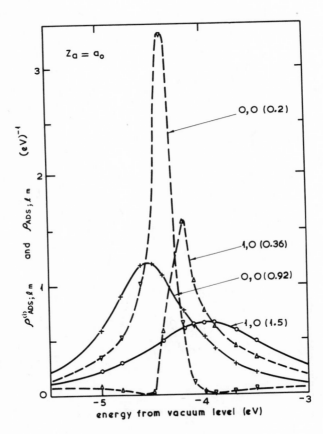

FIGURE 7.5

Solid lines; the partial components $\rho^{(i)}_{ADS;0,0}$ and $\rho^{(i)}_{ADS;1,0}$ for the *isolated* s and p resonances. The parameter values used in the calculation of $\rho^{(i)}_{ADS;0,0}$ are $v_{a;l} = -10.6$ eV, $a_0 = 1.8$ Å; $E_s^0 = -4.48$ eV (from the vacuum level). The parameter values used in the calculation of $\rho^{(i)}_{ADS;1,0}$ are $v_{a;l} = -18.0$ eV, $a_0 = 1.8$ Å; $E_p^0 = -4.02$ eV (from the vacuum level). Broken lines: the partial components $\rho_{ADS;0,0}$ and $\rho_{ADS;1,0}$ when the s and p resonances are allowed to interact through the metal. The parameter values used in the calculation are $v_{a;0} = v_{a;1} = -10.6$ eV, $v_{a;1} = -18.0$ eV, $a_0 = 1.8$ Å. The notation is the same as in Fig. 7.4.

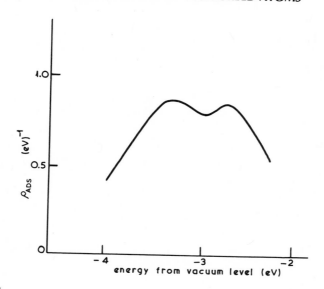

FIGURE 7.6

Diagram illustrating the possibility of fine structure in an s-type adatom resonance, induced by structure in the surface density of states of the substrate.

two states through the metal can lead to an adatom density of states very much different from the sum of the corresponding noninteracting terms. This is demonstrated in Fig. 7.5, which shows the effect of a p state, lying 0.5 eV above an s state in the free atom ($Z_a = \infty$), on the s-type resonance and vice versa. The partial densities $\rho_{\text{ADS};1 \pm 1}(E)$ (not shown in the figure) are not affected by the presence of the s-type resonance, as expected from symmetry considerations. On the other hand, the partial densities $\rho_{\text{ADS};00}(E)$ and $\rho_{\text{ADS};10}(E)$ denoted by the broken curves in the figure, are significantly different from the corresponding quantities (solid curves in the figure) of the isolated (noninteracting) s and p resonances. Note, in particular, that $\rho_{\text{ADS};00}(E)$ has been shifted upwards and is much narrower than the corresponding isolated s-type resonance. This suggests that an empty p-type resonance above the Fermi level, may have a dramatic effect on an s-type resonance lying at, or immediately below, the Fermi level.

We must emphasize that these results were obtained under the assumption that the potential inside the metal is constant, i.e., that the electrons are free there. For example, $G_{00;00}(E)$ for a real metal may be far from constant and could even have a complex pole in the vicinity of E_s^0, which in turn might produce some fine structure in $\rho_{\text{ADS}}(E)$ in the manner indicated schematically in Fig. 7.6 (to be compared with Fig. 7.3). We shall come back to this point when we discuss the energy distribution of the field-emitted electrons from the neighborhood of the adatom.

7.3. ENERGY DISTRIBUTION OF ELECTRONS FIELD-EMITTED FROM THE NEIGHBORHOOD OF A SINGLE ADATOM

We assume that the potential inside the metal ($z < 0$) is not affected by the application of the electric field, and approximate the potential field on the vacuum side of the interface ($z > 0$) by

$$V = V_R^F + U_a^F(r_a), \qquad 0 < z < z_0 = 2a_0 \qquad (7.38)$$
$$V_R^F = V_R - eFa_0 \qquad (7.39a)$$
$$U_a^F(r_a) = 0, \qquad r_a > 0 \qquad (7.39b)$$
$$V = V_s(z) - eFz, \qquad z > z_0 \qquad (7.40)$$

where V_s is the zero field potential barrier defined by Eq. (7.2) and F denotes the magnitude of the applied field. The second term in Eq. (7.39a) is the average value of the electric field term in the region $0 < z < z_0$. We assume that

$$E_F < V_R^F < E_F + \phi \qquad (7.41)$$

We note that the adatom potential in the presence of an applied field denoted by U_a^F, may be different from its zero field expression given by U_a, and that, in principle, it must be calculated self-consistently for any given value of the field. Consider, for example, a half-occupied s-type resonance (with its center at E_F when $F = 0$). As the field increases and the resonance is pulled down in energy [because of the second term in Eq. (7.39a)] the electronic charge in the adatom sphere will increase, and therefore the adatom potential will change accordingly. A simple model calculation of the resonance level as a function of applied field has been given by Bennett and Falicov (1966). They found that the position of the resonance level (its center) relative to its zero field value is mostly determined, about 90% of it, by the field term ($-eFa_0$), which implies that other changes in the adatom potential brought about by the application of the field cancel each other to a very large degree. In what follows we assume, for the sake of simplicity, that $U_a^F = U_a$. It is also worth noting that the second term in Eq. (7.39a) implies that the electric field lines penetrate the adatom terminating on the metal–vacuum interface at $z = 0$. Of course, this is an approximation, and in some instances (e.g., for metallic adsorbates) may lead to an overestimation of the corresponding energy shift.

The energy distribution of the field-emitted electrons from a site of area A (a subregion of a single crystal plane on a field emission tip) containing a single adatom (situated somewhere near the center of the site) is given by (see Section 6.3)

$$j(E) = -\left(\frac{\hbar}{2m\pi^3}\right) f(E) \iint d^2q_\parallel \, \mathrm{Im}\langle \chi'_{q_\parallel} | G_{m-a}(E) | \chi'_{q_\parallel} \rangle \, \frac{|T'_{s(E,q_\parallel)}|^2}{|1 + R'_s(E,q_\parallel)|^2}$$

(7.42)

$j(E) \, dE$ is the number of electrons emitted from area A with energy between E and $E + dE$, per unit time, divided by A.

$$\langle \chi'_{q_\parallel} | G_{m-a}(E) | \chi'_{q_\parallel} \rangle = \iint \chi'^*_{q_\parallel}(\mathbf{r}) G_{m-a}(E,\mathbf{r},\mathbf{r}') \chi'_{q_\parallel}(\mathbf{r}') \, d^3r \, d^3r'$$

(7.43)

where $G_{m-a}(E,\mathbf{r},\mathbf{r}')$ is the Green's function for the metal–adatom complex when the field is applied to the surface, and

$$\chi'_{q_\parallel}(\mathbf{r}) = \frac{1}{\sqrt{VA}} \exp(i\mathbf{q}_\parallel \cdot \mathbf{r}_\parallel) \, \delta(z - z_0)$$

(7.44)

$T'_s(E,\mathbf{q}_\parallel)$ and $R'_s(E,\mathbf{q}_\parallel)$ are defined as follows. Let an electron of energy E and transverse (parallel to the surface) wave vector $\mathbf{q}_\parallel$, be incident on a barrier, defined by,

$$\begin{aligned} V &= 0, & z &< z_0 \\ &= V_s(z) - eFz, & z &> z_0 \end{aligned}$$

(7.45)

from the left. The wave function which describes the scattering of the electron by the above barrier can be written in the following form:

$$\psi = \exp(i\mathbf{q}_\parallel \cdot \mathbf{r}_\parallel) \left\{ \exp\left[i\left(\frac{2mE}{\hbar^2} - q_\parallel^2\right)^{1/2} (z - z_0) \right] \right.$$

$$\left. + R'_s(E,\mathbf{q}_\parallel) \exp\left[-i\left(\frac{2mE}{\hbar^2} - q_\parallel^2\right)^{1/2} (z - z_0) \right] \right\}$$

(7.46a)

for $z < z_0$

$$\psi = \exp(i\mathbf{q}_\parallel \cdot \mathbf{r}_\parallel) \frac{T'_s(E,\mathbf{q}_\parallel)}{[\lambda_{q_\parallel}(z)]^{1/2}} \exp\left[i \int_{b_{q_\parallel}}^{z} \lambda_{q_\parallel}(z) dz \right]$$

(7.46b)

for $z \gg b_{q_{\parallel}}$

$$\lambda_{q_{\parallel}}(z) \equiv \{(2m/\hbar^2) [E - V_s(z) + eFz) - q_{\parallel}^2]\}^{1/2} \tag{7.47}$$

$b_{q_{\parallel}}$ in Eq. (7.46) denotes the larger of the two roots (classical turning points) of the equation $\lambda_{q_{\parallel}}(z) = 0$. Equation (7.46) defines $R'_s(E, \mathbf{q}_{\parallel})$ and $T'_s(E, \mathbf{q}_{\parallel})$, which can de determined, for any given surface barrier $V_s(z)$, by solving numerically the appropriate Schrödinger equation.

For the purpose of calculating the matrix element defined by Eq. (7.43), we replace the potential field on the vacuum side of the interface $(z > 0)$ by [note the analogy with Eq. (7.7)]

$$V = V_R^F + U_a(r_a) \tag{7.48}$$

V_R^F is given by Eq. (7.39a) and we have assumed that $U_a^F(r_a) = U_a(r_a)$ [see comments following Eq. (7.41)].

One can easily show, by repeating the arguments leading to Eq. (5.69), that

$$\langle \chi'_{\mathbf{q}_{\parallel}} | G_{m-a}(E) | \chi'_{\mathbf{q}_{\parallel}} \rangle = \alpha'_{q_{\parallel}} [1 + R^{m-a}(\mathbf{q}_{\parallel}, \mathbf{q}_{\parallel})] \tag{7.49}$$

$$\alpha'_{q_{\parallel}} \equiv - \left(\frac{im}{\hbar^2} \right) \Big/ K_z^+(q_{\parallel})$$

$$= - \left(\frac{m}{\hbar^2} \right) \left[\frac{2m}{\hbar^2} (V_R^F - E) + q_{\parallel}^2 \right]^{-1/2} \tag{7.50}$$

Equation (7.49) is obtained from Eq. (5.69) by putting $\mathbf{R}_s = 0$, and replacing $\mathbf{R}^c$ by $\mathbf{R}^{m-a}$. $\mathbf{R}^{m-a}$, which describes the "reflection" of the electron by the potential field to the left of $z = z_0$, is defined by

$$\int G_0(E, \mathbf{r}, \mathbf{r}'') T_{m-a}(\mathbf{r}'', \mathbf{r}') \exp[i\mathbf{K}^-(\mathbf{q}_{\parallel}) \cdot (\mathbf{r}' - \mathbf{z}_0)] \, d^3r'' \, d^3r'$$
$$= \sum_{\mathbf{q}'_{\parallel}} R^{m-a} (\mathbf{q}'_{\parallel}, \mathbf{q}_{\parallel}) \exp[i\mathbf{K}^+(\mathbf{q}'_{\parallel}) \cdot (\mathbf{r} - \mathbf{z}_0)], \qquad z > z_0 \tag{7.51}$$

We note that throughout the present section $\mathbf{K}^\pm(\mathbf{q}_{\parallel})$ are given by Eq. (7.11) with V_R replaced by $V_R^F \cdot \mathbf{R}_s = 0$ because the wave $\exp[i\mathbf{K}^+(\mathbf{q}_{\parallel}) \cdot (\mathbf{r} - \mathbf{z}_0)]$ is not reflected by the constant potential field in the region $z > z_0$. We note that $R^{m-a}(\mathbf{q}_{\parallel}, \mathbf{q}'_{\parallel})$ is a semicontinuous matrix with indices $\mathbf{q}$ given by

$$\mathbf{q}_{\parallel} = (q_x, q_y) = (2\pi/\sqrt{A})(n_x, n_y) \tag{7.52}$$

Although $R^{m-a}(\mathbf{q}_\|, \mathbf{q}'_\|)$ is defined for any pair $\mathbf{q}_\|, \mathbf{q}'_\|$, for the purpose of evaluating $j(E)$ as defined by Eq. (7.42) we need $R^{m-a}(\mathbf{q}_\|, \mathbf{q}_\|)$ only for small values of $|\mathbf{q}_\||$. We assume, as in the case of emission from a clean metal surface (see Section 5.2), that only values of $\mathbf{q}_\|$ in the central region of the SBZ of the *metal substrate* will contribute significantly to $j(E)$. Therefore, in what follows we shall assume that $\mathbf{q}_\|$ in Eq. (7.49) lies in this region.

Using the fact that the adatom potential U_a does not overlap with the potential field in the region $z < 0+$, we can expand $G_0 T_{m-a}$ in Eq. (7.51) as follows (the notation follows the convention introduced in Section 5.2.2 whereby the arguments of G_0, T_m, and T_a and the integral signs are omitted):

$$
\begin{aligned}
G_0 T_{m-a} = & \; G_0 T_m + [G_0 T_a + \cdots] + G_0 T_m [G_0 T_a + \cdots] \\
& + [G_0 T_a + \cdots] G_0 T_m \\
& + G_0 T_m [G_0 T_a + \cdots] G_0 T_m
\end{aligned}
\tag{7.53}
$$

$$
\begin{aligned}
[G_0 T_a + \cdots] \equiv & \; G_0 T_a + G_0 T_a G_0 T_m G_0 T_a \\
& + G_0 T_a G_0 T_m G_0 T_a G_0 T_m G_0 T_a + \cdots
\end{aligned}
\tag{7.54}
$$

We note that all higher-order terms in Eq. (7.54) begin and end with T_a. The matrix T_a describes the scattering of the electron by the adatom potential U_a. We have

$$
\begin{aligned}
\int G_0(E, \mathbf{r}, \mathbf{r}'') T_a(\mathbf{r}'', \mathbf{r}') j_l(q r'_a) Y_{lm}(\Omega'_a) \; d^3 r'' \; d^3 r' \\
= S_l^a h_l^{(1)}(q r_a) Y_{lm}(\Omega_a), \qquad r_a > 0
\end{aligned}
\tag{7.55}
$$

The integration in Eq. (7.55) is over the volume of the adatom sphere $[T_a(\mathbf{r}'', \mathbf{r}') = 0$ if r''_a and/or r'_a is greater than $a_0]$. We note that q, throughout this section, is given by Eq. (7.8) with V_R replaced by V_R^F. The matrix T_m describes the scattering of the electrons by the potential field in the region $z < 0+$, which includes the potential field inside the metal *and* the step in the barrier at $z = 0$. We have

$$
\begin{aligned}
\int G_0(E, \mathbf{r}, \mathbf{r}'') T_m(\mathbf{r}'', \mathbf{r}') \exp[i\mathbf{K}^-(\mathbf{k}_\| + \mathbf{g}) \cdot \mathbf{r}'] \; d^3 r'' \; d^3 r' \\
= \sum_{\mathbf{g}'} R_{\mathbf{g}'\mathbf{g}}(E, \mathbf{k}_\|) \exp[i\mathbf{K}^+(\mathbf{k}_\| + \mathbf{g}') \cdot \mathbf{r}], \qquad z > 0
\end{aligned}
\tag{7.56}
$$

We note that $T_m(\mathbf{r}'', \mathbf{r}') = 0$ if $\mathbf{r}''$ and/or $\mathbf{r}'$ lies on the right of the interface at $z = 0$. $R_{\mathbf{g}'\mathbf{g}}(E, k_\|)$ are the matrix elements defined by Eq. (7.13). We remember that $\mathbf{k}_\|$ in the above equation is a reduced wave vector in the SBZ of the metal substrate. Using Eqs. (7.55), (7.56) and (3.58), and (7.10) (the latter two equations are used for expanding planes waves into spherical waves around $\mathbf{R}_a$ and spherical waves around $\mathbf{R}_a$ into plane waves, respectively), one can show that

$$\int G_0(\mathbf{r},\mathbf{r}'') T_m(\mathbf{r}'',\mathbf{r}') \exp[i\mathbf{K}^-(\mathbf{q}_{\parallel}) \cdot (\mathbf{r}' - \mathbf{z}_0)] \, d^3r'' \, d^3r'$$

$$= \sum_{\substack{\mathbf{k}_{\parallel} \\ (SBZ)}} \sum_{\mathbf{g}'} R^{(m)}(\mathbf{k}_{\parallel} + \mathbf{g}',\mathbf{q}_{\parallel}) \exp[i\mathbf{K}^+(\mathbf{k}_{\parallel} + \mathbf{g}') \cdot (\mathbf{r} - \mathbf{z}_0)], \qquad z > z_0 \tag{7.57}$$

$$R^{(m)}(\mathbf{k}_{\parallel} + \mathbf{g}',\mathbf{q}_{\parallel}) \equiv \exp[-i\mathbf{K}^-(\mathbf{q}_{\parallel}) \cdot \mathbf{z}_0 + i\mathbf{K}^+(\mathbf{k}_{\parallel} + \mathbf{g}') \cdot \mathbf{z}_0] R_{\mathbf{g}'\mathbf{0}}(E,\mathbf{k}_{\parallel}) \, \delta_{\mathbf{k}_{\parallel}\mathbf{q}_{\parallel}} \tag{7.58}$$

$$\int [G_0(\mathbf{r},) T_a(,\mathbf{r}') + \cdots] \exp[i\mathbf{K}^-(\mathbf{q}_{\parallel}) \cdot (\mathbf{r}' - \mathbf{z}_0)] \, d^3r'$$

$$= \sum_{\substack{\mathbf{k}_{\parallel} \\ (SBZ)}} \sum_{\mathbf{g}'} R^{(1)}(\mathbf{k}_{\parallel} + \mathbf{g}',\mathbf{q}_{\parallel}) \exp[i\mathbf{K}^+(\mathbf{k}_{\parallel} + \mathbf{g}') \cdot (\mathbf{r} - \mathbf{z}_0)], \qquad z > z_0 \tag{7.59}$$

$$R^{(1)}(\mathbf{k} + \mathbf{g}',\mathbf{q}_{\parallel}) \equiv \left(\frac{2\pi}{A}\right) \frac{\exp[i\mathbf{K}^+(\mathbf{k}_{\parallel} + \mathbf{g}') \cdot (\mathbf{z}_0 - \mathbf{R}_a)]}{q[q^2 - (\mathbf{k}_{\parallel} + \mathbf{g}')^2]^{1/2}}$$

$$\times \sum_{\substack{lm \\ l'm'}} (i)^{-l} Y_{lm}(\Omega_{\mathbf{K}+(\mathbf{k}_{\parallel}+\mathbf{g}')}) \Pi_{lm;l'm'}(E) A_{l'm'}(\mathbf{q}_{\parallel}) \tag{7.60}$$

where $\mathbf{\Pi}(E)$ is the matrix defined by Eq. (7.21).

$$A_{l'm'}(\mathbf{q}_{\parallel}) \equiv 4\pi i^{l'}(-1)^{m'} Y_{l'-m'}(\Omega_{\mathbf{K}-(\mathbf{q}_{\parallel})}) \exp[-i\mathbf{K}^-(\mathbf{q}_{\parallel}) \cdot (\mathbf{z}_0 - \mathbf{R}_a)] \tag{7.61}$$

$$\int G_0(\mathbf{r},) T_m[G_0 T_a(,\mathbf{r}') + \cdots] \exp[i\mathbf{K}^-(\mathbf{q}_{\parallel}) \cdot (\mathbf{r}' - \mathbf{z}_0)] \, d^3r'$$

$$= \sum_{\substack{\mathbf{k}_{\parallel} \\ (SBZ)}} \sum_{\mathbf{g}'} R^{(2)}(\mathbf{k}_{\parallel} + \mathbf{g}',\mathbf{q}_{\parallel}) \exp[i\mathbf{K}^+(\mathbf{k}_{\parallel} + \mathbf{g}') \cdot (\mathbf{r} - \mathbf{z}_0)], \qquad z > z_0 \tag{7.62}$$

where

$$R^{(2)}(\mathbf{k}_{\parallel} + \mathbf{g}',\mathbf{q}_{\parallel}) = \left(\frac{2\pi}{A}\right) \sum_{\mathbf{g}} \frac{\exp[-i\mathbf{K}^-(\mathbf{k}_{\parallel} + \mathbf{g}) \cdot \mathbf{R}_a]}{q[q^2 - (\mathbf{k}_{\parallel} + \mathbf{g})^2]^{1/2}} R_{\mathbf{g}'\mathbf{g}}(E,\mathbf{k}_{\parallel})$$

$$\times \sum_{\substack{l'm' \\ lm}} (i)^{-l} Y_{lm}(\Omega_{\mathbf{K}-(\mathbf{k}_{\parallel}+\mathbf{g})}) \Pi_{lm;l'm'}(E) A_{l'm'}(\mathbf{q}_{\parallel}) \exp[i\mathbf{K}^+(\mathbf{k}_{\parallel} + \mathbf{g}') \cdot \mathbf{z}_0] \tag{7.63}$$

$$\int [G_0(\mathbf{r},) T_a + \cdots] G_0 T_m(,\mathbf{r}') \exp[i\mathbf{K}^-(\mathbf{q}_{\parallel}) \cdot (\mathbf{r}' - \mathbf{z}_0)] \, d^3r'$$

$$= \sum_{\substack{\mathbf{k}_{\parallel} \\ (SBZ)}} \sum_{\mathbf{g}'} R^{(3)}(\mathbf{k}_{\parallel} + \mathbf{g}',\mathbf{q}_{\parallel}) \exp[i\mathbf{K}^+(\mathbf{k}_{\parallel} + \mathbf{g}') \cdot (\mathbf{r} - \mathbf{z}_0)], \qquad z > z_0 \tag{7.64}$$

where $R^{(3)}(\mathbf{k}_{\parallel} + \mathbf{g}',\mathbf{q}_{\parallel})$ is given by Eq. (7.60) with $A_{l'm'}(\mathbf{q}_{\parallel})$ replaced by

$$B_{l'm'}(\mathbf{q}_{\parallel}) \equiv 4\pi \sum_{\mathbf{g}} R_{\mathbf{g}\mathbf{0}}(E,\mathbf{q}_{\parallel}) \exp[-i\mathbf{K}^-(\mathbf{q}_{\parallel}) \cdot \mathbf{z}_0 + i\mathbf{K}^+(\mathbf{q}_{\parallel} + \mathbf{g}) \cdot \mathbf{R}_a]$$

$$\times i^{l'}(-1)^{m'} Y_{l'-m'}(\Omega_{\mathbf{K}+(\mathbf{q}_{\parallel}+\mathbf{g})}) \tag{7.65}$$

$$\int G_0(\mathbf{r},) T_m[G_0 T_a + \cdots] G_0 T_m(,\mathbf{r}') \exp[i\mathbf{K}^-(\mathbf{q}_{\parallel}) \cdot (\mathbf{r}' - \mathbf{z}_0)] \, d^3r'$$

$$= \sum_{\substack{\mathbf{k}_{\parallel} \\ (SBZ)}} \sum_{\mathbf{g}'} R^{(4)}(\mathbf{k}_{\parallel} + \mathbf{g}',\mathbf{q}_{\parallel}) \exp[i\mathbf{K}^+(\mathbf{k}_{\parallel} + \mathbf{g}') \cdot (\mathbf{r} - \mathbf{z}_0)], \qquad z > z_0 \tag{7.66}$$

where $R^{(4)}(\mathbf{k}_{\parallel} + \mathbf{g}', \mathbf{q}_{\parallel})$ is given by Eq. (7.63) with $A_{l'm'}(\mathbf{q}_{\parallel})$ replaced by $B_{l'm'}(\mathbf{q}_{\parallel})$.

Using Eqs. (7.51), (7.53), and (7.57)–(7.66) we finally obtain

$$R^{m-a}(\mathbf{q}_{\parallel},\mathbf{q}_{\parallel}) = R^{(m)}(\mathbf{q}_{\parallel},\mathbf{q}_{\parallel}) + R^{(1)}(\mathbf{q}_{\parallel},\mathbf{q}_{\parallel}) + R^{(2)}(\mathbf{q}_{\parallel},\mathbf{q}_{\parallel})$$
$$+ R^{(3)}(\mathbf{q}_{\parallel},\mathbf{q}_{\parallel}) + R^{(4)}(\mathbf{q}_{\parallel},\mathbf{q}_{\parallel}) \qquad (7.67)$$

In what follows we assume that there exists an isolated s-type adatom resonance in the region of the Fermi level, in which case $\Pi(E)$ is given by Eq. (7.26), and the formulas for $R^{(i)}(\mathbf{q}_{\parallel},\mathbf{q}_{\parallel})$, $i = 1,2,3,4$ simplify as follows [we put $\mathbf{z}_0 = (0,0,2a_0)$ and $\mathbf{R}_a = (0,0,a_0)$]

$$R^{(1)}(\mathbf{q}_{\parallel},\mathbf{q}_{\parallel}) = -\left(\frac{2\pi}{A}\right) \frac{\exp[-2a_0|\mathbf{K}_z^+(\mathbf{q}_{\parallel})|]}{|qK_z^+(\mathbf{q}_{\parallel})|} \left[\frac{\alpha_s}{E - E_s^0 - \alpha_s G_{00;00}(E)}\right]$$
$$(7.68)$$

$$R^{(2)}(\mathbf{q}_{\parallel},\mathbf{q}_{\parallel}) = -\left(\frac{2\pi}{A}\right) \sum_{\mathbf{g}} \frac{\exp[-a_0|K_z^+(\mathbf{q}_{\parallel} + \mathbf{g})|]}{|qK_z^+(\mathbf{q}_{\parallel} + \mathbf{g})|} R_{0\mathbf{g}}(E,\mathbf{q}_{\parallel})$$
$$\times \exp[-3a_0|K_z^+(\mathbf{q}_{\parallel})|] \frac{\alpha_s}{E - E_s^0 - \alpha_s G_{00;00}(E)} \qquad (7.69\text{a})$$

$$\simeq -\left(\frac{2\pi}{A}\right) \frac{\exp[-4a_0|K_z^+(\mathbf{q}_{\parallel})|]}{|qK_z^+(\mathbf{q}_{\parallel})|} R_{\mathbf{oo}}(E,\mathbf{q}_{\parallel}) \frac{\alpha_s}{E - E_s^0 - \alpha_s G_{00;00}(E)}$$
$$(7.69\text{b})$$

$$R^{(3)}(\mathbf{q}_{\parallel},\mathbf{q}_{\parallel}) = -\left(\frac{2\pi}{A}\right) \frac{\exp[-3a_0|K_z^+(\mathbf{q}_{\parallel})|]}{|qK_z^+(\mathbf{q}_{\parallel})|}$$
$$\times \sum_{\mathbf{g}} R_{\mathbf{g0}}(E,\mathbf{q}_{\parallel}) \exp[-a_0|K_z^+(\mathbf{q}_{\parallel} + \mathbf{g})|] \qquad (7.70\text{a})$$
$$\times \frac{\alpha_s}{E - E_s^0 - \alpha_s G_{00;00}(E)}$$

$$\simeq -\left(\frac{2\pi}{A}\right) \frac{\exp[-4a_0|K_z^+(\mathbf{q}_{\parallel})|]}{|qK^+q_{\parallel})|} R_{\mathbf{oo}}(E,\mathbf{q}_{\parallel})$$
$$\times \frac{\alpha_s}{E - E_s^0 - \alpha_s G_{00;00}(E)} \qquad (7.70\text{b})$$

$$R^{(4)}(\mathbf{q}_{\parallel},\mathbf{q}_{\parallel}) = -\left(\frac{2\pi}{A}\right) \exp[-4a_0|K_z^+(\mathbf{q}_{\parallel})|] \left(\sum_{\mathbf{g}} \sum_{\mathbf{g}'} R_{\mathbf{0g}}(E,\mathbf{q}_{\parallel})\right.$$
$$\times \frac{\exp\{-a_0[|K_z^+(\mathbf{q}_{\parallel} + \mathbf{g})| + |K_z^+(\mathbf{q}_{\parallel} + \mathbf{g}')|]\}}{|qK_z^+(\mathbf{q}_{\parallel} + \mathbf{g})|} R_{\mathbf{g'0}}(E,\mathbf{q}_{\parallel})\right)$$
$$\times \frac{\alpha_s}{E - E_s^0 - \alpha_s G_{00;00}(E)} \qquad (7.71\text{a})$$

$$\simeq -\left(\frac{2\pi}{A}\right) \frac{\exp[-6a_0|K_z^+(\mathbf{q}_{\parallel})|]}{|qK_z^+(\mathbf{q}_{\parallel})|} [R_{\mathbf{oo}}(E,\mathbf{q}_{\parallel})]^2$$
$$\times \frac{\alpha_s}{E - E_s^0 - \alpha_s G_{00;00}(E)} \qquad (7.71\text{b})$$

The energy distribution of the emitted electrons is, according to Eqs. (7.42), (7.49), and (7.67), given by

$$j(E) = j_m(E) + \Delta j(E) \tag{7.72}$$

where

$$j_m(E) = -\left(\frac{\hbar}{2m\pi^3}\right)f(E) \int\int d^2q_\parallel \alpha'_{q_\parallel} \, \text{Im}[R^{(m)}(\mathbf{q}_\parallel,\mathbf{q}_\parallel)]\frac{|T'_s(E,\mathbf{q}_\parallel)|^2}{|1 + R'_s(E,\mathbf{q}_\parallel)|^2}$$

$$\simeq -\left(\frac{\hbar}{2m\pi^3}\right)f(E) \int\int d^2q_\parallel \alpha'_{q_\parallel} \text{Im}[R_{\mathbf{oo}}(E,\mathbf{q}_\parallel)]\frac{\exp[-4a_0|K_z^+(\mathbf{q}_\parallel)|]|T'_s(E,\mathbf{q}_\parallel)|^2}{|1 + R'_s(E,\mathbf{q}_\parallel)|^2} \tag{7.73}$$

may, for practical purposes, be identified with the energy distribution from the clean metal surface for the same applied field, and

$$\Delta j(E) = -\left(\frac{\hbar}{2m\pi^3}\right)f(E) \int\int d^2q_\parallel \alpha'_{q_\parallel} \, \text{Im} \, [R^{(1)}(\mathbf{q}_\parallel,\mathbf{q}_\parallel) + R^{(2)}(\mathbf{q}_\parallel,\mathbf{q}_\parallel)$$

$$+ R^{(3)}(\mathbf{q}_\parallel,\mathbf{q}_\parallel) + R^{(4)}(\mathbf{q}_\parallel,\mathbf{q}_\parallel)]\frac{|T'_s(E,\mathbf{q}_\parallel)|^2}{|1 + R'_s(E,\mathbf{q}_\parallel)|^2} \tag{7.74}$$

represents a correction term due to the adsorbed atom. We note that only small values of $|\mathbf{q}_\parallel|$ (within the central region of the SBZ of the metal substrate) contribute to the integrals in Eqs. (7.73) and (7.74). We note that also $\Delta j(E)$ is not necessarily positive at every value of the energy.

The (total) emitted current density, i.e., the total current emitted from the given site divided by the area A of this site, is given by

$$J = J_m + \Delta J \tag{7.75}$$
$$J_m = \int j_m(E) \, dE \tag{7.76}$$
$$\Delta J = \int \Delta j(E) \, dE \tag{7.77}$$

It is obvious from Eqs. (7.68)–(7.71) that ΔJ, the contribution of the emitted current density associated with the adsorbed atom, is inversely proportional to A (the area of the site). In practice, A is the area sampled by the probe hole in the anode of the field emission diode (see Section 2.3). Depending on the experimental setup, A could be anything between 200 Å^2 and 10^4 Å^2, but as we have seen in Section 2.3 it cannot be determined very accurately. Obviously, if A is very large, $\Delta J \ll J_m$ and a single adatom will have no significant (measurable) effect on either the current density or the energy distribution of the emitted electrons. Figure 7.7 (Clark and Young, 1968) shows the increase in the current emitted from an area of 250 Å^2 or so ($\sim$ 30 surface atoms) when a Sr atom is adsorbed upon this surface. The curve shown in this figure was obtained as follows. An appropriate

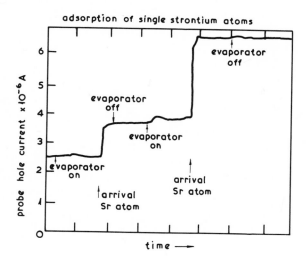

FIGURE 7.7

Probe hole current as a function of time as the strontium source is switched on and off. When a strontium atom arrives on the surface being viewed a step increase in the current takes place. (From Gadzuk and Plummer, 1973.)

source of Sr (a small coil of Mo wire coated with $SrCO_3$) was placed inside the vacuum system of the field emission diode. When heated, to approximately 1200 K, this source produces Sr atoms. Before activating the source in the above manner, the probe current corresponding to the clean surface was measured and was seen to remain constant with time. The total energy distribution from the clean surface could be obtained at this stage. Then the evaporator (source) was turned on. When a Sr atom arrived on the surface being viewed, the current changed in a steplike fashion as can be seen from Fig. 7.7. One can obtain the total energy distribution of the emitted electrons, when a single adatom sits on the surface under consideration, by switching off the evaporator immediately after the arrival of the first adatom has been registered in the above manner. This experiment of Clark and Young and similar experiments by Plummer and Young (1970) proved that the change in the probe current after adsorption of a single atom can be quite large, at least for certain adsorbates (in the Plummer and Young experiments these were Sr, Ba, and Ga) and that, therefore, a spectroscopy of singly adsorbed atoms based on field emission is possible in such cases. Plummer and Young (1970) measured, as well, the energy distribution from various single-crystal planes of tungsten before and after the adsorption of single Sr, Ba, and Ca atoms on these planes. We show some of their results in Figs. 7.8–7.11. We emphasize that the quantity shown in these figures

$$\mathcal{R}(E) = j(E)/j_m(E) \qquad (7.78)$$

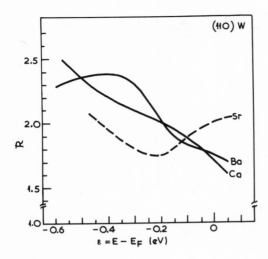

FIGURE 7.8

Experimental $\mathscr{R}$ factor for Ba, Sr, and Ca on W(110). (From Plummer and Young, 1970.)

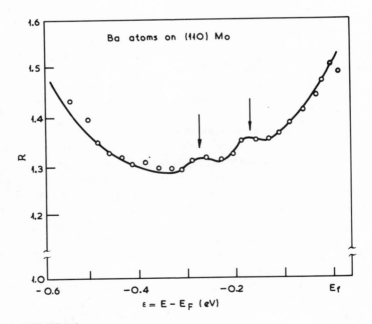

FIGURE 7.9

Experimental $\mathscr{R}$ factor for Ba on Mo(110). (From Gadzuk and Plummer, 1973.)

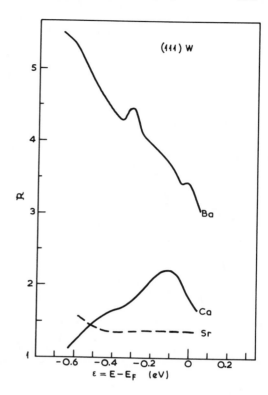

FIGURE 7.10

Experimental $\mathcal{R}$ factor for Ba, Ca, and Sr on W(111). (From Plummer and Young, 1970.)

is the ratio of $j(E)$ (the measured energy distribution after adsorption of a single adatom) to $j_m(E)$ (the measured energy distribution from the clean metal surface). In general it is better if $j(E)$ is divided by a calculated free-electron distribution $j_0(E)$, rather than $j_m(E)$, because the latter can have structure of its own and can, therefore, lead to structure in $\mathcal{R}(E)$ which has nothing to do with the adsorbed atom. On the other hand, the energy distribution from the clean W(110), Mo(110), W(111), and W(013) surfaces in the energy region -0.7 eV $< E - E_F < 0$ is not significantly different from the free-electron distribution and, therefore, we may assume that to a first approximation

$$\mathcal{R}(E) \simeq R(E) + \text{const} \tag{7.79}$$

where

$$R(E) = j(E)/j_0(E) \tag{7.80}$$

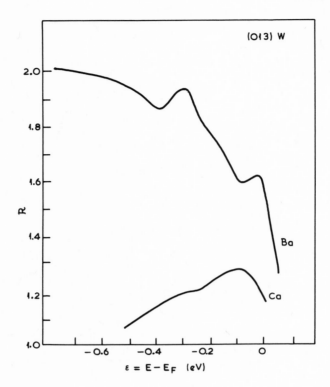

FIGURE 7.11

Experimental $\mathcal{R}$ factor for Ba and Ca on W(013). (From Plummer and Young, 1970.)

in interpreting the experimental results shown in Figs. 7.8–7.11. The question arises as to whether one can understand the structure of these curves on the basis of the formulas given in the present section which, in turn, are based on an independent electron picture of the metal–adatom complex. Because Ba, Sr, and Ca are electropositive adsorbates with a partially empty s-type atomic resonance in the region of the Fermi level, we can use Eqs. (7.68)–(7.71) for at least a qualitative analysis of the data shown in Figs. 7.8–7.11, assuming to begin with that it is possible to explain these data within the independent electron approximation. Substituting Eq. (7.72) into Eq. (7.80), and noting that

$$j_m(E) \propto j_0(E) \tag{7.81}$$

in the present case, we obtain

$$\mathcal{R}(E) \simeq \text{const} + \Delta j(E)/j_0(E) \tag{7.82}$$

Let us, for the sake of simplicity, drop the term proportional to $R^{(4)}(\mathbf{q}_\parallel,\mathbf{q}_\parallel)$ in the integrand of Eq. (7.74) since this is much smaller than the other three terms in this integrand, and let us also replace $K_z^+(\mathbf{q}_\parallel)$ by $K_z^+(\mathbf{q}_\parallel = 0) = q$ in Eqs. (7.68)–(7.70), in which case we obtain

$$
\Delta j(E) \simeq -\frac{f(E)}{\hbar \pi^2}\left(\frac{1}{A|q|^3}\right) \iint d^2 q_\parallel \left\{ \exp(-2a_0|q|) \right.
$$
$$
\times \operatorname{Im}\left[\frac{\alpha_s}{E - E_s^0 - \alpha_s G_{00;00}(E)}\right]
$$
$$
\left. + \exp(-4a_0|q|) \operatorname{Im}\left[\frac{2\alpha_s R_{OO}(E,\mathbf{q}_\parallel)}{E - E_s^0 - \alpha_s G_{00;00}(E)}\right]\right\} \frac{|T_s'(E,\mathbf{q}_\parallel)|^2}{|1 + R_s'(E,\mathbf{q}_\parallel)|^2}
$$

$$(7.83)$$

where

$$
|q| = [(2m/\hbar^2)(V_R^F - E)]^{1/2} \tag{7.84}
$$

and

$$
G_{00;00}(E) = -\left(\frac{1}{2\pi}\right) \iint\limits_{SBZ} d^2 k_\parallel \sum_{g'} \sum_{g}
$$
$$
\times \frac{\exp\{-a_0[|K_z^+(\mathbf{k}_\parallel + \mathbf{g})| + |K_z^+(\mathbf{k}_\parallel + \mathbf{g}')|]\}}{|q K_z^+(\mathbf{k}_\parallel + \mathbf{g})|} R_{g'g}(E,\mathbf{k}_\parallel) \quad (7.85)
$$

In a first approximation we can drop the terms corresponding to $g,g' \neq 0$ and write $G_{00;00}(E)$ as follows:

$$
G_{00;00}(E) = g_r(E) - i g_i(E) \tag{7.85a}
$$
$$
g_r(E) \simeq - \iint\limits_{SBZ} \frac{\exp\{-2a_0[(2m/\hbar^2)(V_R^F - E) + k_\parallel^2]^{1/2}\}}{2\pi |q|^2}
$$
$$
\times \operatorname{Re}[R_{OO}(E,\mathbf{k}_\parallel)]d^2 k_\parallel \tag{7.85b}
$$
$$
g_i(E) \simeq \iint\limits_{SBZ} \frac{\exp\{-2a_0[(2m/\hbar^2)(V_R^F - E) + k_\parallel^2]^{1/2}\}}{2\pi |q|^2}
$$
$$
\times \operatorname{Im}[R_{OO}(E,\mathbf{k}_\parallel)]d^2 k_\parallel \tag{7.85c}
$$

We note that $g_i(E)$ is a positive number. In order to see that, observe that $j_m(E)$, given by Eq. (7.73), is a positive quantity and that, since $(-\alpha_{q_\parallel}') > 0$,

$\text{Im}[R_{OO}(E,\mathbf{k}_\parallel)]$ must also be a positive quantity. $g_i(E)$ is obviously related to the partial density of states (at $z = 0$) of the clean metal substrate [$U_a = 0$ in Eq. (7.48)] given by

$$-\left(\frac{1}{\pi}\right) \iint_{\text{SBZ}} \text{Im}[\langle \chi_{\mathbf{q}_\parallel} | G_m(E) | \chi_{\mathbf{q}_\parallel}\rangle] \, d^2q_\parallel$$

$$= \frac{1}{\pi} \iint_{\text{SBZ}} |\alpha'_{q_\parallel}| \text{Im}[R_{OO}(E,\mathbf{q}_\parallel)] \, d^2q_\parallel \qquad (7.86)$$

where $\chi_{\mathbf{q}_\parallel}$ is the function defined by Eq. (5.37).

Using Eqs. (7.82)–(7.85a) we find that, to a first approximation,

$$\mathcal{R}(E) \simeq \mathcal{A}(E) + \mathcal{B}(E) + \text{const} \qquad (7.87)$$

$$\mathcal{A}(E) \simeq \frac{\exp(2a_0|q|)}{A|q|^2} \frac{\alpha_s^2 g_i(E)}{[E - E_s^0 - \alpha_s g_r(E)]^2 + \alpha_s^2 g_i^2(E)} \qquad (7.88)$$

$$\mathcal{B}(E) \simeq \left(\frac{1}{A|q|^2}\right) \frac{2\alpha_s\{-b_1(E)[E - E_s^0 - \alpha_s g_r(E)] + b_2(E)g_i(E)\alpha_s\}}{[E - E_s^0 - \alpha_s g_r(E)]^2 + \alpha_s^2 g_i^2(E)}$$

$$\qquad (7.89)$$

$$b_1(E) \equiv \frac{1}{(\text{SBZ})_0} \iint_{(\text{SBZ})_0} \text{Im}[R_{OO}(E,\mathbf{k}_\parallel)] \, d^2k_\parallel \qquad (7.90a)$$

$$b_2(E) \equiv \frac{1}{(\text{SBZ})_0} \iint_{(\text{SBZ})_0} \text{Re}[R_{OO}(E,\mathbf{k}_\parallel)] \, d^2k_\parallel \qquad (7.90b)$$

where $(\text{SBZ})_0$ denotes the area of the central region of the SBZ of the metal substrate. The fact that $j_m(E)$ for the surfaces considered in Figs. 7.8–7.11 is free-electron-like in the energy region $-0.7 \text{ eV} < E - E_F < 0$, suggests that $b_1(E)$ and $b_2(E)$ are smooth functions of the energy in this region [this follows from Eq. (7.73)]. Let us, for the sake of simplicity, replace $b_1(E)$ and $b_2(E)$ in Eq. (7.89) by constants, $\bar{b}_1$ and $\bar{b}_2$, respectively, in which case

$$\mathcal{B}(E) \simeq \left(\frac{1}{A|q|^2}\right) \frac{2\alpha_s\{\bar{b}_2\alpha_s g_i(E) - \bar{b}_1 [E - E_s^0 - \alpha_s g_r(E)]\}}{[E - E_s^0 - \alpha_s g_r(E)]^2 + \alpha_s^2 g_i^2(E)} \qquad (7.91)$$

We emphasize that $g_i(E)$ and $g_r(E)$, given by Eq. (7.85), may have structure, e.g., a peak, in the given energy region even when $b_1(E)$ and $b_2(E)$ have none. [Note that the integrals in Eqs. (7.85b) and (7.85c) extend over the entire SBZ whereas the integrals in Eqs. (7.90) extend only over the central SBZ.] For example, a surface resonance in the outer region of the SBZ may lead to a peak in $g_i(E)$ but not in $b_1(E)$. Let us first consider the case when $g_r(E)$ and $g_i(E)$ are structureless. Replacing these functions by constants, $\bar{g}_r$ and $\bar{g}_i$, respectively, in Eqs. (7.88) and (7.91), we obtain

$$\mathcal{A}(E) \simeq \frac{\exp(2a_0|q|)}{A|q|^2}\left[\frac{|\alpha_s|(\Gamma_s/2)}{(E-E_s)^2 + (\Gamma_s/2)^2}\right] \tag{7.92}$$

$$\mathcal{B}(E) \simeq \left(\frac{2}{A|q|^2}\right)\frac{\bar{b}_2\bar{g}_i\alpha_s^2 - \bar{b}_1\alpha_s(E-E_s)}{(E-E_s)^2 + (\Gamma_s/2)^2} \tag{7.93}$$

where

$$E_s = E_s^0 + \alpha_s\bar{g}_r \tag{7.94}$$
$$\Gamma_s = 2\bar{g}_i|\alpha_s| \tag{7.95}$$

The variation of $[\mathcal{A}(E) + \mathcal{B}(E)]$ with energy is shown, schematically, by the solid curve in Fig. 7.12. The broken curve shows $\mathcal{A}(E)$. The peak heights have been normalized in both cases. The observed $\mathcal{R}(E)$ will, of course, depend on the

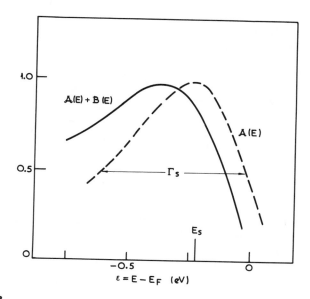

FIGURE 7.12

Variation with energy of $\mathcal{A}(E)$ and of $\mathcal{R}(E) = \mathcal{A}(E) + \mathcal{B}(E)$; $\mathcal{A}(E)$ and $\mathcal{B}(E)$ as defined by Eqs. (7.92) and (7.93), respectively. The peak heights have been normalized in both curves.

position of E_s relative to E_F and on the magnitude of Γ_s. Suppose we are looking at $\mathcal{R}(E)$ in the energy region -0.7 eV $< E - E_F < 0$, as in Figs. 7.8–7.11. If E_s lies outside this region, say $E_s = E_F - 0.8$ eV, and $\Gamma_s \simeq 1.0$ eV, $\mathcal{R}(E)$ will look live curve (a) in Fig. 7.13. If E_s lies further down in energy, curve (b) might be obtained. On the other hand, if E_s lies nearer the Fermi level, say at $E_s = E_F - 0.2$ eV and $\Gamma_s \simeq 0.4$ eV, $\mathcal{R}(E)$ will look like curve (c) in Fig. 7.13. In this way we can interpret the Ba on W(110) curve in Fig. 7.8 by assuming an s-type resonance centered at $E_s = E_F - 0.4$ eV. The Ca on W(110) is compatible with an s-type resonance centered further down in energy. The Sr on W(110) curve in the same figure cannot be explained in this simple manner. The Ca on W(111) curve shown in Fig. 7.10 [similarly the Ca on W(013) curve shown in Fig. 7.11]

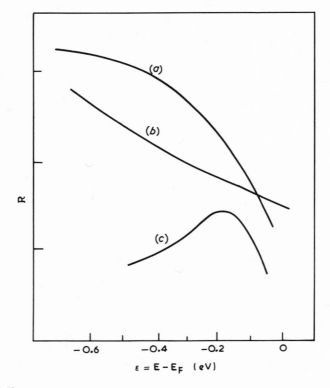

FIGURE 7.13

Diagram illustrating possible shapes of enhancement factor, according to Eqs. (7.87), (7.92), and (7.93). The three different curves correspond to different values of E_s and Γ_s (see text).

could also be explained as due to an s-type resonance centered at $E_s \simeq E_F - 0.2$ eV, provided one can understand why the Ca on W(110) resonance lies well below E_F compared to the Ca on W(111) [or the Ca on W(013)]. One must also explain the fact that in the latter case this resonance is much narrower than on W(110). These differences imply that the shift in the energy level of the valence state of the electron (in the atom) and its broadening depend critically on the electronic structure of the surface (crystal plane) under consideration. On the other hand it is possible that our assumption of an isolated s-type resonance does not apply equally well for every plane, i.e., an empty adatom resonance may play an important role when the atom is adsorbed on some planes but not when the atom is adsorbed on other planes, depending on its position relative to the Fermi level. We note that an empty p-type resonance *just* above E_F is more likely when Ca is adsorbed on the W(111) and W(013) planes, than on the W(110) plane, because of the smaller work function of these planes, and that this resonance interacting with the s resonance below E_F can lead to an upward shift and a narrowing of the latter (see, e.g., Fig. 7.5).

It is obvious that the sharp peaks in the $\mathscr{R}(E)$ curves of Ba on Mo(110), Ba on W(111), and Ba on W(103) shown in Figs. 7.9, 7.10, and 7.11, respectively, cannot be explained on the basis of Eqs. (7.92) and (7.93). Inelastic tunneling (see Section 6.4) is not likely to produce measurable structure at very low coverage (a single adatom in the present case). What then could be the origin of these sharp peaks? Of particular importance is the fact that Ba on W(110) does not lead to fine structure of this kind, whereas Ba on Mo(110) does, which suggests that the origin of this structure lies in the metal substrate. Let us consider what the effect on $\mathscr{R}(E)$ might be of a pronounced structure in $G_{00;00}(E)$ due to a pole (in the complex energy plane but with a small imaginary part) somewhere within the energy region under consideration. [Such a pole could be the result of a sharp surface resonance on the clean surface in the outer region of the SBZ—see comments following Eq. (7.91).] Let us assume that

$$G_{00;00}(E) \simeq \bar{g}_r - i\bar{g}_i + \frac{\alpha_{ss}}{E - \epsilon_{ss} + i\gamma_{ss}/2} \tag{7.96}$$

so that $g_r(E)$ and $g_i(E)$, as defined by Eqs. (7.85), are given by

$$g_r(E) \simeq \bar{g}_r + \frac{\alpha_{ss}(E - \epsilon_{ss})}{(E - \epsilon_{ss})^2 + (\gamma_{ss}/2)^2} \tag{7.97a}$$

$$g_i(E) \simeq \bar{g}_i + \frac{\alpha_{ss}(\gamma_{ss}/2)}{(E - \epsilon_{ss})^2 + (\gamma_{ss}/2)^2} \tag{7.97b}$$

We further assume that $E_s - \Gamma_s/2 < \epsilon_{ss} < E_s + \Gamma_s/2$ and $\gamma_{ss} \ll \Gamma_s$. Substituting Eq. (7.97) into Eq. (7.88) and Eq. (7.91) we obtain

$$\mathcal{A}(E) \simeq \frac{\exp(2a_0|q|)}{A|q|^2} \left\{ \left[|\alpha_s|(\Gamma_s/2) + \frac{\alpha_s^2\alpha_{ss}(\gamma_{ss}/2)}{(E - \epsilon_{ss})^2 + (\gamma_{ss}/2)^2} \right] \right.$$
$$\times \left[\left(E - E_s - \frac{\alpha_s\alpha_{ss}(E - \epsilon_{ss})}{(E - \epsilon_{ss})^2 + (\gamma_{ss}/2)^2} \right)^2 \right.$$
$$\left. \left. + \left(\frac{\Gamma_s}{2} + \frac{\alpha_s\alpha_{ss}(\gamma_{ss}/2)}{(E - \epsilon_{ss})^2 + (\gamma_{ss}/2)^2} \right)^2 \right]^{-1} \right\}$$

(7.98)

and a similar expression for $\mathcal{B}(E)$, which differ from the corresponding expressions given by Eqs. (7.92) and (7.93), only in a narrow region of width γ_{ss} around $E = \epsilon_{ss}$. Consequently, $\mathcal{R}(E)$ will look very much the same as before, e.g., like (a) or (b) in Fig. 7.13, except for a sharp structure, e.g., a peak, of width γ_{ss} at $E = \epsilon_{ss}$. It is clear that there may be more than one pole of $G_{00;00}(E)$ in the energy region $E_s - \Gamma_s/2 < E < E_s + \Gamma_s/2$, in which case $g_r(E)$ and $g_i(E)$ will contain additional terms, similar to the second one in Eqs. (7.97a) and (7.97b), which in turn would lead to additional structure in the $\mathcal{R}(E)$ curve. We believe that such might be the origin of the sharp peaks in the enhancement factors of Ba on Mo(110) in Fig. 7.9 and of Ba on W(111) and W(013) shown in Figs. 7.10 and 7.11. The similarity between the last two curves implies, according to this interpretation, a corresponding similarity in the electronic structure of the W(111) and W(013) clean surfaces. On the other hand, the absence of a similar structure in the enhancement factor of Ca on these planes can, possibly, be explained if we assume, as earlier suggested, that in the case of Ca on W(111) and W(013) we are not dealing with an isolated s-type resonance, and that, therefore, Eqs. (7.88) and (7.91) are not applicable. It is possible, in other words, that a p-type resonance just above E_F might lead to the elimination of the fine structure in $\mathcal{R}(E)$, producing at the same time a resonance which is much narrower than an isolated s-type resonance would be.

Impressed by the remarkable similarity between the enhancement factors of Ba on W(111) and Ba on W(013), Gadzuk (1970) suggested that the fine structure of these curves has its physical origin in the atom itself. Since an independent-electron model of the atom cannot produce such a structure he suggested that correlation effects (which are responsible for the multiplet structure of the excitation spectrum of the free atom) are also responsible for the observed structure in $\mathcal{R}(E)$. While one cannot exclude the possibility that electron correlation within the atom might be important, it seems to us that a direct correlation of the structure in $\mathcal{R}(E)$ with the excitation spectrum of the free atom is not possible and that Gadzuk's attempt to do so cannot be justified theoretically. A calculation of the effect of electron correlation on the electronic structure of the metal–adatom complex and on the energy distribution of the field-emitted electrons from such a system is an extremely difficult problem, and, at present, we do not have a theory for evaluating these effects. We believe that, at this stage, it would be more useful to calculate, using a formalism such as the one described in the present chapter, the energy distribution of the emitted electrons for realistic models of the metal–adatom (one-electron) potential, and we hope that such a calculation will be carried out in the near future. It might turn out that we can understand the observed structure in $\mathcal{R}(E)$ in the manner indicated in the present chapter and that we do not need to involve many-body effects. On the other hand, the opposite might be true. Whatever the outcome, the results of such a calculation are bound to be very interesting. It is also desirable to have the experiments of Plummer and Young (Figs. 7.8–7.11) repeated using a different apparatus and to obtain, if possible $R(E)$ over a wider energy region [see comments following Eq. (7.78)].

CHAPTER EIGHT

FIELD EMISSION FROM SEMICONDUCTOR SURFACES

8.1. INTRODUCTION

The preparation of high-quality semiconductor field emission tips appears to be more difficult than that of metal emitters. Although total energy distributions of the emitted electrons from clean and well-defined crystallographic planes have been obtained (see Sections 8.5 and 8.6), the amount of such data is still very small. This is unfortunate because, as we hope to demonstrate in the present chapter, a lot of interesting physics can be learned from such studies. Obviously, the energy distribution of the emitted electrons can provide us with information on the surface density of states, as in the case of metal emitters. In the case of semiconductors, however, and in contrast to the situation in metal emitters, it cannot always be assumed that the occupation of the one-electron states under conditions appropriate to a field emission experiment is the same as in equilibrium (zero emitted current). Thus a new dimension is introduced in the field emission process which leads, or may lead, to new phenomena which do not occur in field emission from metal surfaces (see Sections 8.6 and 8.7). Even in the absence of nonequilibrium effects (i.e., at sufficiently low emitted currents) the theory is not as straightforward as for metal emitters, because of field penetration into the semiconductor which leads to band bending at the surface.

8.2. BAND BENDING IN THE ZERO EMITTED CURRENT APPROXIMATION

Let us assume that the semiconductor crystal occupies the half-space from $z = -\infty$ to $z = 0$ and that the semiconductor–vacuum interface (the xy plane) is parallel to a crystallographic plane. Let us further assume that an electric field (appropriate to a field emission experiment) is applied normal to the surface. We shall specify this field by its magnitude, denoted by F_0, on the vacuum side of the interface. [This model, entirely satisfactory in the case of field emission from a given plane on a metallic field emission tip, may not be as good as in the case of a plane on a semiconducting tip because of field penetration (see comments following Eq. (8.18) and Section 8.7). We use it, nevertheless, because of its simplicity and because, at the present time, there is no theory based on a more realistic geometrical model of the emitter. It should be good enough for at least a semi-quantitative analysis of relevant experimental data.] The field penetrates into the semiconductor leading to band bending as shown, schematically, in Fig. 8.1. $E_c(z)$ and $E_v(z)$ denote the energy at the bottom of the conduction band and valence band, respectively. Band bending implies that, away from the surface, the average (over a unit cell) density of states is given by

$$\rho(E;z) \simeq \rho_\infty(E - E_c(z)) \tag{8.1}$$

where $\rho_\infty(E)$ is the density of states in the unperturbed infinite crystal with the energy measured from the bottom of the conduction band. Equation (8.1) is, of course, not valid in the surface region ($z_s < z$) which consists of a few (in practice three or four) atomic layers. There, and for the same reasons as for a metal, the density of states may be very much different from the bulk density of states. The validity of Eq. (8.1) in the rest of the space charge region (see Fig. 8.1) rests on the fact that

$$E_c(z) = \text{const} - eV(z) \tag{8.2}$$

where $V(z)$, the electrostatic potential due to field penetration, varies so slowly with z that it is practically constant over a distance of the order of the lattice constant. The constancy of the Fermi level in Fig. 8.1 implies an equilibrium (Fermi–Dirac) distribution of the electrons inside the semiconductor. This, the zero-emitted-current approximation, applies in the limit of very small emitted current densities when the influence of the current flowing through the sample can be neglected. When this is the case, $E_c(z)$, corresponding to a given applied field,

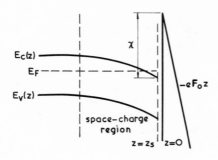

FIGURE 8.1

Band bending due to field penetration.

may be determined by solving Poisson's equation

$$\frac{d^2 E_c}{dz^2} = \left(\frac{4\pi e}{K}\right) \bar{\nu}(z) \tag{8.3}$$

where K is the dielectric constant of the semiconductor and $\bar{\nu}(z)$ the charge density (averaged over a unit cell), with boundary conditions

$$\frac{dE_c}{dz} \to 0 \qquad \text{as } z \to -\infty \tag{8.4a}$$

$$\left.\frac{dE_c}{dz}\right|_{z=z_s} = -\left(\frac{e}{K}\right) F_s = -\left(\frac{e}{K}\right)(F_0 + 4\pi Q_s) \tag{8.4b}$$

F_0 is the *magnitude* of the applied field (appropriate to a field emission experiment) on the vacuum side of the interface, and Q_s is the charge per unit area in the surface region, for the given field. $(-F_s/4\pi)$ equals the total charge per unit area in the region $z < z_s$. The charge density at $z < z_s$ is given by

$$\bar{\nu}(z) = e[p(z) - n(z) + N_d^+(z) - N_a^-(z)] \tag{8.5}$$

where $n(z)$ is the concentration of electrons in the conduction band, $p(z)$ the concentration of holes in the valence band, $N_d^+(z)$ the concentration of ionized donor impurities, and $N_a^-(z)$ the concentration of ionized acceptor impurities. We have

$$n(z) = \int_{E_c(z)}^{\infty} \rho_\infty(E - E_c(z)) f(E) \, dE \tag{8.6a}$$

$$p(z) = \int_{-\infty}^{E_c(z)-E_g} \rho_\infty(E - E_c(z))[1 - f(E)] \, dE \tag{8.6b}$$

where $f(E)$ denotes, as usual, the Fermi–Dirac distribution function, and $E_g \equiv E_c(z) - E_v(z)$ is the energy gap, which is, of course, independent of z.

$$N_d^+(z) = N_d \left\{ 1 + 2 \exp\left[\frac{E_F - E_c(z) - \epsilon_d}{k_B T}\right]\right\}^{-1} \tag{8.7a}$$

$$N_a^-(z) = N_a \left\{ 1 + 2 \exp\left[-\frac{E_F - E_c(z) - \epsilon_a}{k_B T}\right]\right\}^{-1} \tag{8.7b}$$

where N_d and N_a are the concentrations of the donor and acceptor impurity atoms, and ϵ_d and ϵ_a are the energy levels, measured from the bottom of the conduction band, of the corresponding (localized around these atoms) electron states. $\rho_\infty(E)$ is known from calculations of the energy-band structure of the bulk semiconductor. Similarly N_d, N_a, ϵ_d, ϵ_a are assumed known, and so is the Fermi level E_F. The position of the latter relative to $E_c(z \to -\infty)$ is determined from the condition $\bar{\nu}(z \to -\infty) = 0$. We see that $\bar{\nu}(z)$, as defined by Eqs. (8.5)–(8.7), is, for a given sample, a unique function of $E_c(z)$. It is obvious that without an analytic expression for this function, the solution of Eq. (8.3) will be extremely laborious. It is equally obvious that in order to obtain such an expression we must approximate ρ_∞ in Eq. (8.6) by a relatively simple function. Fortunately, the integral in Eq. (8.6a) depends, to a very good degree of approximation, only on the density of states in the vicinity of E_c and, similarly, the integral in Eq. (8.6b) is determined, to a very good degree of approximation, by the density of states near E_v. It is generally assumed that (see, e.g., any textbook on semiconductor physics)

$$\rho_\infty(E - E_c) \simeq \frac{1}{2\pi^2}\left(\frac{2m_n}{\hbar^2}\right)^{3/2}(E - E_c)^{1/2}, \qquad E \geq E_c \qquad (8.8a)$$

$$\simeq \frac{1}{2\pi^2}\left(\frac{2m_p}{\hbar^2}\right)^{3/2}(E_c - E_g - E)^{1/2}, \qquad E \leq E_c - E_g \qquad (8.8b)$$

where m_n and m_p, the effective masses of electrons and holes, respectively, are determined empirically so that the above expressions give correctly the density of states near E_c and E_v, respectively. Substituting Eq. (8.8) into Eq. (8.6) one obtains

$$n(z) \simeq A^c F_{1/2}\left[\frac{E_F - E_c(z)}{k_B T}\right] \qquad (8.9a)$$

$$p(z) \simeq \alpha A_c F_{1/2}\left[-\frac{E_F - E_c(z) + E_g}{k_B T}\right] \qquad (8.9b)$$

where $A^c \equiv (1/2\pi^2)(2m_n k_B T/\hbar^2)^{3/2}$, $\alpha \equiv (m_p/m_n)^{3/2}$, and $F_j(y)$ is the Fermi integral of order j defined by

$$F_j(y) = \int_0^\infty \frac{x^j\,dx}{1 + \exp(x - y)} \qquad (8.10)$$

With $\bar{\nu}(z)$ given by Eqs. (8.5), (8.7), and (8.9) a solution of Eq. (8.3) with boundary conditions given by Eq. (8.4) can be obtained without much difficulty. For a given specimen one obtains a set of solutions corresponding to different values of F_s [as defined by Eq. (8.4b)]. We denote these solutions by

$$E_c(z) = E_c(z; F_s) \qquad (8.11)$$

The difference between the Fermi level and the bottom of the conduction band at the surface, defined by

$$\theta_s^b(F_s) = E_F - E_c(z_s; F_s) \tag{8.12}$$

is of particular importance. We note that

$$\theta_\infty^b \equiv E_F - E_c(z \to -\infty) \tag{8.13}$$

is independent of the applied field [see comments following Eq. (8.7)]. Seiwatz and Green (1958) established the following relation between F_s and θ_s^b:

$$-F_s = \frac{k_B T K}{e L_D} \left\{ \frac{N_d}{n_i} \ln \left[\frac{1 + (1/2) \exp(W_{d,i} - u_s)}{1 + (1/2) \exp(W_{d,i} - u_b)} \right] \right. $$
$$+ \frac{N_a}{n_i} \ln \left[\frac{1 + (1/2) \exp(u_s - W_{a,i})}{1 + (1/2) \exp(u_b - W_{a,i})} \right]$$
$$+ \frac{(2/3)}{F_{1/2}(W_{i,c})} [F_{3/2}(u_s - W_{c,i}) - F_{3/2}(u_b - W_{c,i})]$$
$$\left. - \frac{(2/3)}{F_{1/2}(W_{v,i})} [F_{3/2}(W_{v,i} - u_b) - F_{3/2}(W_{v,i} - u_s)] \right\}^{1/2} \tag{8.14}$$

where

$$u_s \equiv u_b + (\theta_s^b - \theta_\infty^b)/k_B T \tag{8.15}$$
$$u_b \equiv (E_F - E_i)/k_B T \tag{8.16}$$
$$W_{\mu,\nu} \equiv (E_\mu - E_\nu)/k_B T \tag{8.17}$$

E_v, E_c, E_i, E_a, and E_d are parameters of the infinite crystal. They denote (respectively) the energy at the top of the valence band, the energy at the bottom of the conduction band, the Fermi level in the intrinsic (pure) semiconductor at the given temperature, the energy level of the acceptor impurity states, and the energy level of the donor states. $F_j(y)$ is the integral defined by Eq. (8.10). n_i is the concentration of electrons (in the conduction band) of the intrinsic semiconductor and

$$L_D \equiv (k_B T K / 8 \pi n_i e^2)^{1/2} \tag{8.18}$$

One can show that if the left-hand side of Eq. (8.14) is positive (negative) then $u_s < u_b$ ($u_s > u_b$). The sign of the square root is taken accordingly. Simpler expressions, approximations to Eq. (8.14), valid in certain ranges of impurity concentration, temperature, and applied field, have been derived by a number of authors (Kingston and Neustadter, 1955; Dousmanis and Duncan, 1958; Tsong, 1979).

It is evident from various calculations based on these formulas (see, e.g., Tsong, 1979) that, for the applied fields used in a typical field emission experiment, most of the variation of the field and charge concentration occurs within 100 Å or so from the surface, which is approximately one order smaller than the radius of a typical field-emitter. This means that the surface of the emitter is approximately an equipotential and that the one-dimensional approximation to the potential field [Eq. (8.3)] is reasonably good. However, there may be situations when field penetration is considerably larger. This is, for example, the case when current saturation occurs. In that case, however, the internal flow of current plays a critical role and Eq. (8.14) (or any other equation based on equilibrium considerations) is not valid anyway (see Section 8.7).

In order to find the value of F_s corresponding to a given value of F_0 (the value of the applied field on the vacuum side of the interface), one must first evaluate Q_s, the charge per unit area in the surface region, which appears in Eq. (8.4b). It turns out [see Eqs. (8.21)–(8.26) and the comments at the end of this section] that for a given value of F_0, Q_s is a function of θ_s^b and therefore through Eq. (8.14) of F_s. We have

$$Q_s = Q_s(\theta_s^b(F_s);F_0) \tag{8.19a}$$

and approximately

$$Q_s \simeq Q_s(\theta_s^b(F_s)) \tag{8.19b}$$

$\theta_s^b(F_s)$ is given by Eq. (8.14) or some equivalent expression. On the other hand, according to Eq. (8.4b)

$$Q_s = (F_s - F_0)/4\pi \tag{8.20}$$

A plot of Q_s versus F_s, for a given value of F_0, is, according to Eq. (8.20), a straight line. The point where this straight line crosses the plot of Q_s versus F_s, obtained from Eq. (8.19), determines the values of F_s, Q_s, and θ_s^b [through Eq. (8.14)] for the given value of F_0.

The problem of evaluating $Q_s(F_s;F_0)$, as defined by Eq. (8.19), is a difficult one. The charge density in the surface region ($z_s < z$) is given by

$$\nu(\mathbf{r}) = e[\nu_{\text{ion}}(\mathbf{r}) - \nu_{\text{el}}(\mathbf{r})] \tag{8.21}$$

The first term in the above equation represents the contribution to the charge density from the ions (the nuclei and the inner-shell electrons) in the surface region. [It may be assumed that the orbitals of the inner-shell electrons are the same as in the free atom.] The electronic term is given by

$$\nu_{el}(\mathbf{r}) = \int_{-\infty}^{\theta_s^b} \rho(\epsilon_s;\mathbf{r}) \, d\epsilon_s \tag{8.22}$$

where $\rho(\epsilon_s;\mathbf{r})$ is the local density of electron states (excluding the contribution from inner-shell orbitals already taken into account by ν_{ion}), and

$$\epsilon_s \equiv E - E_c(z_s;F_s) \tag{8.23}$$

In writing Eq. (8.22) we have assumed, for the sake of clarity, an absolute zero temperature. The extension to finite temperatures is trivial. The charge per unit area in the surface region is given by

$$Q_s = Q_{ion} + Q_{el}(\theta_s^b;F_0) \tag{8.24}$$

$$Q_{ion} = e \int_{z>z_s} \left[\frac{1}{A_0} \iint_{A_0} \nu_{ion}(\mathbf{r}) \, d^2r_\| \right] dz \tag{8.25}$$

$$Q_{el}(\theta_s^b;F_0) = -e \int_{z>z_s} \left[\frac{1}{A_0} \iint_{A_0} \nu_{el}(\mathbf{r}) \, d^2r_\| \right] dz \tag{8.26}$$

where A_0 is the area of the surface unit cell. We note that $\rho(\epsilon_s;\mathbf{r})$ may depend, to some degree, on the applied field F_0 and this, in turn, leads to the dependence of Q_{el} on F_0 indicated in Eq. (8.26). We emphasize that the potential energy field

$$V(\mathbf{r}) = V_c(\mathbf{r}) + V_{xc}(\mathbf{r}) \tag{8.27}$$

in the surface region ($z > z_s$) must be self-consistent with $\nu_{el}(\mathbf{r})$ as defined by Eq. (8.22), i.e., $V_c(\mathbf{r})$, the electrostatic potential energy, must satisfy Poisson's equation with the charge density given by Eq. (8.21), the boundary condition

$$\frac{dV_c}{dz} \simeq -eF_0 \qquad \text{for } z \gg 0 \tag{8.28}$$

and must lead to a value of dV_c/dz at z_s (average over the surface unit cell) equal to

$$\frac{dV_c}{dz}\bigg|_{z=z_s} = -\left(\frac{e}{K}\right)F_s \tag{8.29}$$

where F_s is such that $\theta_s^b(F_s)$, determined from Eq. (8.14) equals the input value of this quantity appearing in Eq. (8.22). Similarly, $V_{xc}(\mathbf{r})$, the exchange part of the potential must be given in terms of $\nu_{ep}(\mathbf{r})$ in the manner described in Section 3.2. It is also assumed that $V(\mathbf{r})$ joins smoothly to the bulk potential at $z = z_s$. It is obvious that solving the Schrödinger equation for a given input potential to obtain the one-electron states of the semi-infinite crystal and $\rho(\epsilon_s,\mathbf{r})$ in the surface region, and the subsequent iteration to self-consistency is a very difficult job. Moreover, because of the boundary condition [Eq. (8.28)], the entire calculation must be repeated for different values of F_0, if it turns out that $\rho(\epsilon_s,\mathbf{r})$ depends strongly on the applied field. However, if $\rho(\epsilon_s,\mathbf{r})$ is, at least approximately, independent of the applied field and of the degree of band bending, we can calculate self-consistently $\rho(\epsilon_s,\mathbf{r})$ for $F_0 = 0$ and $F_s = 0$ [the latter condition corresponding to zero band bending implies a neutral surface region ($Q_s = 0$) and is therefore possible only for a specific value of the Fermi level E_F (see Section 8.3)] and use it thereafter for any values of F_0 and F_s. We note that even when $\rho(\epsilon_s,\mathbf{r})$ does not depend on F_s, Q_s does so through θ_s^b, as can be seen from Eqs. (8.22)–(8.26). The numerical evaluation, however, of the function $Q_s(\theta_s^b(F_s))$, when $\rho(\epsilon_s,\mathbf{r})$ is known, is relatively simple. Therefore, the self-consistency problem for nonzero applied fields is reduced to that of determining F_s from Eqs. (8.19b) and (8.20).

8.3. ELECTRONIC STRUCTURE OF SEMICONDUCTOR SURFACES

When no external field is applied to the surface ($F_0 = 0$) and no band bending occurs ($F_s = 0$), the one-electron states of the semi-infinite semiconductor can be described in the same way as for a metal. We note, however, that reconstruction of the top one or two surface layers is much more common among semiconductor surfaces (because of the strong directional bonding between the atoms) than it is for metals (see, e.g., Harrison, 1976). The area of the unit cell and the symmetry of the reconstructed surface can be obtained from LEED experiments but the determination of the position of the surface atoms, within the unit cell, based on a mathematical analysis of the LEED intensity curves (magnitude of diffraction amplitudes versus energy curves) is extremely difficult, if not impossible with the means presently available to LEED analysis. [The large unit cell of the reconstructed surface demands that a very large number of **g** vectors (reciprocal vectors of the reconstructed surface) be included in the calculation of the amplitude of the diffracted beams (see Section 4.4.2) and this makes the calculation of these amplitudes prohibitively long even for today's fast computers.] In some cases, however, a reasonable model of the reconstructed surface can be obtained in an indirect way, e.g., by calculating the energy spectrum of the surface states for two or three plausible models of the surface structure (plausible in the sense that they are compatible with LEED data and general bonding considerations) and comparing the results for the different models with an "experimental spectrum" obtained from angle-resolved photoemission measurements.

Appelbaum, Baraff, and Haman (1975; 1976) calculated (self-consistently) the surface density of states for the unreconstructed Si(100) surface, shown in Fig. 8.2a, and for two plausible models of the corresponding reconstructed surface. They came to the conclusion that the model of the reconstructed surface which fits the available experimental data better is the (2 × 1) surface shown in Fig. 8.2b. This is known as the pairing model (Schlier and Farnsworth, 1959) of the reconstructed Si(100) surface. We need not explain here the method used by Appelbaum *et al.* (1976) to calculate, self-consistently, the surface potential and the corresponding one-electron states. It is worth noting, however, that their method is not based on the muffin-tin approximation of the potential and that the transition from a potential depending on (x,y,z) inside the metal to a surface potential depending only on z for $z > z_a$ is a smooth one. (In practice z_a, measured from the center of the last (top) layer of atoms, is equal approximately to 2 Å.) There are approximations in their calculation, especially in the way the electron wave functions in the surface region $(z > z_s)$ are joined to approximate Bloch waves of the unperturbed (infinite) crystal in the region $z < z_s$, and also in the way the density of states is obtained from calculated wave functions at a limited number

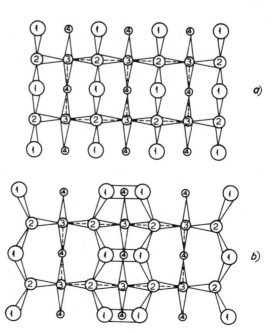

FIGURE 8.2

(a) and (b) show, respectively, the top four planes of atoms of the unreconstructed and of the reconstructed (pairing model) Si(100) surface. The circles labeled by the same integer denote atoms in the same plane with higher numbers indicating increasing distance from the vacuum. In (a) the interplane distance is $\frac{1}{4}a$ where $a = 5.43$ Å. The nearest-neighbor distance in a plane is $a/\sqrt{2}$ and the bulk bond length is $(\frac{3}{16})^{1/2} a = 2.35$ Å. In (b) rows of surface atoms move toward each other so that all bonds have the bulk bond length.

of points in the SBZ, but there is no reason to doubt the essential correctness of their results. In Fig. 8.4. we show the bands of surface states on the unreconstructed Si(100) surface, as calculated by Appelbaum *et al.* (1975), along the symmetry directions of the SBZ (shown in Fig. 8.3). We note that the top of the valence band (shaded area in Fig. 8.4a) occurs at $E = 0.06884$ a.u. on the energy axis of Fig. 8.4. The bottom of the conduction occurs at $E \simeq 0.11$ a.u. on the same axis ($E_g = 0.04$ a.u. $= 1.11$ eV). Following Appelbaum *et al.* we refer to the surface band shown in Fig. 8.4a as the lower gap surface state (LGSS) and to the one shown in Fig. 8.4b as the upper gap surface state (UGSS). Because the formation of the unreconstructed Si(100) surface results in the breaking of two bonds per surface atom (see Fig. 8.2a) it is reasonable to assume that the formation of the surface removes from the valence band (in the surface region) the equivalent of two electron states per (surface) unit cell. At the same time, because of the hybridizational nature of the energy gap, the equivalent of two electrons states per unit cell is removed from the conduction band (in the surface region). This is consistent with the appearance of two bands of surface states, as shown in Fig. 8.4, each band extending over the entire SBZ and therefore containing two surface states per unit cell. (We note that around Γ the surface states in the LGSS band become surface resonances, for around Γ there exist bulk states of the same E and $\mathbf{k}_\parallel$. For practical purposes these resonances are equivalent to true surface states and are treated as such by Appelbaum *et al.*) When there is no externally applied field the electronic charge in the surface bands must be equal to that missing from the valence band in order that overall charge neutrality is maintained. When no band bending occurs overall charge neutrality is equivalent to charge neutrality in the surface region ($Q_s = 0$). In this case the Fermi level (at absolute zero temperature) must be such that there are two electrons per unit cell in the surface bands. [When there is no band bending $E_c(z)$ and, therefore, ($E_F - E_c$) is constant throughout the crystal and is determined by the concentration of donor and/ or acceptor impurities. It follows that the value of ($E_F - E_c$) determined by the degree of doping must equal the value of ($E_F - E_c$) determined by charge neu-

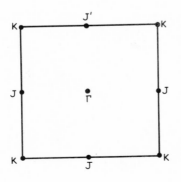

FIGURE 8.3
SBZ of the (unreconstructed) Si(100) surface.

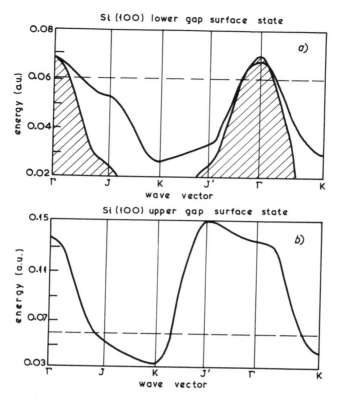

FIGURE 8.4

(a) the LGSS energy along the principal symmetry directions of the SBZ. The shaded area represents the continuum of bulk energy levels near the top of the valence band. (b) The UGSS energy along the principal symmetry directions of the SBZ. (From Appelbaum *et al.*, 1975.)

trality in the surface region ($Q_s = 0$) if long-range band bending is to be avoided.] If the UGSS band in Fig. 8.4 did not overlap the LGSS, charge neutrality could be achieved by filling the LGSS completely and leaving the UGSS completely empty, i.e., by putting E_F in between the two surface bands. We would have in that case a semiconducting surface. As it is (Fig. 8.4) the two surface bands overlap, as indicated by the (approximate) position of the Fermi level (broken line in the figure). Besides the two bands shown in Fig. 8.4 corresponding to broken bonds, there are isolated pockets of surfaces states at K at energies near the bottom of the valence but these have no significant effect on the charge density in the surface region (see, e.g., Appelbaum *et al.*, 1975) and need not concern us here.

The surface states near the energy gap ($E_v < E < E_c$) on the reconstructed

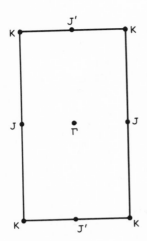

FIGURE 8.5

SBZ of the reconstructed (pairing model) Si(100) surface.

surface described by Fig. 8.2b are shown in Fig. 8.6 (Appelbaum, Baraff, and Hamann, 1976) along the symmetry direction of the corresponding SBZ, shown in Fig. 8.5. (We note that the point group symmetry is the same as that of the ideal surface.) The states in the π_b band shown in this figure derive from even (bonding) combinations, and those in the π_a band from odd (antibonding) combination, of the broken bond surface states in the LGSS band of the unreconstructed surface. We note that the enlarged (by a factor of 2) unit cell of the reconstructed surface would in the absence of distortion lead to two degenerate such states for a given $\mathbf{k}_\parallel$ in the new (reduced by a factor of 2) SBZ. In a similar manner the UGSS band of the ideal surface gives rise to a σ_a band at higher

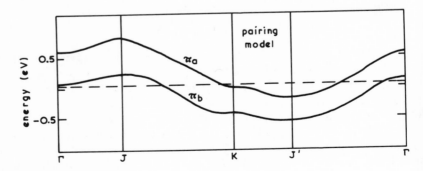

FIGURE 8.6

Bands of surface states for the pairing model of the reconstructed Si(100) surface. The energy is measured from the valence band maximum. (From Appelbaum *et al.*, 1976.)

energies (well above the bottom of the conduction band) and to a bonding σ_b band at lower energies among the valence band. A detailed description of the surface states and resonances on the reconstructed Si(100) surface (in some of these the charge density is maximum between the first and second layers of atoms) can be found in the paper of Appelbaum *et al.* Charge neutrality of the surface region ($Q_s = 0$), and therefore overall neutrality (when no band bending occurs), is achieved by putting two electrons (per unit surface cell) in the bands shown in Fig. 8.6, which in turn determines the position of the Fermi level (shown by a broken line in the figure). Chadi (1979) has shown that this, almost semimetallic surface according to Fig. 8.6, can be transformed into a semiconducting surface by a relatively small adjustment of the position of the atoms in the top layers. The corresponding surface band structure is shown in Fig. 8.7. The upper band in this figure is completely empty and the lower band completely occupied. Angle-resolved photoemission measuresment (Himpsel and Eastman, 1979) show that the Si(100) surface is in fact a semiconducting surface. A brief review of theoretical and experimental results on Si(100) and other semiconductor surfaces can be found in the article of Inglesfield and Holland (1981).

The local density of states in the surface region of a semi-infinite semiconductor can be written, as for a metal (see Section 4.3), as the sum of two terms

$$\rho(\epsilon_s, \mathbf{r}) = \rho_b(\epsilon_s, \mathbf{r}) + \rho_s(\epsilon_s, \mathbf{r}) \tag{8.30}$$

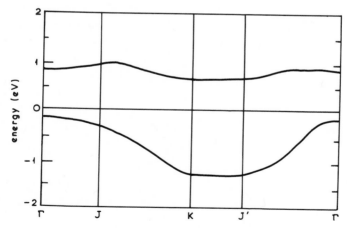

FIGURE 8.7

Surface states on reconstructed Si(100) surface. The position of the atoms in the top layer are slightly different from those in the pairing model (Fig. 8.2b). The surface states below the maximum of the valence band (zero of energy) are occupied and those above it are empty. (From Chadi, 1979.)

$\rho_b(\epsilon_s,\mathbf{r})$, the contribution to ρ from the bulk states, is given by (this equation is exact when no band bending occurs)

$$\rho_b(\epsilon_s,\mathbf{r}) = \frac{2V}{4\pi^3} \iint\limits_{\text{SBZ}} d^2k_{\parallel} \left[\sum_{\alpha} \left(\frac{\partial k_{\alpha z}}{\partial \epsilon_s}\right) |\Psi^B(\mathbf{r})|^2\right]_{\epsilon_s^B=\epsilon_s} \qquad (8.31)$$

where $B = (\epsilon_s^B,\mathbf{k}_{\parallel},\alpha)$ characterizes the Bth eigenstate in the manner described in Section 4.2. The energy is measured from the bottom of the conduction band [see Eq. (8.23)]; when $\epsilon_s^B > 0$ ($\epsilon_s^B < -E_g$) the eigenenergy lies in the conduction band (valence band). We note that the SBZ in Eq. (8.31) is that of the reconstructed surface.

We denote the contribution to ρ_b from the conduction band (valence band) by $\rho_n(\rho_p)$, i.e.,

$$\begin{aligned}\rho_b(\epsilon_s,\mathbf{r}) &= \rho_n(\epsilon_s,\mathbf{r}), & \epsilon_s &> 0 \\ &= 0, & -E_g &< \epsilon_s < 0 \\ &= \rho_p(\epsilon_s,\mathbf{r}), & \epsilon_s &< -E_g\end{aligned} \qquad (8.32)$$

The second term in Eq. (8.30), representing the contribution to $\rho(\epsilon_s,\mathbf{r})$ from surface states is given by [see Eq. (4.29)]

$$\rho_s(\epsilon_s,\mathbf{r}) = \frac{A}{2\pi^2} \iint\limits_{\text{SBZ}} \sum_{\mu} \delta(\epsilon_s - \epsilon_\mu(\mathbf{k}_{\parallel})) |\Psi^S(\mathbf{r})|^2 d^2k_{\parallel} \qquad (8.33)$$

where $S \equiv (\mathbf{k}_{\parallel},\mu)$ are the quantum numbers which characterize the surface state, μ being a (surface) band index. The energy $\epsilon_\mu(\mathbf{k}_{\parallel})$ of this state is measured relative to $E_c(z_s)$, and Ψ^S is assumed normalized according to Eq. (4.23).

In what follows we refer to the surface states, obtained with zero band bending and zero applied field, which are occupied at zero temperature, as donor surface states, and to those which are empty at zero temperature, as acceptor surface states. For example, in the case of the Si(100) surface bands shown in Fig. 8.6, all the surface states above E_F (broken line in the figure) are acceptor states and those below E_F are donor states. It is clear that a donor state is "neutral" when occupied in the same way that an occupied electron state in the valence band is "neutral"; when empty it contributes a positive charge e to the surface region in the same way that a hole in the valence band carries a positive charge e. Similarly, an acceptor surface state is neutral when empty; it contributes a negative charge $(-e)$ to the surface region when occupied in the same way that an electron in the conduction band carries a negative charge $(-e)$. We denote the density of donor [acceptor] surface states by $\rho_{s;a}(\epsilon_s)[\rho_{s;a}(\epsilon_s)]$. From Eq. (8.33) We have

$$\rho_{s;a}(\epsilon_s) = \frac{1}{2\pi^2} \iint_{SBZ} \sum_{\mu} \delta(\epsilon_s - \epsilon_{\mu;a}(\mathbf{k}_{\|})) \, d^2k_{\|} \tag{8.34}$$

where $\epsilon_{\mu;a}(\mathbf{k}_{\|})$ is the energy of the μth state at $\mathbf{k}_{\|}$ (of the acceptor type). $\rho_{s;d}(\epsilon_s)$ is given by the same expression with a replaced by d.

Let us now assume that overpopulating the surface region (e.g., by allowing additional electrons into acceptor surface states), or depopulating the surface region (e.g., by removing electrons from donor surface states) does not lead to any significant change in the surface density of states itself. Then, and as long as the above assumption remains valid, the net charge in the surface region (per unit area) is given by

$$Q_s(\theta_s^b) = -e \int \rho_{s;a}(\epsilon_s) f(\epsilon_s,\theta_s^b) \, d\epsilon_s + e \int \rho_{s;d}(\epsilon_s)[1 - f(\epsilon_s,\theta_s^b)] \, d\epsilon_s$$
$$+e \int_{-\infty}^{-E_g} \overline{\rho}_p(\epsilon_s)[1 - f(\epsilon_s,\theta_s^b)] \, d\epsilon^s - e \int_0^{\infty} \overline{\rho}_n(\epsilon_s) f(\epsilon_s,\theta_s^b) \, d\epsilon_s \tag{8.35}$$

whatever the value of the applied field

$$\overline{\rho}_n(\epsilon_s) \equiv \int_{z>z_s} \left[\frac{1}{A_0} \iint_{A_0} \rho_n(\epsilon_s,\mathbf{r}) \, d^2r_{\|} \right] dz \tag{8.36a}$$

$$\tag{8.36b}$$

$$dz \, \overline{\rho}_p(\epsilon_s) \equiv \int_{z>z_s} \left[\frac{1}{A_0} \iint_{A_0} \rho_p(\epsilon_s,\mathbf{r}) \, d^2r_{\|} \right] dz$$

where $\rho_n(\epsilon_s,\mathbf{r})$ and $\rho_p(\epsilon_s,\mathbf{r})$ are defined by Eq. (8.32), and

$$f(\epsilon_s,\theta_s^b) = \left[1 + \exp\left(\frac{\epsilon_s - \theta_s^b}{k_B T} \right) \right]^{-1} \tag{8.37}$$

The last two terms in Eq. (8.35) are comparatively very small quantities, and therefore

$$Q_s(\theta_s^b) \simeq -e\int\rho_{s;a}(\epsilon_s)f(\epsilon,\theta_s^b)d\epsilon_s + e\int\rho_{s;d}(\epsilon_s)[1 - f(\epsilon,\theta_s^b)] \, d\epsilon_s \tag{8.38}$$

The above equation provides us with a relatively simple expression for the quantity θ_s^b [introduced, in the first instance, by Eq. (8.19b]. We note that, in general, band bending occurs and therefore Q_s is different from zero even when $F_0 = 0$.

Obviously, overpopulation (similarly depopulation) of the surface region beyond a certain limit may lead to significant changes in the density of surface states, in which case the above method of determining Q_s ceases to be valid. It is reasonable to assume that this limit is reached when Q_s becomes comparable to the electronic charge per unit area in surface states when $Q_s = 0$. The case when most of the excess charge Q_s in the surface states is induced by the application of an electric field F_0 is of particular importance in relation to field emission from the given surface. In this case we obtain an upper bound for Q_s from Eq. (8.20), by putting $F_s = 0$. For a typical value of the applied field, say $F_0 = 0.4 \text{ V/Å}$, this leads to $Q_s \leq 0.2 \times 10^{14}$ electrons per cm^2. In the case of the reconstructed Si(100) surface this corresponds to $Q_s < 0.12$ electrons per unit surface cell, which is much smaller than the two electrons per unit cell residing in the π-bands (Fig. 8.6) when $Q_s = 0$, and, therefore, the above method of calculating Q_s [based on Eq. 8.38)] should be valid. It is also worth noting that the band bending, measured by (see Fig. 8.1)

$$\Delta E_c \equiv E_c(-\infty) - E_c(z_s) \tag{8.39}$$

required to produce that much excess charge in the surface bonds is of the order of

$$\Delta E_c \simeq 0.04 \text{ eV} \tag{8.40}$$

which is much less than the width of these bands.

8.4. FIELD EMISSION IN THE ZERO INTERNAL CURRENT APPROXIMATION

In the zero (internal) current approximation we assume that there is no (resistive) voltage drop inside the semiconductor. Figure 8.8 shows (schematically) a possible energy level diagram for an n-type semiconductor. In this example the conduction and valence band bend upwards at the surface. ΔE_c [defined by Eq. (8.39)] is relatively large and negative when $F_0 = 0$ to compensate for electronic charge in an acceptor surface band (shown by the shaded area in the figure). For the given electric field the bending is reduced but not eliminated, and the conduction band remains nondegenerate ($\theta_s^b < 0$). Figure 8.9 shows an energy diagram corresponding to a degenerate at the surface conduction band ($\theta_s^b > 0$). In this example the density of acceptor surface states is smaller and/or the applied field larger leading to a degenerate conduction band at the surface. The constancy of the Fermi level from $z = -\infty$ right up to the semiconductor–vacuum interface, implies that the occupation of the bulk states is practically the same as for equilibrium. The same need not, however, apply to the surface band. In other words,

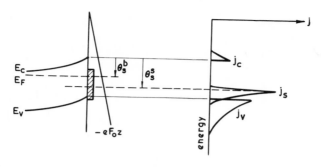

FIGURE 8.8

Schematic description of field emission from an n-type semiconductor when the conduction band is nondegenerate at the surface ($\theta_s^b < 0$). The shaded area denotes a band of surface states of the acceptor type.

whereas electrons emitted from the valence band or the conduction band are automatically replaced by other electrons from the interior of the crystal, this may not be so for electrons emitted from surface states. If that is the case, the electrons in the surface band cannot be assumed to be in thermal equilibrium with the electrons in the bulk bands, although they can be in "pseudoequilibrium" among themselves, in the sense that the occupation of the surface states is determined by a Fermi–Dirac distribution function given by Eq. (8.37) with θ_s^b replaced by θ_s^s, an effective (or pseudo-) Fermi level given by

$$\theta_s^s = \theta_s^b(F_s) - \Delta\theta_s^s \qquad (8.41)$$

One way to calculate θ_s^s is described in Section 8.6. We note that a nonzero $\Delta\theta_s^s$ distinguishes the "zero-internal-current approximation" from the "zero-emitted-

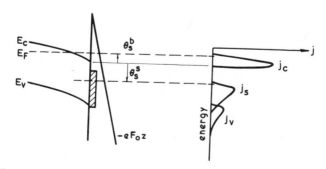

FIGURE 8.9

Schematic description of field emission from an n-type semiconductor when the conduction band is degenerate at the surface ($\theta_s^b > 0$). The shaded area denotes a band of surface states of the acceptor type.

current approximation" ($\Delta\theta_s^s = 0$). (The magnitude of θ_s^s has been exaggerated in Figs. 8.8 and 8.9.)

Let us assume that the plane $z = 0$ lies sufficiently away (2 Å or so) from the center of the last (top) layer of atoms, so that

$$V = V(z) \qquad \text{for } z > 0 \tag{8.42}$$

According to classical electrostatics

$$V(z) \simeq \chi - \frac{(K-1)}{(K+1)} \frac{e^2}{4(z+z_i)} - eF_0(z+z_i) \qquad \text{as } z \to \infty \tag{8.43}$$

χ, the electron affinity, is the energy difference between the vacuum level and the bottom of the conduction band at the surface (the latter energy equals zero by definition). K is the dielectric constant of the semiconductor and F_0 is the magnitude of the applied field (see Fig. 8.1). $z = -z_i$ defines the image plane which lies somewhere between the last plane of atoms and the plane $z = 0$. (z_i is by definition positive; we assume that $0 < z_i < 1$ Å.) Our assumption (see Section 8.2) that the local density of states in the surface region is, to a good approximation independent of the applied field and of band bending, implies that the electronic wave functions in the surface region are also independent (to a good approximation) of the applied field and of band bending. The latter may shift the entire energy spectrum by ΔE_c [see Eq. (8.39)] but does not change significantly the wave functions. [We avoid an explicit reference to the above-mentioned energy shift by measuring all energies relative to $E_c(z_s)$.] Assuming, without any loss of generality, that a slot of zero potential of infinitesimal thickness exists at $z = 0$, we can write the wave functions there in the following form:

$$\Psi^B = (2V)^{-1/2} \sum_{\mathbf{g}} [A_{\mathbf{g}}^{B+} \exp(i\mathbf{K}_{\mathbf{g}}^+ \cdot \mathbf{r}) + A_{\mathbf{g}}^{B-} \exp(i\mathbf{K}_{\mathbf{g}}^- \cdot \mathbf{r})], \quad z \simeq 0 \tag{8.44}$$

$$\Psi^S = A^{-1/2} \sum_{\mathbf{g}} [A_{\mathbf{g}}^{S+} \exp(i\mathbf{K}_{\mathbf{g}}^+ \cdot \mathbf{r}) + A_{\mathbf{g}}^{S-} \exp(i\mathbf{K}_{\mathbf{g}}^- \cdot \mathbf{r})], \quad z \simeq 0 \tag{8.45}$$

Ψ^B are the bulk states appearing in Eq. (8.31) and Ψ^S the surface states appearing in Eq. (8.33)

$$\mathbf{K}_{\mathbf{g}}^{\pm} = \left(\mathbf{k}_{\|} + \mathbf{g}, \pm \left[\frac{2m\epsilon_s}{\hbar^2} - (\mathbf{k}_{\|} + \mathbf{g})^2\right]^{1/2}\right) \tag{8.46}$$

where $\mathbf{g}$ are the reciprocal vectors of the (reconstructed) surface.

The (total) energy distribution of the emitted electrons is given by

$$j(\epsilon_s) = j_b(\epsilon_s) + j_s(\epsilon_s) \tag{8.47}$$

where $j_b(\epsilon_s)$ is the contribution from the bulk bands and $j_s(\epsilon_s)$ that from the surface bands. A formal expression for $j_b(\epsilon_s)$ and $j_s(\epsilon_s)$ can be derived in exactly the same way as for a metal surface (Section 5.2). We obtain

$$
j_b(\epsilon_s;\theta_s^b) = \frac{\hbar f(\epsilon_s;\theta_s^b)}{4\pi^3 m} \iint_{\substack{\text{SBZ}}} d^2k_\parallel \left[\sum_\alpha \left(\frac{\partial k_{\alpha z}}{\partial \epsilon_s}\right) \sum_g |A_g^B|^2 \right]_{\epsilon_s^B = \epsilon_s} \tag{8.48}
$$
$$
\times \frac{|T_s(\epsilon_s,\mathbf{k}_\parallel + \mathbf{g})|^2}{|1 + R_s(\epsilon_s,\mathbf{k}_\parallel + \mathbf{g})|^2}
$$
$$
\simeq \frac{\hbar f(\epsilon_s,\theta_s^b)}{4\pi^3 m} \iint_{\substack{\text{central}\\ \text{SBZ}}} d^2k_\parallel \left[\sum_\alpha \left(\frac{\partial k_{\alpha z}}{\partial \epsilon_s}\right) |A_0^B|^2 \right]_{\epsilon_s^B = \epsilon_s} \frac{|T_s(\epsilon_s,\mathbf{k}_\parallel)|^2}{|1 + R_s(\epsilon_s,\mathbf{k}_\parallel)|^2}
$$
$$
\tag{8.49}
$$

A_g^B are defined by Eq. (8.52) and T_s and R_s are defined by Eq. (8.55). The comments made in relation to the sum over α in Eq. (4.19) apply as well here. We note in particular that the integrand in the above equation vanishes identically for those values of $\mathbf{k}_\parallel$ (for given ϵ_s) for which no bulk state exists.

$$
j_s(\epsilon_s;\theta_s^s) = \frac{\hbar f(\epsilon_s,\theta_s^s)}{2\pi^2 m} \iint_{\substack{\text{SBZ}}} d^2k_\parallel \left[\sum_\mu \delta(\epsilon_s - \epsilon_\mu(\mathbf{k}_\parallel)) \sum_g |A_g^S|^2 \right] \tag{8.50}
$$
$$
\times \frac{|T_s(\epsilon_s,\mathbf{k}_\parallel + \mathbf{g})|^2}{|1 + R_s(\epsilon_s,\mathbf{k}_\parallel + \mathbf{g})|^2}
$$
$$
\simeq \frac{\hbar f(\epsilon_s,\theta_s^s)}{2\pi^2 m} \iint_{\substack{\text{central}\\ \text{SBZ}}} d^2k_\parallel \sum_\mu \delta(\epsilon_s - \epsilon_\mu(\mathbf{k}_\parallel)) |A_0^S|^2 \frac{|T_s(\epsilon_s,\mathbf{k}_\parallel)|^2}{|1 + R_s(\epsilon_s,\mathbf{k}_\parallel)|^2}
$$
$$
\tag{8.51}
$$

A_g^M $(M = B,S)$ is given by

$$
A_g^M \equiv A_g^{M+} + A_g^{M-} \tag{8.52}
$$

where $A_g^{M\pm}$ are the coefficients appearing in Eqs. (8.44) and (8.45). We note that, according to Eqs. (8.44)–(8.46) the normalized wave function $\Psi^M(M = B,S)$ at $z = 0$ is given by

$$
\Psi^M(\mathbf{r}_\parallel;z = 0) = \Omega^{-1/2} \sum_g A_g^M \exp[i(\mathbf{k}_\parallel + \mathbf{g}) \cdot \mathbf{r}_\parallel] \tag{8.53}
$$

where $\Omega = (2V)$, A for $M = B,S$, respectively. Self-consistent calculations of the potential and of the electronic wave functions in the surface region, such as those of Appelbaum *et al.* (1975; 1976) give the wave functions in the above form. Therefore, when the results of such calculations are known, the coefficients appearing in Eqs. (8.49) and (8.51) are obtained without further calculation. {We have already pointed out that the calculation of $(\partial k_{az}/\partial \epsilon_s)|A_g^B|^2$, appearing in Eq. (8.49), does not require an explicit evaluation of $(\partial k_{az}/\partial \epsilon_s)$ [see, e.g., comments following Eq. (4.26)].} It is also worth noting that if need be a correction to the zero-field wave functions at $z = 0$ due to field penetration beyond the $z = 0$ plane [implied by the last term in Eq. (8.43)] can be made without much difficulty.

$T(\epsilon_s, \mathbf{q}_\parallel)$ and $R(\epsilon_s; \mathbf{q}_\parallel)$ are defined as follows. Let the wave $\exp(iq_z z)$ where $q_z = (2m\epsilon_s/\hbar^2 - q_\parallel^2)^{1/2}$, be incident, from the left, on the barrier

$$U = V(z), \quad z > 0 \qquad (8.54)$$
$$= 0, \quad z > 0$$

where $V(z)$ is the potential defined by Eq. (8.42). The corresponding solution of the Schrödinger equation has the following asymptotic form:

$$\Psi_{\epsilon_s, \mathbf{q}_\parallel} = \exp(iq_z z) + R_s(\epsilon_s, \mathbf{q}_\parallel) \exp(-iq_z z), \quad z < 0 \qquad (8.55)$$
$$= \frac{T_s(\epsilon_s, \mathbf{q})}{[\lambda_{q_\parallel}(z)]^{1/2}} \exp\left[i \int_{b_{q_\parallel}}^z \lambda_{q_\parallel}(z)\, dz\right], \quad z \gg b_{q_\parallel}$$

where $\lambda_{q_\parallel}(z)$ denotes, as usual, the z component of the classical momentum (divided by $\hbar$) and $b_{q_\parallel}$ is the right classical turning point of the barrier. The above equation defines T_s and R_s for any ϵ_s and $\mathbf{q}_\parallel$.

In the limit of very small emitted current densities (zero-emitted-current approximation) we have

$$\theta_s^s = \theta_s^b \qquad (8.56)$$

and θ_s^b can be calculated for a given value F_0 of the externally applied field in the manner already described in the previous two sections. Substituting this value of θ_s^b in Eqs. (8.49) and (8.51) one obtains the contribution to the energy distribution from the bulk and surface states, respectively. The contribution from bulk states may be further separated into a conduction band contribution denoted by $j_c(\epsilon_s)$, and a valence band contribution denoted by $j_v(\epsilon_s)$. We have

$$j_b(\epsilon_s) = j_c(\epsilon_s), \quad \epsilon_s > 0 \qquad (8.57a)$$
$$= 0, \quad -E_g < \epsilon_s < 0 \qquad (8.57b)$$
$$= j_v(\epsilon_s), \quad \epsilon_s < -E_g \qquad (8.57c)$$

j_c, j_v, and j_s are shown schematically in Fig. 8.8 for a nondegenerate at the surface conduction band, and in Fig. 8.9 for a degenerate at the surface conduction band. (A finite temperature, say $T = 300$ K, is assumed in both cases.) The (total) emitted current density is given by

$$J = J_c + J_v + J_s \tag{8.58}$$

where J_c, J_v, and J_s denote the contribution to the current density from the conduction, valence, and surface bands, respectively. We have

$$J_c = e \int_0^\infty j_c(\epsilon_s) \, d\epsilon_s \tag{8.59a}$$

$$J_v = e \int_{-\infty}^{-E_g} j_v(\epsilon_s) \, d\epsilon_s \tag{8.59b}$$

$$J_s = e \int_{-\infty}^\infty j_s(\epsilon_s) \, d\epsilon_s \tag{8.59c}$$

where e denotes, as usual, the magnitude of the electronic charge. Because each of the integrands vanishes outside a limited energy region, replacing the limits of the integration by infinities, as we have done, does not lead to any error.

APPROXIMATE FORMULAS FOR THE TUNNELING CURRENT

A calculation of the energy distribution and of the emitted current density from a semiconductor surface based on the equations of the preceding section has not been performed as yet. The few calculations that have been made so far are based on approximate expressions for j_b and j_s obtained by a straightforward extension of the free-electron theory of electron emission from metals. Formulas of this kind for j_v and j_c have been derived by Fischer (1962) and by Stratton (1962; 1964), and for j_s by Gadzuk (1972) and Modinos (1974). We obtain Stratton's formulas for j_v and j_c from Eq. (8.49), if we assume that

$$\hbar \left(\frac{\partial k_{\alpha z}}{\partial \epsilon_s} \right) \frac{(\hbar/m) \, | \, T_s(\epsilon_s, k_{\parallel}) \, |^2}{| \, 1 + R_s(\epsilon_s, k_{\parallel}) \, |^2} \, | A_0^B |^2 \simeq D(w) \tag{8.60}$$

where $D(w)$ is the probability of electron tunneling through the image potential barrier defined by Eq. (8.43) evaluated in the WKB approximation,

$$D(w) \simeq \exp[-Q(w)] \tag{8.61}$$

$$Q(w) = 2 \left(\frac{2m}{\hbar^2} \right)^{1/2} \int_{z1}^{z2} \left[\chi - \frac{(K-1)}{(K+1)} \frac{e^2}{4z} - eF_0z - w \right]^{1/2} dz \tag{8.62}$$

$$w \equiv \epsilon_s - \hbar^2 k_{\parallel}^2 / 2m \tag{8.63}$$

z_1 (z_2) is the smaller (larger) root of the integrand in Eq. (8.62). This equation implies that the image approximation to the potential barrier, Eq. (8.43), is valid close to the semiconductor–vacuum interface. Substituting Eq. (8.60) into Eq. (8.49) we obtain

$$j_b(\epsilon_s;\theta_s^b) \simeq \frac{f(\epsilon_s,\theta_s^b)}{4\pi^3\hbar} \iint_{\substack{\text{central}\\\text{SBZ}}} d^2k_\parallel \sum_\alpha D(\epsilon_s - \hbar^2 k_\parallel^2/2m) \qquad (8.64)$$

The sum over α in the above equation takes into account the fact that for given ϵ_s and for a given value of $\mathbf{k}_\parallel$, there may be none (in which case the integrand vanishes identically), one, or more than one bulk state contributing to the tunneling current. It is seen from Eq. (8.64) that the αth band will contribute to j_b if the projection of the constant energy surface $\epsilon_{s\alpha}(\mathbf{k}) = \epsilon_s$ onto the surface overlaps with the central region of the SBZ. We note that, apart from the topological effects introduced by the sum over α, Eq. (8.64) is identical with Eq. (1.82) which gives the Fowler–Nordheim TED from a metal surface. We have seen that the Fowler–Nordheim theory fails to reproduce the fine structure in the energy distribution, of the electrons emitted from a metal surface, and can not be relied upon for a calculation of the *magnitude* of the emitted current density. Obviously, the same applies in relation to electron emission from semiconductor surfaces. Even in the absence of surface resonances Eq. (8.64) cannot be relied upon for an exact quantitative analysis of $j_b(\epsilon_s)$. We have seen, for example, that the presence of surface states in the energy gap implies a deficiency in the surface density of states of the valence band and the conduction band (see Section 8.3). This fact, which is taken into account by the value of $|A_0^B|$ in Eq. (8.49), is disregarded in Eq. (8.64), which, in this respect, overestimates the contribution to the tunneling current from the bulk bands. Approximating the surface potential by an image barrier at small distances from the semiconductor–vacuum interface may lead to further overestimation of the tunneling current. On the other hand, if a surface resonance exists at a given $\mathbf{k}_\parallel$, $(\partial k_{\alpha z}/\partial\epsilon_s)|A^B{}_0|^2$ in Eq. (8.49) peaks at the corresponding resonance energy as described in Section 8.4. The corresponding "excess" contribution to the tunneling current resulting from such resonances is missing in Eq. (8.64). [In some cases, for example the surface resonance around Γ in Fig. 8.4a, resonances are obviously an "extension" of a band of true surface states. In such cases we can take the "excess" contribution to local density of states and to the tunneling current, resulting from such resonances, approximately into account by treating these resonances as true surface states. Throughout this chapter we assume for the sake of simplicity that all surface resonances can be treated in this way.]

In spite of the obvious limitations of Eq. (8.64), a theory of field emission from semiconductors based on this equation, with the contribution to the tunneling current from surface states (and resonances) taken into account separately, can be quite useful, in the same way that the Fowler–Nordheim theory is useful in the study of field emission from metal surfaces. We remember that the Fowler–Nordheim theory, though it fails to predict the fine structure in the energy distribution, it predicts correctly, in most cases, the overall shape and the full width at half-maximum of this distribution; and though it cannot be relied upon for an estimate of the magnitude of the emitted current density, it describes correctly the relationship between this quantity and the externally applied field. We expect, therefore, that a theory based on Eq. (8.64) will describe correctly, in most cases, the overall shape of the energy distribution from the conduction and valence bands of the semiconductor, and the functional dependence of J_c and J_v (the contributions to the emitted current density from the conduction and valence band, respectively) on the applied field.

The integral $Q(w)$, defined by Eq. (8.62), has the same form as that defined by Eq. (1.33) and can be evaluated in the same way. If we expand $Q(w)$ in a Taylor series around $w = w'$, we obtain {note the analogy with the corresponding expression for metal surfaces [Eqs. (1.83)–(1.86)]}

$$-Q(w) = -b(w') + c(w')(w - w') \tag{8.65a}$$

$$b(w') = \frac{4}{3}\left(\frac{2m}{\hbar^2}\right)^{1/2} \frac{(\chi - w')^{3/2}}{eF_0} v(y_{w'}) \tag{8.65b}$$

$$c(w') = 2\left(\frac{2m}{\hbar^2}\right)^{1/2} \frac{(\chi - w')^{1/2}}{eF_0} t(y_{w'}) \tag{8.65c}$$

$$y_{w'} = (e^3 F_0)^{1/2} \left(\frac{K-1}{K+1}\right)^{1/2} \bigg/ (\chi - w') \tag{8.65d}$$

v and t are the functions tabulated in Table 1.1. Stratton (1962, 1964) derived explicit formulas for j_c, j_v, J_c, and J_v, for the case when there is only a single conduction band [a single term replaces the sum over α in Eq. (8.64)] described by

$$\epsilon_n(\mathbf{k}) \equiv \frac{\hbar^2 k^2}{2m_n} \tag{8.66a}$$

$$m_n = r_n m \tag{8.66b}$$

and a single valence band described by

$$\bar{\epsilon}_p(\mathbf{k}) \equiv \frac{\hbar^2 k^2}{2m_p} \tag{8.67a}$$

$$m_p = r_p m \tag{8.67b}$$

Where $\bar{\epsilon}$ measures the energy from the top of the valence band positively down-wards, i.e.,

$$\bar{\epsilon} = -E_g - \epsilon, \qquad \epsilon < -E_g \tag{8.68}$$

where ϵ is the energy measured from the bottom of the conduction band. Stratton's formulas are summarized below:

$$j_c(\epsilon_s,\theta_s^b) \simeq \left(\frac{m}{2\pi^2\hbar^3}\right)\frac{e^{-(b_1+c_1\theta_s^b)}}{c_1}f(\epsilon_s,\theta_s^b)e^{c_1\epsilon_s}[1 - e^{-r_nc_1\epsilon_s}] \qquad \text{if } \theta_s^b > 0 \tag{8.69}$$

$$b_1 \equiv b(\theta_s^b), \qquad c_1 \equiv c(\theta_s^b) \tag{8.69a}$$

$$j_c(\epsilon_s,\theta_s^b) \simeq \left(\frac{m}{2\pi^2\hbar^3}\right)\frac{e^{-b0}}{c_0}f(\epsilon_s,\theta_s^b)e^{c_0\epsilon_s}[1 - e^{-r_nc_0\epsilon_s}] \qquad \text{if } \theta_s^b < 0 \tag{8.70}$$

$$b_0 = b(0), \qquad c_0 \equiv c(0) \tag{8.70a}$$

Equation (8.69), which gives the energy distribution of the electrons emitted from the conduction band when the latter is degenerate at the surface, follows from Eq. (8.64) and (8.61) when use is made of Eq. (8.66) and of the expansion described by Eq. (8.65) with $w' \equiv \theta_s^b$. We note that Eq. (8.69) is valid only when $c_1 k_B T$ < 1. (In practice $c_1 \simeq 10$ eV^{-1}.) Equation (8.70), which gives j_c when the conduction band is *not* degenerate at the surface, follows from Eq. (8.64) and (8.61) when use is made of Eq. (8.66) and of the expansion described by Eq. (8.65) with $w' = 0$. Equation (8.70) is valid when $c_0 k_B T < 1$. (In practice $c_0 \simeq 10$ eV^{-1}.) Moreover, because $\theta_s^b < 0$, the Fermi–Dirac distribution function in Eq. (8.70) can be approximated by

$$f(\epsilon_s, \theta_s^b) \simeq e^{\theta_s^b/k_BT}e^{-\epsilon_s/k_BT}, \qquad \text{when } \theta_s^b < 0, \qquad \epsilon_s > 0 \tag{8.71}$$

Theoretical energy distributions [Eq. (8.69) and (8.70) with $r_n = 1$] are shown in Fig. 8.10 for a degenerate and in Fig. 8.11 for a nondegenerate (at the surface) conduction band (Arthur, 1964).

Substituting Eqs. (8.70) and (8.71) into Eq. (8.59a), one finds that the contribution to the emitted current density from a nondegenerate (at the surface) conduction band is given by

$$J_c \simeq \left(\frac{em_n}{2\pi^2\hbar^3}\right)(k_BT)^2 e^{\theta_s^b/k_BT}e^{-b0}, \qquad \theta_s^b < 0 \tag{8.72}$$

The contribution to the emitted current density from a degenerate (at the surface) conduction band is calculated from Eq. (8.59a) with $j_c(\epsilon_s)$ given by Eq. (8.69). In

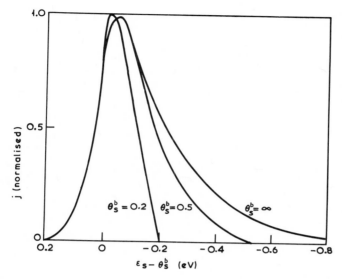

FIGURE 8.10

Theoretical energy distributions of field-emitted electrons from a degenerate (at the surface) conduction band ($\theta_s^b > 0$). Obtained from Eq. (8.69) with $r_n = 1$. $F_0 = 0.45$ V/Å. (From Arthur, 1964.)

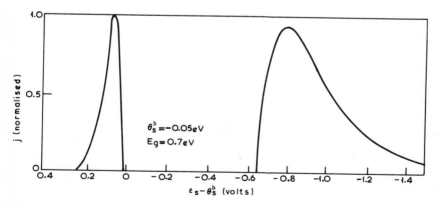

FIGURE 8.11

Theoretical energy distribution of field-emitted electrons from nondegenerate conduction band and valence band. Obtained from Eqs. (8.70) with $r_n = 1$ and Eq. (8.75) with $r_p = 1$. $F_0 = 0.45$ V/Å. (From Arthur, 1964.)

the limit $T \to 0$, when $f(\epsilon_s, \theta_s^b) \simeq 1$ (0) for $\epsilon_s < \theta_s^b (\epsilon_s > \theta_s^b)$, we obtain

$$J_c = \left(\frac{em}{2\pi^2\hbar^3}\right) \frac{e^{-b_1}}{c_1^2} \left[1 - e^{-c_1\theta_s^b} + \frac{e^{-c_1\theta_s^b}}{1-r_n} - \frac{e^{-r_n c_1\theta_s^b}}{1-r_n}\right], \quad \theta_s^b > 0, \qquad r_n \neq 1$$

$$\text{(8.73a)}$$

$$= \left(\frac{em}{2\pi^2\hbar^3}\right) \frac{e^{-b_1}}{c_1^2} [1 - e^{-c_1\theta_s^b} - c_1\theta_s^b e^{-c_1\theta_s^b}], \qquad \theta_s^b > 0, \qquad r_n = 1 \quad \text{(8.73b)}$$

When $c_1\theta_s^b \gg 1$, the expression in the square brackets equals unit, and the above equation becomes formally identical with the Fowler–Nordheim expression given by Eq. (1.57). We note, however, that the work function

$$\phi = \chi - \theta_s^b \tag{8.74}$$

appearing in b_1 and c_1 [see Eqs. (8.69a) and (8.65)] depends through θ_s^b on the applied field. We note also that the numerical values of the functions t and v (for given ϕ and F_0) are slightly different from the corresponding ones for metals because of the finite value of K in Eq. (8.65d). It is interesting to consider a plot $\ln(J_c/F_0^2)$ versus $1/F_0$. If $c_1\theta_s^b \gg 1$ over the entire range of the applied field, this plot will be a straight line, for it is reasonable to assume that θ_s^b becomes independent of the applied field above a certain level of degeneration. If, however, at the lower applied fields the surface is nondegenerate ($\theta_s^b < 0$), or if the level of degeneration is small ($c_1\theta_s^b < 1$), this plot will not be a straight line. Its slope will be larger at the lower applied fields, decreasing as the field increases, becoming eventually (when $c_1\theta_s^b \gg 1$) a straight line.

The energy distribution of the electrons emitted from the valence band when the latter is nondegenerate at the surface (this is practically always the case) is given by

$$j_v(\bar{\epsilon}_s) \simeq \left(\frac{m}{2\pi^2\hbar^3}\right) \frac{e^{-b_{v0}}}{c_{v0}} f(-E_g - \bar{\epsilon}_s, \theta_s^b) e^{-c_{v0}\bar{\epsilon}_s} [1 - e^{-r_p c_{v0}\bar{\epsilon}_s}] \tag{8.75}$$

where $\bar{\epsilon}_s$ measures the energy from the top of the valence band (at the surface) positively downwards, and

$$b_{v0} = b(-Eg), \qquad c_{v0} = c(-E_g) \tag{8.75a}$$

Equation (8.75) follows from Eqs. (8.64) when use is made of Eq. (8.67) and of the expansion described by Eq. (8.65) with $w' = -E_g$. In the present case

$$f(-E_g - \bar{\epsilon}_s, \theta_s^b) \simeq 1 - \exp[-(E_g + \theta_s^b)/k_BT] \exp(-\bar{\epsilon}_s/k_BT) \tag{8.76}$$

A theoretical energy distribution [Eq. (8.75)] with $r_p = 1$ is shown in Fig. 8.11 (Arthur, 1964). Substituting Eqs. (8.75) and (8.76) into Eq. (8.59b) we obtain the following formula for the valence band contribution to the current density:

$$J_v = \left(\frac{em}{2\pi^2\hbar^3}\right)\frac{e^{-b_{v0}}}{c_{v0}^2}\left\{\frac{1}{1+r_p} - (c_{v0}k_BT)^2 \exp[-(E_g + \theta_s^b)/k_BT]\right\} \quad (8.77)$$

It has been assumed, in deriving the above equation that $c_{v0}k_BT < 1$ ($c_{v0} \simeq 10$ eV^{-1} and, therefore, at room temperature $c_{v0}k_BT \simeq 1/4$). Because of the smallness of the second term in the brackets of Eq. (8.77), J_v is to a very good approximation independent of the temperature. It is also evident from Eqs. (8.75a) and (8.65) that a plot of $\ln(J_v/F_0^2)$ versus $1/F_0$ is a straight line whose slope is proportional to $(\chi + E_g)^{3/2}$.

An approximate formula for the energy distribution of the electrons emitted from surface states can be obtained as follows. We assume that only one band of surface states contributes significantly to the field emission current. [There may be additional bands lying well below the top of the valence band which do not contribute to the tunneling current. These bands are of the donor type, i.e., they are neutral when occupied (see Section 8.3).] We further assume that the sth surface band, which contributes to the tunneling current, is a circular band, i.e.,

$$\epsilon_{s;a}(\mathbf{k}_\parallel) = \epsilon_s^0 + \hbar^2 k_\parallel^2/2m^* \quad (8.78)$$

The corresponding density of states, as defined by Eq. (8.34), is given by

$$\rho_{s;a} = \frac{m^*}{\pi\hbar^2} \equiv N_s, \qquad \epsilon_s^0 < \epsilon_s < \epsilon_s^0 + \Gamma_s \quad (8.79)$$

where ϵ_s^0 denotes the energy at the bottom of the surface band (measured from the bottom of the conduction band at the surface). The corresponding energy level diagram is shown schematically in Fig. 8.12. This model was originally suggested by Handler (1960) as an approximation to the density of surface states on a clean germanium surface. [If we assume that the distribution of surface states on, say, the Ge(100) surface is not very different from that shown in Fig. 8.6 for the Si(100) surface we can say that the $s;\alpha$ ($s;d$) bands in Fig. 8.12 represent approximately the π_a (π_b) of Fig. 8.6.] Finally, by analogy to Eq. (8.60) we put

$$\frac{|T_s(\epsilon_s,\mathbf{k}_\parallel)|^2}{|1 + R_s(\epsilon_s,\mathbf{k}_\parallel)|^2}|A_0^s|^2 \simeq D(w)\overline{Q}^2 \quad (8.80)$$

in the integrand of Eq. (8.51). $D(w)$ is given by Eq. (8.61) and $\overline{Q}$ is an effective wave vector to be treated as an adjustable parameter. Substituting Eq. (8.80) into

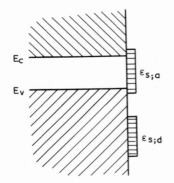

E_C

$\varepsilon_{s;a}$

E_V

$\varepsilon_{s;d}$ **FIGURE 8.12**

Bands of surface states on a clean germanium surface according to the Handler model.

Eq. (8.51) and using Eq. (8.78) we obtain (Modinos, 1974)

$$j_s(\epsilon_s,\theta_s^s) \simeq \left(\frac{\hbar}{2\pi m}\right) N_s\overline{Q}^2 f(\epsilon_s,\theta_s^s)D\left(\epsilon_s - \frac{m^*}{m}(\epsilon_s - \epsilon_s^0)\right) \qquad (8.81a)$$

$$D\left(\epsilon_s - \frac{m^*}{m}(\epsilon_s - \epsilon_s^0)\right) = \exp\left[-b\left(\epsilon_s - \frac{m^*}{m}(\epsilon_s - \epsilon_s^0)\right)\right] \qquad (8.81b)$$

Eq. (8.81b) follows from Eq. (8.61) and Eq. (8.65) with $w = w' = \epsilon_s - (m^*/m)(\epsilon_s - \epsilon_s^0)$. N_s is the (constant) density of states defined by Eq. (8.79). The current density emitted from the surface band is given by

$$J_s(\theta_s^s,F_0) = e\int_{\epsilon_s^0}^{\epsilon_s^0 + \Gamma_s} j_s(\epsilon_s,\theta_s^s)\, d\epsilon_s \qquad (8.82)$$

When J_s is sufficiently small, $\theta_s^s = \theta_s^b$ and, therefore, once the value of θ_s^b has been determined (in the manner described in Sections 8.2 and 8.3), the value of J_s can be obtained by a numerical evaluation of the above integral. In principle, however, θ_s^s may be different from θ_s^b even in the zero-internal-current approximation and these two parameters must be calculated self-consistently (see Section 8.6).

It would seem from Eqs. (8.73) and (8.77) that $J_c > J_v$ when the conduction band is degenerate at the surface, a situation demonstrated schematically in Fig. 8.9. We must remember, however, that the above equations apply when the conduction and valence bands are described by Eqs. (8.66) and (8.67), respectively. The shape of constant energy surfaces in the vicinity of the bottom of the conduction band is of particular importance. It may happen that the electron states at the bottom of the conduction band correspond to $\mathbf{k}_\parallel$ values outside the central region of the SBZ, in which case J_c could be much small than J_v [it is obvious that Eq. (8.66) does not apply in this case and therefore Eq. (8.73) does not apply either]. In germanium, for example, a constant energy surface near the bottom of the conduction band consists of ellipsoids centered away from the origin along the ⟨111⟩ directions, as shown in Fig. 8.13. It is clear from this figure that the components of the **k** vector parallel to the (100) surface of the corresponding states do not lie in the central region of the SBZ of the unreconstructed surface. (This may or may not be so for a reconstructed surface.) Because of the transverse energy (∼2.4 eV) associated with these states the normal energy of an electron at the bottom of the conduction band is considerably smaller than that of an electron with $k_{\parallel} \simeq 0$ at the top of the valence band. Accordingly the tunneling probability, which diminishes rapidly as the normal energy decreases [see, e.g., Eq. (8.61)],

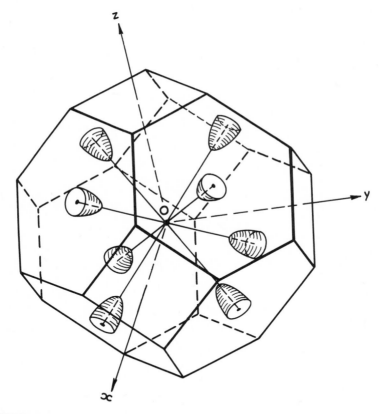

FIGURE 8.13

The Brillouin zone of germanium. The ellipsoids along the (111) directions show a constant energy surface near the minimum of the conduction band.

will be larger for electrons at the top of the valence band leading to $J_v \gg J_c$, even when the conduction band is degenerate at the surface. It is worth remembering, however, that this argument does not necessarily apply to a reconstructed Ge(100) surface.

8.5. FIELD EMISSION FROM GERMANIUM (100)

Shepherd and Peria (1973) were able to obtain reproducible energy distributions from the center of the (100) plane of a germanium (40 Ω cm, n type) after careful preparation and cleaning of its surface. This involved sputter-cleaning and field evaporation followed by annealing of the emitter by heating to approximately 400°C for 5 sec. Figure 8.14 shows samples of energy distributions from the clean annealed surface at various values of the applied voltage. It is obvious that the energy distribution has two peaks—a high-energy peak between 0.6 and 0.7 eV below the Fermi level, followed by a second peak (or shoulder) at a lower energy—and that the ratio of the amplitude of the high-energy peak to the amplitude of the low-energy peak decreases as the applied field increases. [The position

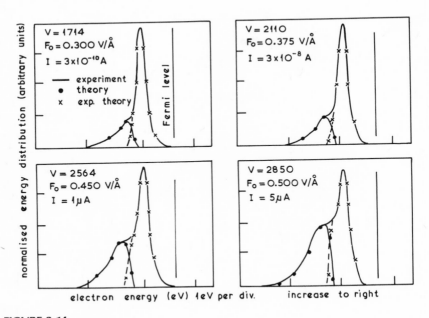

FIGURE 8.14

Energy distributions of field-emitted electrons from Ge(100) at various values of the applied field. (From Shepherd and Peria, 1973.)

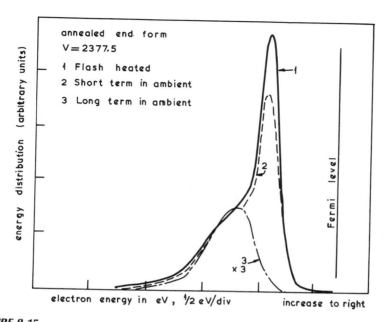

FIGURE 8.15

Effect of adsorption on the energy distribution of field-emitted electrons from Ge(100). (From Shepherd and Peria, 1973.)

of the Fermi level on the retarding potential axis (emitter collector bias) is independent of the specific properties of the emitter (for a given collector) when there is no internal (ohmic) voltage drop and can therefore be determined from a tungsten emitter. When an internal voltage drop does occur, the position of the Fermi level on the retarding potential axis is displaced by that amount which can be determined experimentally.] The following experimental observations show that the high-energy peak originates from a band of surface states: (a) The amplitude of this peak is reduced with adsorption; this is demonstrated in Fig. 8.15. (b) Sputtering of the emitter caused disorder and removed the high-energy peak. (c) The intensity of this peak would oscillate with time as the tip was being slowly field-evaporated, indicating that the band of surface states does not exist when the (100) plane is too small. (d) The high-energy peak is *extremely* sensitive to the orientation of the emitter, which indicates that the band of surface states exists only within the central region of the (100) plane. The low-energy peak, in contrast to the high-energy one, is not particularly sensitive to the condition of the surface, which indicates that it originates from bulk states. Since this peak is part of a distribution which extends to energies well below the high-energy (surface-band) peak, it must originate from the valence band. We note that a surface band over-lapping the valence band is also consistent with available experimental data on Ge(100) from other sources (Handler, 1960; Frankl, 1967; Inglesfield and Holland, 1981). Shepherd and Peria (1973) calculated a theoretical energy distribution from the valence band of Ge, using Eq. (8.64). They took into account the fact that there are three valence bands ($\alpha = 1,2,3$) in Ge which (near the top of the band) can be described approximately by

$$\bar{\epsilon}_{p1,2}(\mathbf{k}) = \frac{\hbar^2 k^2}{2m_{p1,2}} \tag{8.83a}$$

$$\bar{\epsilon}_{p3}(\mathbf{k}) = \delta\bar{\epsilon}_{p3} + \frac{\hbar^2 k^2}{2m_{p3}} \tag{8.83b}$$

Here, as in Eq. (8.67), the energy is measured positively downward from the top of the valence band. $\delta\bar{\epsilon}_{p3} = 0.3$ eV is the separation between the upper and the lower valence band maxima, $m_{p1} = m/4.8$, $m_{p2} = m/21.4$, $m_{p3} = m/13.1$ are effective masses. The tunneling probability $D(w)$ was calculated on the basis of Eqs. (8.61)–(8.63) with $(\chi + E_g) = 4.77$ eV and $K = 16$. In order to obtain a better fit to the experimental data they multiplied the integrand in Eq. (8.64) with a correction factor

$$t_c = [1 - 4(\epsilon_{\parallel}/\lambda E_{ps})^{1/2}]^{3/2} \qquad \text{if } \epsilon_{\parallel} < \lambda E_{ps}/16$$
$$= 0 \qquad \text{otherwise}$$

where $\epsilon_{\parallel} = \hbar^2 k_{\parallel}^2/2m$, $E_{ps} = 0.7$ eV is the "passing energy" of the analyzer and λ is an adjustable parameter. Though the need for the above correction factor was attributed by Shepherd and Peria to instrumental reasons, it could be attributed, at least partly, to the inadequacy of Eq. (8.64) (see comments following that equation). The local field factor $\beta = F_0/V$, V being the applied voltage, was an additional parameter. The procedure followed by Shepherd and Peria was to fit the energy distribution at the lowest possible value of the applied voltage by varying β, for a given value of λ, and then to use the value of β determined in this way to calculate the energy distribution at the higher applied voltages and compare them with the experimental distributions at these voltages. The process was repeated for different values of λ (leading in turn to different values of β) until a good fit was obtained at all applied voltages. In this way both λ and the field factor β were determined uniquely for a given set of effective masses. It was found that a variation in the values of the latter had a significant effect on the calculated distribution and that the agreement between theory and experiment was best when the expressions given by Eq. (8.83), which apply to the (100) direction, were used. Figure 8.14 shows the best fit to the data (dots on solid line) obtained with $\lambda =$

3. {We note that for a nondegenerate (at the surface) valence band $f(\epsilon_s, \theta_s^b)$ is practically equal to unity [see Eq. (8.76)] and, therefore, the shape of the energy distribution is practically independent of θ_s^b.} The result of subtracting the theoretical valence band distribution (dots on solid line) from the experimental distribution (solid line) is also shown in Fig. 8.14 ($\times$ on solid or dashed line). It is obvious that the latter quantity which represents the energy distribution of the electrons emitted from the surface band (band of surface states/resonances), broadens as the applied field increases, though not as much as the valence band distribution.

The position of the Fermi level in Fig. 8.14 implies that the conduction band is degenerate at the surface for the values of the applied field used in the experiment. However, no emission was observed from the conduction band. Shepherd and Peria explained this using the argument already described at the end of the previous section, namely, that no electron states exist at the bottom of the conduction band with sufficiently small $k_{\parallel}$ to contribute significantly to the tunneling current. [This argument does not necessarily apply to a reconstructed Ge(100) surface and, therefore, cannot be accepted without reservations.]

Figure 8.16 (Shepherd and Peria, 1973) shows Fowler–Nordheim plots from the clean annealed surface that gave the energy distributions shown in Fig. 8.14. The probe–hole data were obtained by integrating the experimental energy distribution. For instrumental reasons, the probe–hole current was known only within a scaling factor G of the order of 5×10^6. A theoretical plot for the valence band contribution to the current density (obtained by numerical integration of the energy distribution) is also shown in Fig. 8.16. We note that in spite of the fact that the probe–hole current comes not from one band of states but from four bands (the surface band and the three components of the valence band) the FN plot is a straight line whose slope is not significantly different from that of a single valence band with a work function $\phi = \chi + E_g \simeq 4.77$ eV.

Shepherd and Peria produced an estimate of the density of surface states assuming (a) that the conduction band is degenerate at the surface at all applied fields and (b) that *all* states in the surface band (assumed to be of the acceptor type) are occupied at all applied fields. The positions on the retarding potential axis of the high-energy threshold of the energy distribution from the surface band and of the surface state peak were determined experimentally within $\pm 2k_B T$ and $\pm k_B T$, respectively. These are shown in Fig. 8.17 as a function of the applied field. The rapid excursions of the peak and threshold at large fields was attributed to an Ohmic (IR) voltage drop in the emitter which was subtracted out to give the corrected points denoted by $\triangle$ and $\circ$ in the figure. [At sufficiently high currents the shift in the energy distribution along the retarding potential axis varies linearly with the emitted current and the slope is equal to the total resistance of the emitter. With the resistance of the emitter obtained in this way one can calculate the IR voltage drop at the lower emitted currents as well.] At the lower applied field ($F_0 = 0.3$ V/Å) the valence band was located ($0.77 \pm k_B T$) eV below the Fermi

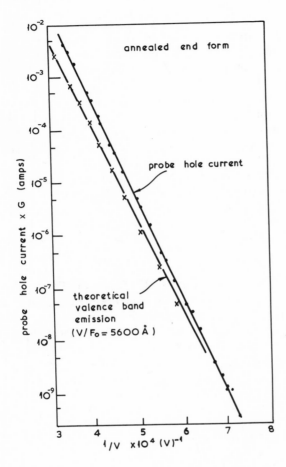

FIGURE 8.16

Fowler–Nordheim plots for the clean annealed Ge(100) surface; and a theoretical plot for the valence band contribution to the emitted current. (From Shepherd and Peria, 1973.)

level which implies that for this field $\theta_s^b = (4 \pm 1)\, k_B T$. [We remember that θ_s^b, defined by Eq. (8.12), gives the Fermi level relative to the bottom of the conduction band at the surface.] We have seen that the value of θ_s^b uniquely determines the value of the electric field F_s [defined by Eq. (8.4b)], through Eq. (8.14). That equation simplifies considerably when the surface band is filled at all fields, the conduction band is degenerate at the surface, and the bulk is practically intrinsic. From the experimental value of θ_s^b, for $F_0 = 0.3\ \text{V/Å}$, after determining the corresponding value of F_s, and substituting F_0 and F_s into Eq. (8.20) they found that the electronic charge in the fully occupied (acceptor) surface band is given by

$$Q_s = (6.3 \pm 2.4) \times 10^{12} \text{ electrons/cm}^2 \qquad (8.84)$$

Using the above value of Q_s (which by assumption does not depend on the applied field) they estimated the increase ($4k_BT$) in the value of θ_s^b corresponding to an increase of F_0 from 0.3 V/Å to 0.5 V/Å. Since the observed shift in θ_s^b when F_0 was increased from 0.3 to 0.5 V/Å was no more than $3k_BT$ this leaves at least $1k_BT$ unaccounted for. This may be, as the above authors suggested, due to the uncertainty of the measured values. The other possibility suggested by them, namely, that the surface band is always full but the density of surface states as such is increased when the applied field is increased, does not seem very probable. It would require an increase of 40% in the density of surface states to keep the change in θ_s^b to $3k_BT$. Because the surface band is fully occupied, according to Shepherd and Peria, Eq. (8.84) gives the total number of states in the band. Approximating this band by a rectangular one [Eq. (8.79)] with a width Γ_s equal to the full width at half-maximum of the surface state peak ($\Gamma_s \sim 0.2$ eV for $F_0 = 0.3$ V/Å) gives an average density of states $N_s \simeq 3 \times 10^{13}$ states/cm^2 eV. This is significantly smaller than other estimates of this quantity [8×10^{13} states/cm^2 eV according to Handler and Portnoy (1959); 3×10^{14} states/cm^2 eV according to Margoninski (1963)]. We should also point out that the assumption of a fully occupied surface band (whatever the value of the applied field) makes it very difficult to explain the increase in the width of the surface state peak in the energy

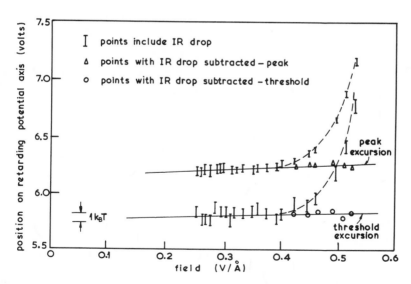

FIGURE 8.17

Positions on the retarding potential axis of the high-energy threshold and peak of the field-emission energy distribution from the surface band of a Ge(100) surface as a function of applied field. (From Shepherd and Peria, 1973.)

distribution of the emitted electrons. This increase, evident in Fig. 8.14, is discussed in more detail by Shepherd and Peria (1973).

It has been assumed, in the above analysis, that $\theta_s^s = \theta_s^b$, i.e., that the electrons in the surface band are in thermal equilibrium with the electrons in the bulk states of the semiconductor whatever the value of the current emitted from the surface band. In other words, we have assumed that an electron emitted from the surface band is automatically replaced by an electron flowing into this band from either the conduction band or the valence band of the semiconductor. This is not necessarily so in practice.

8.6. REPLENISHMENT OF THE SURFACE STATES

We assume for the sake of simplicity, that we have only one surface band, that it is of the acceptor type and that it is described by Eq. (8.79). Under steady state conditions the current tunneling out from the surface band is equal to a *net* current of electrons *from* the valence and conduction bands *into* the surface band. The flow of electrons with energy between ϵ_s and $\epsilon_s + d\epsilon_s$ from the surface band to the valence bands is approximately by

$$v_p S p_s N_s f(\epsilon_s, \theta_s^s)\, d\epsilon_s \tag{8.85}$$

N_s is the density of states defined by Eq. (8.79), and $f(\epsilon_s, \theta_s^s)$ determines the occupation of these states under conditions appropriate to a field emission experiment, as explained at the beginning of Section 8.4. [The assumption that the electrons in the surface band are in thermal equilibrium among themselves is justified when the collision time (relaxation time) of the scattering mechanism which leads to this equilibrium is less than the lifetime of an electron with respect to tunneling (this is the time that elapses before an electron tunnels out of the surface band into vacuum).] p_s in Eq. (8.85) denotes the concentration of holes (in the valence band) at the surface. When the valence band is nondegenerate throughout the crystal (including the surface) we obtain

$$p_s = p_\infty \exp[(\theta_\infty^b - \theta_s^b)/k_B T] \tag{8.86}$$

where p_∞ is the hole concentration deep inside the semiconductor ($z \rightarrow -\infty$), and θ_s^b and θ_∞^b are the quantities defined by Eqs. (8.12) and (8.13), respectively. Of the remaining factors in formula (8.85), v_p denotes a velocity which when mulitplied by the concentration gives the number of holes crossing a unit area from one direction. It is given by

$$v_p \simeq (k_B T/2\pi m_p) \tag{8.87}$$

where m_p is the effective mass of the holes defined by Eq. (8.8b). Finally S denotes an average recombination cross section. Besides the flow of electrons given by Eq. (8.85) there will be a flow of electrons in the reverse direction, from the valence band to surface states in the energy interval between ϵ_s and $\epsilon_s + d\epsilon_s$. This is given by

$$g_t v_p S N_s [1 - f(\epsilon_s, \theta_s^s)] \, d\epsilon_s \qquad (8.88)$$

g_t is determined by the requirement that when there is equilibrium between the valence band and the surface band $(\theta_s^b = \theta_s^s)$ the sum of the two terms, given by Eqs. (8.85) and (8.88), respectively, is zero. We obtain

$$g_t = p_s \exp[(\theta_s^b - \epsilon_s)/k_B T] \qquad (8.89)$$

The net current density $J_{v \to s}$ from the (nondegenerate) valence band to the surface band is obtained by subtracting the current given by Eq. (8.85) from the one given by Eq. (8.88), integrating the difference over the width of the surface band, and multiplying the final result with the electronic charge. We obtain

$$
\begin{aligned}
J_{v \to s}(\theta_s^b, \Delta\theta_s^s) = {} & e v_p S p_\infty \exp[(\theta_\infty^b - \theta_s^b)/k_B T] N_s k_B T \\
& \times \ln\{1 + \exp[(\theta_s^b - \Delta\theta_s^s - \epsilon_s^0)/k_B T]\}[\exp(\Delta\theta_s^s/k_B T) - 1]
\end{aligned}
\qquad (8.90)
$$

[In deriving the above equation we assumed that $\theta_s^s = \theta_s^b - \Delta\theta_s^s$, the effective Fermi level in the surface band, lies sufficiently below the top of the surface band at $\epsilon_s = \epsilon_s^0 + \Gamma_s$, so that $f(\epsilon_s^0 + \Gamma_s, \theta_s^s) = 0$.] In a very similar manner we can calculate the net current density flowing from the conduction band into the surface states. We obtain

$$
\begin{aligned}
J_{c \to s}(\theta_s^b, \Delta\theta_s^s) = {} & e v_n S n_\infty \exp[(\theta_s^b - \theta_\infty^b)/k_B T] \\
& \times \{N_s \Gamma_s - N_s k_B T \ln[1 + \exp((\theta_s^b - \Delta\theta_s^s - \epsilon_s^0)/k_B T)]\} \\
& \times [1 - \exp(-\Delta\theta_s^s/k_B T)]
\end{aligned}
\qquad (8.91)
$$

$N_s\Gamma_s$ equals the total number of states (per unit area) in the surface band, and

$$v_n = (k_B T/2\pi m_n) \qquad (8.92)$$

where m_n is the effective mass of the electrons as defined by Eq. (8.8a). In deriving Eq. (8.91) we have assumed that S has the same value in Eq. (8.91) as in Eq. (8.90). This is of the order of $S \simeq 10^{-17} \text{ cm}^2$ (Modinos, 1974).

The net total current density into the surface band is given by

$$J_{vc \to s}(\theta_s^b, \Delta\theta_s^s) = J_{c \to s}(\theta_s^b, \Delta\theta_s^s) + J_{v \to c}(\theta_s^b, \Delta\theta_s^s) \qquad (8.93)$$

Under steady state conditions this must be equal to the current density emitted from the surface band into the vacuum. We calculate the latter quantity, denoted by $J_s(\theta_s^b - \Delta\theta_s^s, F_0)$, from Eqs. (8.81) and (8.82). We have

$$J_s(\theta_s^b - \Delta\theta_s^s, F_0) = J_{vc \to s}(\theta^{bs}, \Delta\theta_s^s) \tag{8.94a}$$

which when written down explicitly reads as follows:

$$
\begin{aligned}
(e\hbar/2m)N_s(\overline{Q}^2/\pi) \int_{\epsilon_s^0}^{\mathcal{E}+\Gamma_s} & \left[1 + \exp\left(\frac{\epsilon_s - \theta_s^b - \Delta\theta_s^s}{k_B T} \right) \right]^{-1} \\
\times \exp & \left\{ -\frac{4}{3}\left(\frac{2m}{\hbar^2}\right)^{1/2} \frac{[\chi - \epsilon_s + (m^*/m)(\epsilon_s - \epsilon_s^0)]^{3/2} \, v(y)}{eF_0} \right\} d\epsilon_s \\
= ev_p S p_\infty \exp & \left[\frac{(\theta_\infty^b - \theta_s^b)}{k_B T} \right] N_s k_B T \ln \left[1 + \exp\left(\frac{\theta_s^b - \Delta\theta_s^s - \epsilon_s^0}{k_B T} \right) \right] \\
\times & \left[\exp\left(\frac{\Delta\theta_s^s}{k_B T} \right) - 1 \right] + ev_n S n_\infty \exp \left[\frac{(\theta_s^b - \theta_\infty^b)}{k_B T} \right] \\
\times & \left\{ N_s \Gamma_s - N_s k_B T \ln \left[1 + \exp\left(\frac{\theta_s^b - \Delta\theta_s^s - \epsilon_s^0}{k_B T} \right) \right] \right. \\
\times & \left. \left[1 - \exp\left(\frac{\Delta\theta_s^s}{k_B T} \right) \right] \right\} \\
v(y), \quad y & = (e^3 F_0)^{1/2} \frac{(K-1)/(K+1)}{\chi - \epsilon_s + (m^*/m)(\epsilon_s - \epsilon_s^0)}
\end{aligned}
\tag{8.94b}
$$

is obtained from Table 1.1.

When $\Delta\theta_s^s \neq 0$, the net charge in the surface region is obtained from Eq. (8.38) with θ_s^b replaced by $\theta_s^s = \theta_s^b - \Delta\theta_s^s$. Substituting Eq. (8.79) in Eq. (8.38) (the second term in that equation equals zero in the present case) we obtain

$$Q_s(\theta_s^b - \Delta\theta_s^s) \simeq -eN_s \int_{\epsilon_s^0}^{\mathcal{E}+\Gamma_s} f(\epsilon_s, \theta_s^b - \Delta\theta_s^s) \, d\epsilon_s \tag{8.95}$$

Because the integrand vanishes before the upper limit of the integration is reached [see comments following Eq. (8.90)], the latter can be replaced by $+\infty$ and we obtain

$$Q_s(\theta_s^b - \Delta\theta_s^s) \simeq -eN_s k_B T \ln\{1 + \exp[(\theta_s^b - \Delta\theta_s^s - \epsilon_s^0)/k_B T]\} \tag{8.96}$$

We remember that [Eq. (8.20)]

$$4\pi Q_s = F_s - F_0 \tag{8.97}$$

where F_0 is the magnitude of the applied field on the vacuum side of the interface, and that F_s, defined by Eq. (8.4b), is according to Eq. (8.14) a unique function of θ_s^b. When both the conduction and valence bands are nondegenerate throughout the crystal, the relation between F_s and θ_s^b becomes (Kingston and Neustadter, 1955)

$$-F_s = \left(\frac{k_B T K}{e L_D}\right)\sqrt{2}\left[\left(\frac{\theta_\infty^b - \theta_s^b}{k_B T}\right)\sinh u_b\right.$$
$$\left. - \cosh u_b + \cosh\left(u_b + \frac{\theta_s^b - \theta_\infty^b}{k_B T}\right)\right]^{1/2} \quad (8.98)$$

[The derivation of Eq. (8.98) assumes also, that all donor and acceptor impurity atoms are ionized. For germanium this is valid for temperatures higher than 100 K.] Substituting Eqs. (8.96) and (8.98) into Eq. (8.97) we obtain

$$-F_0 + 4\pi e N_s k_B T \ln\{1 + \exp[(\theta_s^b - \Delta\theta_s^s - \epsilon_s^0)/k_B T]\}$$
$$= \left(\frac{k_B T K}{e L_D}\right)\sqrt{2}\left[\left(\frac{\theta_\infty^b - \theta_s^b}{k_B T}\right)\sinh u_b - \cosh u_b + \cosh\left(u_b + \frac{\theta_s^b - \theta_\infty^b}{k_B T}\right)\right]^{1/2}$$
$$(8.99)$$

For a given field F_0, θ_s^b, and $\Delta\theta_s^s$ are uniquely determined by solving simultaneously Eqs. (8.94) and (8.99), as follows. We plot $eV_s \equiv (\theta_s^b - \theta_\infty^b)$ versus $\Delta\theta_s^s$ using Eq. (8.99). [θ_∞^b is known for the given emitter (see Eq. (8.13); V_s is the electrostatic potential at the surface relative to its value deep inside the crystal.] Naturally, as $\Delta\theta_s^s$ increases (the effective Fermi level in the surface band goes down relative to the Fermi level in the bulk bands) there are fewer electrons in the surface band and, therefore, the portion of the applied field compensated by the charge in the space charge region of the semiconductor must go up, which leads to an increase in V_s. This is shown schematically by the solid line in Fig. 8.18. For the same applied field we plot eV_s versus $\Delta\theta_s^s$ using Eq. (8.94). It turns out that for a given value of $\Delta\theta_s^s$ there are two solutions represented by the broken and dotted curves, respectively, in the figure. Above a certain value of $\Delta\theta_s^s$ no solution exists. We note that the solid line, obtained from Eq. (8.99), crosses only one of these two curves [it can be the upper (broken) curve, or the lower (dotted) curve]. The crossing point determines the self-consistent solutions for V_s and $\Delta\theta_s^s$ for the given field. Once $\theta_s^b = \theta_\infty^b + eV_s$ and $\Delta\theta_s^s$ have been determined in the above manner, they can be used in the appropriate formulas for the evaluation or the current density and total energy distribution of the emitted electrons. A numerical calculation of these quantities based on the Handler model of the Ge surface, described by Fig. 8.12 and Eq. (8.79), has been performed by the author (Modinos, 1974). The valence band contribution to the emitted current was evaluated approximately using Eqs. (8.75) and (8.76) with $m_p = 0.3m$. The conduction

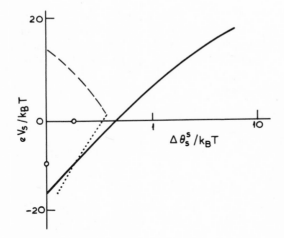

FIGURE 8.18

Solid line: eV_s as a function of $\Delta\theta^s_s$ obtained from Eq. (8.99) for a given applied field. Broken and dotted lines: the upper and lower solutions for eV_s as a function of $\Delta\theta^s_s$ obtained from Eq. (8.94) for the same applied field. The circles denote the self-consistent values of $eV_s = \theta^b_s - \theta^b_\infty$ and $\Delta\theta^s_s = \theta^b_s - \theta^s_s$ used in the calculation of the emitted current.

band remained nondegenerate at the surface for the values of the applied field ($F_0 < 0.4$ V/Å) used in the calculation, and its contribution to the tunneling current, calculated from Eq. (8.72) with $m_n = 0.55m$, was very small (less than 1% of the total current for $F_0 = 0.4$ V/Å and much less than that for smaller fields). We note that Eq. (8.72) overestimates the emitted current from the conduction band when applied to an ideal (unreconstructed) Ge(100) surface. For that surface the contribution of the conduction band to the emitted current is entirely negligible. The parameters which describe the surface band [only the acceptor band contributes to the (excess) charge density in the surface region and to the field emission current in the model under consideration] are as follows: the density of states N_s, the bottom edge of the surface band ϵ^0_s, and its width Γ_s. The values of these and other parameters used in the calculation are shown in Table 8.1. The values for N_s and ϵ^0_s are in agreement with independent estimates of these quantities (see the

TABLE 8.1

Properties of Germanium

Bulk properties of Ge at $T = 300$ K	
$m_n = 0.55$ m	$K = 16$
$m_p = 0.3$ m	$E_g = 0.67$ eV
$n_i = 1.5 \times 10^{13}$ cm^{-3}	$\chi = 4$ eV

Parameters relating to the acceptor surface band on Ge(100)	
$N_s = 10^{14}$ eV^{-1} cm^{-2}	$S = 10^{-17}$ cm^2
$\epsilon^0_s = -E_g - 0.05$ eV	$\overline{Q}^2/\pi = 0.01$ Å^{-2}

discussion at the end of Section 8.5). The exact value of Γ_s, as long as $\Gamma_s > E_g$, does not affect the final results in any significant way, because the effective Fermi level in the surface lies, always, sufficiently below the top energy of this band. $\overline{Q}$, the effective wave vector introduced by Eq. (8.80), and the recombination cross section S were treated as adjustable parameters. Using the parameter values shown in Table 8.1, we calculated the energy distribution of the field-emitted electrons for different values of u_b corresponding to a p-type ($u_b = -4$), intrinsic ($u_b = 0$) and an n-type ($u_b = 4$) specimen, respectively. The results of the calculation for the p-type emitter ($u_b = -4$) are shown in Fig. 8.19a and those for the n-type emitter ($u_b = 4$) in Fig. 8.19b. We note that the zero on the energy axis gives the position of the top of the valence band deep inside the crystal. The position of the Fermi level relative to this zero is indicated by an arrow. The self-consistent values for V_s and $\Delta\theta_s^x$ for the given value of the applied field are shown in the figure captions. We observe that the total energy distributions for the p-type and the n-type specimen are practically identical. [The displacement in the energy axis corresponds to the different position of the Fermi level in relation to the top of the valence band in the two specimens.] This result is in agreement with the experimental observations of Arthur (1964). We note that for the smaller applied fields ($F < 0.3$ V/Å for the n-type specimen; $F < 0.2$ V/Å for the p-type specimen) $V_s < 0$ which means that the top of the valence band (similarly the bottom of the conduction bands) bends upwards at the surface. At higher values of the applied field, $V_s > 0$ and the bands bend downwards at the surface. The top of the valence band *at the surface* lies at $E = -eV_s$, and that of the bottom of the conduction band at the surface at $E = E_g - eV_s$, on the energy axis of Fig. 8.19. We note that the latter quantity lies (for the fields considered) above E_F, i.e., the conduction band (at the surface) is nondegenerate. The calculated contribution to the emitted current density from the conduction band is noted in the figure caption but, being too small, does not show in Fig. 8.19 where the plot is that of the normalized energy distribution (the total energy distribution divided by the emitted current density). The high-energy peak in the energy distributions shown in Fig. 8.19 derives from the surface band, and the low-energy peak (or shoulder) from the valence band. The variations of the energy distribution with the applied field—namely, the widening of the valence band contribution, the broadening to a lesser degree of the high-energy (surface band) peak, and the decrease in the ratio of the high-energy peak amplitude to the low-energy peak amplitude as the field increases—are in very good agreement (at least at a semi-quantitative level) with the results of Shepherd and Peria (1973) shown in Fig. 8.14. The exact shape of the calculated distribution depends of course on the chosen value of $\overline{Q}$ (for details, see Modinos, 1974; Kerkides, 1976). [One could possibly obtain a very good fit between theory and experiment by varing the value of $\overline{Q}$, and/or treating emission from the valence band in the manner of Shepherd and Peria (see Section 8.5), but that in itself would not be more enlightening. In

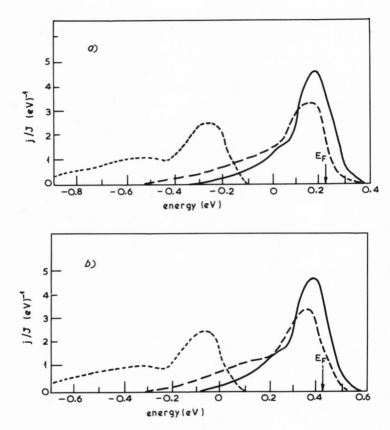

FIGURE 8.19

(a) Theoretical total energy distribution of field-emitted electrons for $u_b = -4$. The values of the parameters are given in Table 8.1. Solid line: $F_0 = 0.2$ V/Å, $eV_s = -6k_BT$, $\Delta\theta_s^s < k_BT$, $J_c = 1.9 \times 10^{-12}$ A/cm^2. $J_s = 1.74 \times 10^{-3}$ A/cm^2, $J_v = 9.98 \times 10^{-4}$ A/cm^2, log $J = -2.56$. Broken line: $F_0 = 0.3$ V/Å, $eV_s = 3.5k_BT$, $\Delta\theta_s^s < k_BT$, $J_c = 1.36 \times 10^{-8}$ A/cm^2, $J_s = 9.14$ A/cm^2, $J_v = 7.04$ A/cm^2, log $J = 1.21$. Dotted line: $F_0 = 0.4$ V/Å, $eV_s = 14k_BT$, $\Delta\theta_s^s = 16.25K_BT$, $J_c = 13.15$ A/cm^2, $J_s = 6.98 \times 10^2$ A/cm^2, $J_v = 6.98 \times 10^2$ A/cm^2, log $J = 3.15$. (b) Theoretical energy distribution of field-emitted electrons for $u_b = 4$. The values of the parameters are given in Table 8.1. Solid line: $F_0 = 0.2$ V/Å, $eV_s = -14k_BT$, $\Delta\theta_s^s < k_BT$, $J_c = 1.9 \times 10^{-12}$ A/cm^2, $J_s = 1.74 \times 10^{-3}$ A/cm^2, $J_v = 9.98 \times 10^{-4}$ A/cm^2. log $J = -2.56$. Broken line: $F_0 = 0.3$ V/Å, $eV_s = -11.5k_BT$, $\Delta\theta_s^s < k_BT$, $J_c = 1.36 \times 10^{-8}$ A/cm^2, $J_s = 9.14$ A/cm^2, $J_v = 7.04$ A/cm^2, log $J = 1.21$. Dotted line: $F_0 = 0.4$ V/Å, $eV_s = 6k_BT$, $\Delta\theta_s^s = 16.25k_BT$, $J_c = 13.5$ A/cm^2, $J_s = 6.98 \times 10^2$ A/cm^2, $J_v = 6.98 \times 10^2$ A/cm^2, log $J = 3.15$.

a proper quantitative calculation emission from the surface band and from the bulk bands must be obtained using the formalism described in Section 8.4.] The theoretical results depicted in Fig. 8.19 differ, or appear to differ, from the findings of Shepherd and Peria (1973) in one respect. For the applied fields used in their experiments the high-energy peak is found somewhere between 0.6 and 0.7 eV below E_F (the Fermi level in the bulk of the semiconductor). The present calculation shows that for the *smaller* applied fields the high-energy peak lies immediately below E_F, in agreement with the earlier experimental results of Arthur (1964).

For larger current densities (corresponding to higher applied fields) our model calculation predicts a shift of the total energy distribution towards lower energies. This shift, which has nothing to do with any internal voltage drop across the bulk of the semiconductor (assumed to be zero in the present calculation), is evident in Fig. 8.19. The nature of this shift and its dependence on the magnitude of S, the recombination cross section appearing in Eqs. (8.90) and (8.91), is shown more clearly in Fig. 8.20. This figure shows the position of the high-energy peak denoted by E_p, relative to E_F as a function of the applied field, for two different values of the cross section S. It is seen from this figure that a smaller value of S can produce a shift of the total energy distribution towards lower energies, comparable to that seen by Shepherd and Peria, for an applied field equal to $F_0 = 0.3$ V/Å or even less (depending on the value of S). It is worth noting that the position of the peak changes very slowly below or above a "threshold" value of the applied field. It is possible that the displacement of the energy distribution observed by Shepherd and Peria (before the onset of the internal voltage drop) is due to the above mechanism. In this respect, the fact that they did not observe any conduction band emission is significant. We remember that according to their

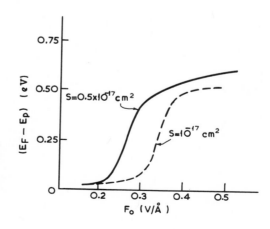

FIGURE 8.20

Position (relative to E_F) of the high-energy (surface band) peak E_p in the energy distribution of the emitted electrons as a function of the applied field for two different values of the cross section S.

analysis, which neglects the dynamics of the replenishment of the surface band, the conduction band is degenerate at the surface. As we have seen (Section 8.5) their explanation of negligible conduction band emission implies an ideal (non-reconstructed) Ge(100) surface, which is most unlikely. The absence of significant emission from the conduction band is perhaps more consistent with our model, which predicts a nondegenerate conduction band for applied fields $F_0 < 0.4$ V/ Å. The values of the (total) emitted current density (J) corresponding to the energy distributions shown in Fig. 8.19 are given in the figure caption. A plot of $\log(J)$ versus $1/F_0$ gives a straight line [for more details see Modinos (1974)], in agreement with the experimental observations.

It is worth remembering that the analysis of the present section applies only as long as the valence and the conduction band are nondegenerate throughout the crystal (including the surface). It has also been assumed that all impurity atoms are ionized. The modification of the formalism necessary to remove the above restrictions is straightforward (Kerkides, 1976). The more general formulas are needed, for example, to calculate the emitted current at higher fields (higher than those shown in Fig. 8.19) when the conduction band becomes degenerate at the surface.

Finally, a brief comment on the internal (supply) current. At low fields (small current densities) when $eV_s \equiv \theta_s^b - \theta_\infty^b$ is negative (see Fig. 8.19) the surface band is practically in thermal equilibrium with the bulk banks ($\Delta\theta_s^s < k_B T$) and the current *into* the surface band comes mostly from the valence band. At higher fields (dotted lines in Fig. 8.19) $eV_s > 0$, and the supply current to the surface band comes from the conduction band. For an n-type emitter there are enough electrons in the conduction band to sustain this current, although this may lead to a small ohmic voltage drop along the emitter. We note, however, that the total emitted current density corresponding to the dotted curves in Fig. 8.19 comes equally from the surface band *and* the valence band even for the n-type specimen. We can think of the electrons emitted from the valence band as holes injected into the valence band at the surface. The hole current towards the interior of the semiconductor contributes as well to the ohmic voltage drop. An ohmic voltage drop implies that the Fermi level inside the semiconductor is replaced by an effective (or pseudo) Fermi level which depends on the position. As long as the internal voltage drop is small (a few volts), and assuming that the effective Fermi level for electrons is the same with that of holes, its only effect is to shift the energy distribution along the retarding potential axis. It is possible, however, for the pseudo-Fermi level in the valence band to lie below that of the conduction band at the surface (and possibly lower than the pseudo-Fermi level in the surface band) when the valence band contribution to the emitted current from an n-type emitter exceeds a certain value. In that case the valence band contribution to the emitted current will be smaller than the one predicted by the present theory (dotted line in Fig. 8.19b).

In the case of a p-type specimen, emission at high fields is further complicated by the onset of saturation phenomena (see next section).

Apart from the work of Shepherd and Peria described in Section 8.5 and that of Arthur (1964) which we have already mentioned, total energy distribution measurements of field-emitted electrons from semiconductor surfaces have been reported (as far as we know) by the following authors. Hughes and White (1969) measured the energy distribution from a GaAs surface. Shcherbakov and Sokolskaya (1963) and Salmon and Braun (1973) measured the energy distribution from a CdS surface. Sykes and Braun (1975) measured the energy distribution from a PbTe surface. Lewis and Fischer (1974) measured the energy distribution from (110), (100), and (111) oriented Si emitters. More recently Rihon (1978a; 1981) measured the energy distribution from the (0001) plane of ZnO. Figure 8.21 shows the energy distribution from this plane at various values of the applied field. These TED curves are typical of wide gap semiconductors (E_g =

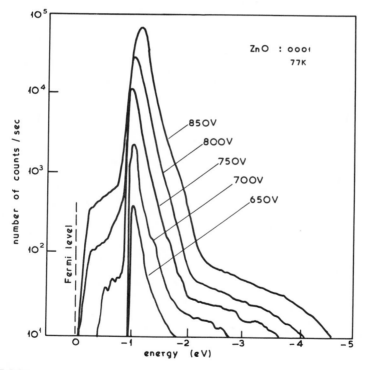

FIGURE 8.21

Energy distribution of field-emitted electrons from the (0001) plane of ZnO, at various values of the applied voltage. (From Rihon, 1978a.)

3.2 eV for ZnO) in the sense that the base width of the distribution ($\sim$3.5 eV or so) is much larger than those of metallic emitters. The peak of the distribution in Fig. 8.2 lies below the Fermi level, which suggests that electrons from surface states contribute most of the tunneling current. Another observation worth noting is that the peak of the distribution does not shift significantly along the energy axis when the total current varies by nearly two orders of magnitude. This suggests that the internal voltage drop is negligible and that band bending, if present, is also very small. It was also found that a Fowler–Nordheim plot of $\ln I$ versus V^{-1} (I is the total current and V the applied voltage as usual) gives a straight line.

8.7. INTERNAL VOLTAGE DROP AND SATURATION EFFECTS

In our theoretical analysis so far we have assumed that the voltage drop IR (I is the total emitted current and R the resistance of the tip) equals zero. We have seen that field penetration leads, in the zero (internal) current approximation, to band bending (shown schematically in Fig. 8.1). It is usually assumed that the only effect of the ohmic voltage drop on the energy level diagram is to bend it (the valence band and the conduction band alike) by an amount equal to the internal voltage drop $\Delta V_i(z)$. The position of the effective Fermi level relative to the bottom of the conduction is the same (in this approximation) as in the zero current limit. The situation is shown schematically in Fig. 8.22. The broken line in this figure shows the effective Fermi level. We assume that any external voltage is applied to the metal support (e.g., the anode voltage is applied between the metal support and the anode), in which case the (total) internal voltage V_i is the difference between the Fermi level in the metal support and the effective Fermi level at the emitting surface. It is made up from the IR voltage drop in the bulk of the

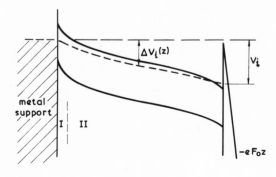

FIGURE 8.22

Internal voltage drop in a semiconductor (n-type) field-emitter. The position of the effective Fermi level (broken line) relative to the bottom of the conduction band is the same (in this approximation) as in the zero-internal-current approximation.

semiconductor (region II in Fig. 8.22) and the voltage drop across the metal–semiconductor junction (region I in the figure). When the metal–semiconductor junction is made into an ohmic contact the contribution to V_i from region I is negligible. The concept of an effective (or pseudo) Fermi level implies that the formulas in the preceding sections which involve the Fermi–Dirac distribution function remain valid, but E_F is now replaced by an effective Fermi level (the same for electrons and holes) which varies *slowly* with position. We denote this quantity by $\zeta(z)$. Our assumption that the internal voltage drop does not change the position of the effective Fermi level *relative* to the bottom of the conduction band means that the occupation of the one-electron states (bulk or surface states) and hence the charge distribution in the space charge region is not affected by the internal voltage drop. It is also evident that under these conditions the energy distribution of the emitted electrons will be shifted downwards by an amount equal to V_i relative to the Fermi level in the metal support, but its shape will not change in any significant way. We have already seen an example of this shift in Fig. 8.17. It turns out that when the effect of the internal voltage drop can be treated in the above manner, V_i is very small (a few volts) compared to the anode voltage (a few thousands volts) and for that reason the field F_0 at the emitting surface for a given V is not affected by the internal voltage drop. It is also obvious that, under the same conditions, the internal voltage drop does not have any significant effect on the current (anode)–voltage relationship.

An extensive experimental study of the (emitted current)–voltage relationship for Ge crystals of different dopings has been reported by Arthur (1965). The field emission tips used in his experiments were prepared from germanium single crystals of 0.001 Ω cm n-, 3 Ω cm n-, 30 Ω cm n-, 46 Ω cm p-, 40 Ω cm p-, 12 Ω cm p-, 0.06 Ω cm p-, and 0.0006 Ω cm p-type. The emission from all n-type crystals resulted in FN plots of log I vs $1/V$ [I is the emitted current from the total emitting area (not a single plane) of the tip and V is the voltage applied between the base of the emitter and the anode] which are always linear. The internal voltage V_i (obtained from the shift in the energy distribution on the retarding potential axis) was small (5 V or less); the emitter resistance was always smaller than 10^7 Ω, in reasonable agreement with the value calculated from the tip geometry and the bulk resistivity. It is evident from these results that emission from n-type Ge crystals conforms with the theoretical ideas described at the beginning of this section. In contrast to these results, emission from the p-type crystals (with the exception of 0.0006 Ω cm samples) produced nonlinear FN plots which were strongly influenced by temperature and illumination. These effects, which were more pronounced with the crystals of higher resistivity, are illustrated in Figs. 8.23 and 8.24. The "light" curve in Fig. 8.23 is typical of the results obtained from p-type crystals. At low fields (region I) log $I = A - B/V$, in accordance with the Fowler–Nordheim law. However, above some particular field, which depends on temperature and illumination, the current becomes almost indepen-

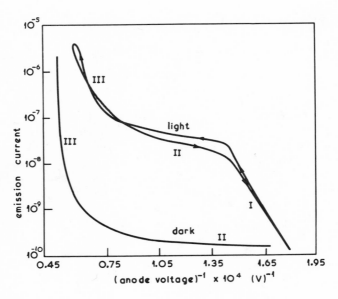

FIGURE 8.23

Field-emitted current as a function of anode voltage for a 40-Ω cm p-type Ge tip at 77 K in the dark and with room lights. (From Arthur, 1965.)

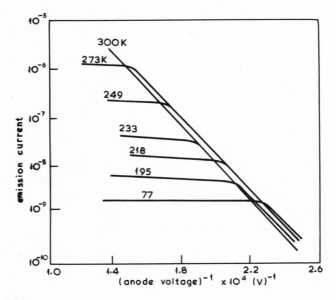

FIGURE 8.24

Field-emitted current as a function of anode voltage for a 40-Ω cm p-type Ge tip at various temperatures in the dark. (From Arthur, 1965.)

dent of the applied voltage (region II is referred to as the saturation region). Finally at higher fields (region III) the current rises sharply once again. The value of the current in the saturation region depends on light intensity (this is obvious from Fig. 8.23) and on the temperature. The latter dependence is demonstrated in Fig. 8.24. Figures 8.23 and 8.24 demonstrate another general result, namely, that increasing the temperature or the light intensity leads to a higher value of the saturation current but never above the linear extension of the low-field data (the latter being practically independent of temperature and light intensity). Another interesting result is demonstrated in Fig. 8.25. This figure shows the internal voltage drop V_i, obtained from the shift of the energy distribution on the retarding (collector) potential axis, as a function of the emitted current. Perhaps the results obtained under darkness are the more significant. We see that in the low-current

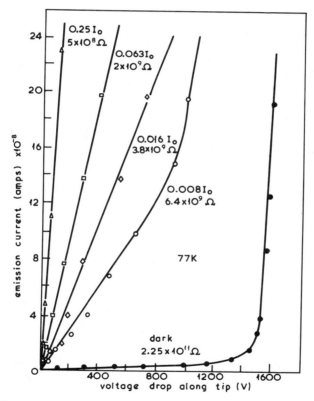

FIGURE 8.25

Field-emitted current versus tip voltage drop for a 40-Ω cm p-type Ge tip at 77 K and various light intensities. (From Arthur, 1965.)

region of this curve (V_i < 1300 V or so), corresponding to the saturation region of the current versus anode voltage curve (V_i in the low-field region of the latter curve is negligible on the scale of Fig. 8.25), V_i varies linearly with the current and that the resistance of the tip equals approximately 2×10^{11} Ω. This value is about 10^5 times greater than the resistance of the given tip estimated from the known bulk resistivity and the geometry of the tip (Arthur, 1965), which suggests that the current in the saturation region is essentially determined by a region of the tip, not far from the emitting surface, which has been depleted of its mobile carriers in the manner indicated schematically in Fig. 8.26. The energy level diagram shown in this figure is appropriate to a p-type emitter when the conduction band is degenerate at the surface. [It is reasonable to assume that, in general, the conduction band will be degenerate above a certain value of the applied field.] The shaded region in Fig. 8.26 constitutes a depletion region between the p-type interior and the n-type surface. In the depletion region the effective Fermi level, shown by the broken line, lies in the middle of the energy gap leading to a minimum concentration of electrons and holes in this region. The situation is the same as that in a $p–n$ junction when it is reverse-biased (electrons flowing from the p side to the n side). In both cases the saturation current, due to thermal generation of electrons on the p side, is practically independent of the voltage across the depletion layer, and strongly dependent on the temperature. (The increase of the saturation current with illumination has also its exact analog in $p–n$ junctions.) The sudden increase in the current at the higher applied fields (see Fig. 8.23) is due to carrier multiplication produced by avalanching in the depletion region (a well-known phenomenon in $p–n$ junctions). The observation by Arthur, that the emitted current in region III (Fig. 8.23) has a very broad energy distribution, indicating heating of the electrons by several eV, is evidence of avalanche breakdown. A direct comparison of the values of the saturation current and breakdown voltage in the field emission experiments and in ordinary $p–n$ junctions is not possible, because the exact shape of the potential and charge distributions is different in the two cases (a theory whcih takes into account the conical shape of the field emission tip is not presently available). Also, leakage along the shank of the emitter and

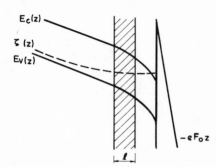

FIGURE 8.26

Schematic description of band bending and of the effective Fermi level in field emission from a p-type semiconductor. The shaded region is depleted of conduction electrons and holes.

electron generation along this surface may be partly responsible, according to Arthur, for the observed large saturation currents (two or three orders of magnitudes larger than in $p–n$ junctions).

Although the voltage drop along the tip is large, of the order of 10^3 (see Fig. 8.25), it is still a small fraction (about a tenth) of the applied (anode) voltage, and cannot by itself explain the nonlinearity of the FN plots shown in Figs. 8.23 and 8.24. In other words, the experimental curves *cannot* be described by a formula

$$I = A \exp[-B/(V - V_i)] \qquad (8.100)$$

where A and B are constants. We should point out at this stage that results similar to those obtained by Arthur from p-type germanium have been reported by a number of other workers for a variety of p-type semiconductors and also for high-resistivity n-type semiconductors with a large energy gap. A list of relevant references can be found in the article of Baskin, Lvov, and Fursey (1971). It appears that in most cases, if not all, a fit of the data by a modified Fowler–Nordheim formula [Eq. (8.100)] is *not* possible. This is demonstrated for p-Si (3 Ω cm) in Fig. 8.27 (Baskin *el al.*, 1971). The experimental results are described by line 1. Line 2 is the experimental curve with the correction due to the internal voltage drop taken into account. Line 4 is the linear extension of the low-field data which obey the Fowler–Nordheim law. The constancy of B in Eq. (8.100) implies that the field factor $\beta = F_0/(V - V_i)$ (F_0 denotes as usual the field at the emitting surface) is independent of the field (and therefore of the voltage). This may not be exactly so. However, by measuring the relative change in the dimensions of the emission pattern one can estimate the correction to the theoretical Fowler–Nordheim plot arising from the variation of β with the applied voltage (Fursey and Egorov, 1969). It turns out that the corrected Fowler–Nordheim curve (line 3 in Fig. 8.27) does not differ significantly from the uncorrected Fowler–Nordheim plot (line 4 in the same figure). It is obvious that for p-type semiconductors our assumptions at the beginning of this section, as to the effect of the internal current on the charge distribution inside the semiconductor and on the emitted current, are not valid beyond the low-field region (region I in Figs. 8.23 and 8.27). This means that we cannot use Eq. (8.14) (the same applies to any other equation based on statistical equilibrium) to relate θ_s^b, the effective Fermi level at the surface relative to the bottom of the conduction band at the surface, to the total charge (per unit area) in the space charge region of the semiconductor or, what amounts to the same thing, the field F_s as defined by Eq. (8.4b).

An approximate theory of saturation effects in field emission from semiconductor surfaces has been developed by Baskin *et al.* (1971) for a planar emitting surface. Their calculation is based on the following assumptions. The semiconductor extends from $z = -\infty$ to the emitting surface at $z = 0$. There are no surface states, and only the conduction band contributes significantly to the emitted

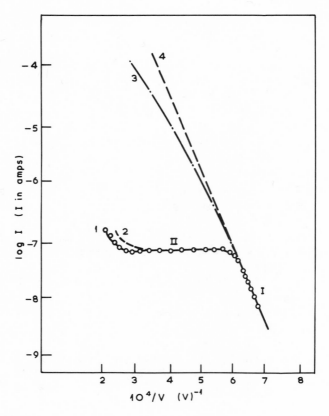

FIGURE 8.27

Emitted current as a function of applied voltage for p-Si (3 Ω cm) in the dark. The experimental results are described by line 1. Line 2 is the experimental curve with the correction due to internal voltage drop taken into account. Line 4 is the linear extension of the low-field data which obey a Fowler–Nordheim relation. Line 3 is a Fowler–Nordheim plot with a correction arising from the variation of β with V. (From Baskin *et al.*, 1971.)

current density. The conduction band contribution to the current density is approximated by Eq. (8.72) when the conduction band is nondegenerate at the surface ($\theta_s^b < 0$) and by Eq. (8.73b) when it is degenerate at the surface ($\theta_s^b > 0$). The effective Fermi level is denoted by $\zeta(z)$, and the difference between it and the bottom of the conduction band $E_c(z)$ is denoted by

$$\theta(z) = \zeta(z) - E_c(z) \tag{8.101}$$

We note that θ_s^b, which appears in Eqs. (8.72) and (8.73b), is given by

$$\theta_s^b = \theta(0) \tag{8.102}$$

In order to evaluate θ_s^b, one needs to know $E_c(z)$ and $\zeta(z)$. These are calculated from the simultaneous solution of

$$\frac{dF}{dz} = \frac{4\pi e}{K}(n - p + N^- - N^+) \equiv -\frac{4\pi}{K}\bar{\nu}(z) \tag{8.103a}$$

$$-eF \equiv dE_c/dz \tag{8.103b}$$

which follows from Poisson's equation, and the equation for the current density given by†

$$-(\mu_n n + \mu_p p)\frac{d\zeta}{dz} = J \tag{8.104}$$

where μ_n and μ_p are the mobilities of electrons and holes, respectively. Under steady state conditions the current density J equals the emitted current density. We note that n, p, N^+, and N^- are given by Eqs. (8.9) and (8.7) with E_F replaced by $\zeta(z)$. (It is assumed that the effective Fermi level is the same for electrons and

---

†The electron contribution to the current density consists of a drift term and a diffusion term. We have

$$J_n = -\mu_n n \frac{\partial E_c}{\partial z} - eD_n \frac{\partial n}{\partial z} \tag{i}$$

when D_n is the diffusion constant for electrons. From Eq. (8.9a) [with E_F replaced by $\zeta(z)$] we obtain

$$\frac{\partial n}{\partial z} = \frac{\partial n}{\partial E_c}\frac{dE_c}{dz} - \frac{\partial n}{\partial E_c}\frac{d\zeta}{dz} \tag{ii}$$

Therefore,

$$J_n = -\mu_n n\frac{dE_c}{dz} - eD_n\frac{\partial n}{\partial E_c}\frac{dE_c}{dz} + eD_n\frac{\partial n}{\partial E_c}\frac{d\zeta}{dz} \tag{iii}$$

At equilibrium $J_n = 0$ and $d\zeta/dz = 0$, which implies that

$$eD_n \partial n/\partial E_c = -\mu_n n \tag{iv}$$

Substituting Eq. (iv) into Eq. (iii) we obtain

$$J_n = -\mu_n n \, d\zeta/dz \tag{v}$$

The hole contribution to the current density is obtained in a similar manner. The sum of the two contributions gives Eq. (8.104).

holes.) The simultaneous Eqs. (8.103) and (8.104) must be solved with the following boundary conditions:

$$\left. \frac{dE_c}{dz} \right|_{z \to -\infty} = \left. \frac{d\zeta}{dz} \right|_{z \to -\infty} = -\frac{J}{\mu_n n_\infty + \mu_p p_\infty} \tag{8.105a}$$

$$\left. \frac{dE_c}{dz} \right|_{z=0} = -\frac{eF_0}{K} \tag{8.105b}$$

where n_∞ and p_∞ are the equilibrium values of the electron and hole concentrations, respectively, deep inside the metal. We note that because there are no surface states in the model under consideration $F_s = F_0$, so that Eq. (8.105b) is the same boundary condition as Eq. (8.4b) of the zero-current approximation. In contrast Eq. (105a) is different from the corresponding boundary condition [Eq. (8.4a)] of the zero-current approximation. The solution of Eqs. (8.103) and (8.104) with the boundary conditions given by Eqs. (8.105) can be reduced to that of the following nonlinear differential equation of the first order (Baskin *et al.*, 1971):

$$\frac{d(eF)}{dy} = -\frac{4\pi e}{K} \frac{k_B T \bar{\nu}(y)}{[eF(y) - J/\mu_n b(y)]} \tag{8.106}$$

with the boundary condition

$$e\mu_n b(y_\infty) F(y_\infty) = J \tag{8.107}$$

where

$$y \equiv \theta(z)/k_B T \tag{8.108a}$$
$$b(y) \equiv n + (\mu_p/\mu_n)p \tag{8.108b}$$

and y_∞ is determined by the condition

$$\bar{\nu}(y_\infty) = 0 \tag{8.108c}$$

A numerical integration of Eq. (8.106), with J treated as a parameter, is straightforward. One determines y_∞ (a negative number) from Eq. (8.108c) and then $F(y_\infty)$ from Eq. (8.107). $F(y)$ for $y > y_\infty$ is obtained in a step by step integration of Eq. (8.106). In that way one obtains a family of curves

$$F = F(y; J) \tag{8.109}$$

corresponding to different values of the current density J. Using Eqs. (8.103b) and (8.105b), we obtain

$$F_0 = F(y_0;J)/K \qquad (8.110)$$

where F_0 is the field on the vacuum side of the emitting surface, and $y_0 = \theta_s^b/k_BT$ [see Eqs. (8.102) and (8.108a)] is the value of y at this surface. On the other hand, we have $J = J_c(\theta_s^b, F_0)$ [given by Eqs. (8.72) and (8.73b)] and, therefore, for a given value of J, F_0 is a function of θ_s^b. We denote this function by

$$F_0 = F_0(\theta_s^b;J) \qquad (8.111)$$

The values of θ_s^b and F_0 corresponding to the chosen value of J are determined graphically from the equation $F_0(\theta_s^b;J) = F(\theta_s^b/k_BT;J)$. The current–field rela-

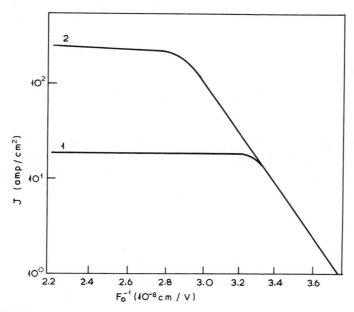

FIGURE 8.28

Emitted current density as a function of applied field for p-Ge with $N_a = 10^{15}$ cm^{-3} at $T = 300$ K. The following relationships between mobility and strength of the internal field F were assumed.
(1) $\mu_n(F) = \mu_n(0)$, $F < 10^2$ V/cm
  $= \mu_n(0)(10^2/F)^{1/2}$, $10^2 < F < 10^4$ V/cm
  $= \mu_n(0)(10^3/F)$, $F \geq 10^4$ V/cm
(2) $\mu_n(F) = \mu_n(0)$, $F < 10^2$ V/cm
  $= \mu_n(0)(10^2/F)^{1/2}$, $F > 10^2$ V/cm
(From Baskin *et al.*, 1971.)

tionship calculated in this manner for p-type Ge is shown in Fig. 8.28 (Baskin *et al.*, 1971). The two curves in this figure correspond to different expressions for the mobility [these are shown in the figure caption; for a justification of these expressions see, e.g., Shockley (1950)]. Baskin *et al.* point out that the experimental values of the saturation current should be larger than the calculated values because of the conical geometry of the real emitter which could lead, according to the above authors, to saturation currents one or even two orders of magnitude greater than those shown in Fig. 8.28. In order to have a clear picture of the depletion region, which is responsible for the saturation in the current density, one must plot the mobile current density b [defined by Eq. (8.108b)] as a function of z. For that purpose one requires, in the first instance, an explicit expression for $y(z)$. Using Eqs. (8.101)–(8.103) and Eqs. (8.108) in Eq. (8.104) one obtains

$$z = k_B T \int_{y0}^{y} \frac{\mu_n b(y) \, dy}{e\mu_n b(y) F(y) - J} \qquad (8.112)$$

Knowing $b(y)$ and having calculated $F(y)$ one can obtain $y(z)$ from the above formula. Baskin *et al.* found that the depletion region (there b is very small) has sharp edges when plotted as a function of z. Figure 8.29 shows their results for the length l of the depletion region (defined as the distance between the points at which the concentration equals two times its minimum value) as a function of the

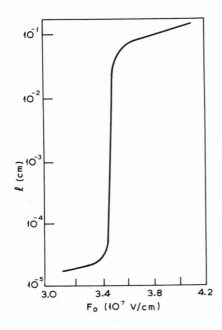

FIGURE 8.29

Extension of the depletion region as a function of applied field for p-Ge with $N_a = 10^{15}$ cm^{-3} at $T = 300$ K. (From Baskin *et al.*, 1971.)

field F_0. They also note that the increase of l with F_0, at constant current, occurs only at the expense of a deeper penetration of the field into the semiconductor. These results for l are in reasonable agreement with measurements of this quantity by Arthur (1965) based on a light probe technique. We note, however, that the calculated penetration of the electric field into the semiconductor (of the order of a few millimeters when current saturation occurs) is large compared to the tip radius, which implies that the one-dimensional theory is not good enough for quantitative estimates when current saturation takes place. Obviously, this theory overestimates both the length of the depletion region and that of overall field penetration. Another result of Baskin et al. is also worth mentioning. They find, as one would expect, that the concentration of conduction band electrons at the surface at first increases slowly as the field F_0 increases, then remains constant over a wide region of F_0 and finally, when saturation sets in, it decreases sharply as the field increases (and the emitted current remains practically constant).

Results similar to those shown in Fig. 8.28 were obtained for n-type silicon. In this case, however, and in contrast to the situation in p-type semiconductors, the value of the saturation current density depends strongly on the concentration of impurities (Baskin et al., 1971).

By construction, the theory of Baskin et al. does not take into account the possibility of avalanche breakdown and, therefore, it tells us nothing about the breakdown region of the current–voltage relationship (region III of the curves in Fig. 8.23).

We have already noted that as a result of the large field penetration (when saturation sets in), the one-dimensional approximation to the potential distribution cannot be used in a quantitative calculation of the current–voltage relationship of a real field-emitter. [The one-dimensional approximation is, however, reasonably good in the low-field, Fowler–Nordheim region (region I of the curves in Figs. 8.22 and 8.25), because in that case most of the variation of the potential occurs within a distance from the emitting surface which is approximately ten times smaller than the tip radius of a typical emitter.] There are, of course, additional approximations: The role of surface states has been neglected. [Their effect on the saturation current is probably small.] It has also been assumed that the effective Fermi level is the same for electrons and holes. This might not be the case in practice. It is also worth remembering that the use of a Fermi–Dirac distribution function [with E_F replaced by $\zeta(z)$] to describe the occupation of the electron states is in itself an approximation. The significance of these approximations was pointed out by Baskin, Lvov, and Fursey (1977). It is clear from their analysis that, while the simple one-dimensional theory (described in the present chapter) is qualitatively correct and good enough, perhaps, for an order-of-magnitude calculation of the observed quantities, a more elaborate theory, which takes into account the shape of the emitter and at least some of the other effects neglected by the simple theory, is required for a proper quantitative analysis of saturation phenomena.

THERMIONIC EMISSION SPECTROSCOPY OF METALS

9.1. BASIC FORMULAS

We know (see Chapter 1) that at sufficiently high temperatures, a small fraction of the electrons in a metal have energies higher than the maximum of the potential barrier at the surface and some of them escape from the metal giving rise to a thermionic current. The potential barrier is shown (schematically) in Fig. 9.1. On the vacuum side of the interface at distances larger than 4 Å or so from the top layer of atoms the electron moves in a potential field described by (see Section 1.3)

$$V(z) \simeq \frac{-e^2}{4z} - eFz + E_F + \phi \qquad (9.1)$$

The maximum of the barrier occurs at z_m, given by Eq. (1.28), and has the value V_{max} given by Eq. (1.29). In a typical thermionic emission experiment the value of the applied electric field is

$$6.5 \times 10^{-5} < F < 6.5 \times 10^{-3} \, \text{V/Å} \qquad (9.2)$$

Therefore, z_m and $\Delta\phi$ [the quantity defined by Eq. (1.76)] are given by

$$240 > z_m > 25 \, \text{Å} \qquad (9.3)$$
$$0.03 < \Delta\phi < 0.3 \, \text{eV} \qquad (9.4)$$

Note that the potential barrier described by Eq. (9.1) decreases smoothly to the right of the barrier maximum (region IV in Fig. 9.1), and therefore the probability that an escaping electron will be reflected backward towards the metal, after leaving the barrier maximum region (region III), is negligible. This means that the detailed shape of the potential in region IV, as long as it is a slowly varying, not reflecting potential, does not affect in any way the emitted current. We assume,

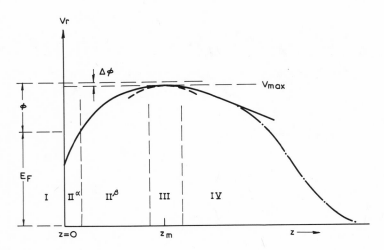

FIGURE 9.1

Surface potential barrier in thermionic emission (schematic). The solid curve represents the actual barrier. The broken curve represents the parabolic approximation to the potential in region III. The dash–dot curve represents a nonreflecting potential in region IV.

without loss of generality, that the potential barrier reduces *slowly* to zero† at some distance from the barrier maximum, in the manner indicated schematically by the dash–dot curve in Fig. 9.1.

A most convenient way of calculating the thermionic current has been described by Herring and Nichols (1949). We imagine that the metal is in thermal equilibrium with an ideal noninteracting electron gas which occupies the half space from the barrier maximum to $z = +\infty$. Using the fact that the electrons on the right of the barrier are free particles and the principle of detailed balancing one can show (see Appendix A) that $j(E)\ dE$ [the number of electrons which, under equilibrium conditions, flow out of the metal per unit area per unit time with (total) energy between E and $E + dE$] is given by

$$j(E) = \frac{2f(E)}{(2\pi)^3 \hbar} \iint\limits_{0 < k_{\parallel} < (2mE/\hbar^2)^{1/2}} [1 - r(E, \mathbf{k}_{\parallel})]\ d^2 k_{\parallel} \qquad (9.5)$$

where $f(E)$ is the Fermi–Dirac distribution function [Eq. (1.5)], and $r(E, \mathbf{k}_{\parallel})$ is the total (elastic plus inelastic) reflection coefficient of a (free) electron incident on

---

†The real part of the constant potential between the muffin-tin spheres inside the metal is taken as the zero of energy.

the metal from the right with energy E and wave vector (parallel to the surface) $\mathbf{k}_{\parallel}$. We note that in deriving Eq. (9.5) we made the assumption that $r(E,\mathbf{k}_{\parallel})$ does not depend on the spin of the electron, but we did take into account spin degeneracy. If we assume that under conditions appropriate to a thermionic emission experiment the energy and momentum distribution of the electrons inside the metal are not significantly affected by the thermally emitted current we can identify $j(E)$, given by Eq. (9.5), with the energy distribution of the thermally emitted electrons. The above assumption is easily satisfied in practice in view of the smallness of the emitted current compared to the very large number of highly mobile electrons available on the metal side of the metal–vacuum interface.

It is obvious from Fig. 9.1 that only electrons with energy near or above the barrier maximum contribute to the thermionic current. For these energies

$$
\begin{aligned}
f(E) &\simeq \exp\left(\frac{E_F - E}{k_B T}\right) \\
&\simeq \exp\left(\frac{E_F - V_{\max}}{k_B T}\right) \exp\left(-\frac{\mathbf{E}}{k_B T}\right)
\end{aligned}
\tag{9.6}
$$

where $\mathbf{E}$ is defined by

$$
\mathbf{E} \equiv E - V_{\max}
\tag{9.7}
$$

Using Eqs. (1.76) and (9.6) we obtain

$$
j(\mathbf{E}) = \frac{1}{4\pi^3 \hbar} \exp\left[-\frac{\phi - (e^3 F)^{1/2}}{k_B T}\right] \exp\left(-\frac{\mathbf{E}}{k_B T}\right) \iint [1 - r(\mathbf{E},\mathbf{k}_{\parallel})]\, d^2 k_{\parallel}
\tag{9.8}
$$

The limits of the $\mathbf{k}_{\parallel}$ integration need not be specified accurately because the integrand vanishes except within a very limited region around $\mathbf{k}_{\parallel} = 0$.

The thermionic (charge) current density is given by

$$
J = e \int_{-\infty}^{+\infty} j(\mathbf{E})\, d\mathbf{E}
\tag{9.9}
$$

Substituting Eq. (9.8) into Eq. (9.9) we obtain

$$
J = A_R \bar{t} T^2 \exp\left[-\frac{\phi - (e^3 F)^{1/2}}{k_B T}\right]
\tag{9.10}
$$

where A_R is the universal constant defined by Eq. (1.75) and $\bar{t}$ is an average transmission coefficient defined by

$$\bar{t}(T,F) = \frac{\hbar^2}{2\pi m} (k_B T)^{-2} \int_{-\infty}^{+\infty} \left\{ \exp\left(-\frac{E}{k_B T}\right) \iint [1 - r(E,k_\parallel)] \, d^2 k_\parallel \right\} dE$$

(9.11)

We note that the integrand in the above equation decreases exponentially in both energy directions, for $E > 0$ because of the exponential term, and for $E < 0$ because of the transmission coefficient $(1 - r)$. The dependence on $\bar{t}$ on the applied field comes through r.

It is worth noting that the original Richardson–Schottky formulas of thermionic emission (see Chapter 1) can be obtained from the above formulas if we assume that

$$r = 0 \qquad \text{for } w > 0$$
$$= 1 \qquad \text{for } w < 0$$

(9.12)

where w is defined by

$$w \equiv E - \hbar^2 k_\parallel^2/2m$$

(9.13)

Equation (9.12) states that an electron with normal energy above the barrier maximum is not reflected. This is approximately true for free-electron metals but not necessarily so for real metal surfaces (see Section 9.7). Equation (9.12) also states that an electron with normal energy below the barrier maximum is entirely reflected. This is not a bad approximation in the limit of very weak applied fields, but it is not a good approximation for the higher applied fields within the thermionic region of F and T [as defined by Eqs. (1.72) and (1.73)]. If we use Eqs. (9.12) and (9.13) in Eq. (9.8) we obtain the following formula for the energy distribution of the thermally emitted electrons:

$$j_0(E) = \frac{m}{2\pi^2 \hbar^3} \exp\left[-\frac{\phi - (e^2 F)^{1/2}}{k_B T}\right] E \exp\left(-\frac{E}{k_B T}\right) \qquad \text{for } E > 0$$
$$= 0 \qquad \text{for } E < 0$$

(9.14)

Replacing $j(E)$ in Eq. (9.9) by the above expression leads one to the original Richardson–Schottky formula for the thermionic current density [Eq. (1.74)]. In order to go beyond the Richardson–Schottky theory it is necessary to calculate in a systematic manner the reflection coefficient $r(E,k_\parallel)$ for realistic models of the metal surface.

In Fig. 9.2 we show the measured, by McRae and Caldwell (1976), total,

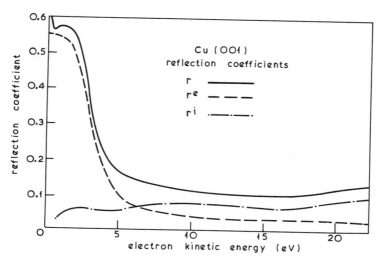

FIGURE 9.2

Measured total, elastic, and inelastic coefficients of reflection of electrons incident normally on Cu(100) at room temperature. (From McRae and Caldwell, 1976.)

elastic, and inelastic coefficients of reflection of low-energy electrons at a copper (100) surface at room temperature. The electrons were incident normally on the surface. The minimum kinetic energy (this equals the total energy of the electron measured from the vacuum level) for which reliable results could be obtained was 0.5 eV, and at this energy the elastic reflection coefficient was $r^e = 0.55$. It is seen from this figure that the inelastic reflection coefficient r^i is very small indeed for kinetic energies smaller than 1 eV, so that for thermal electrons we can put

$$r(\mathbf{E}, \mathbf{k}_\parallel) \simeq r^e(\mathbf{E}, \mathbf{k}_\parallel) \tag{9.15}$$

Incident electrons do, of course, collide (inelastically) with the vibrating ions of the lattice and with other electrons in the metal. It turns out that the effect of the lattice vibrations on r^e is relatively small. We shall see how this can be taken approximately into account in Section 9.4. On the other hand electron–electron collisions have a very significant effect on a beam of thermal electrons incident on the metal with energy in a narrow region, a few $k_B T$, around the barrier maximum. Since $V_{max} - E_F \simeq 4$ eV or so in a typical experiment it is reasonable to assume that the energy (after the collision) of the vast majority of inelastically scattered primary (incident) electrons lies below the barrier maximum at z_m (see Fig. 9.1) and, therefore, these electrons cannot escape from the metal. The same applies to the (secondary) electrons which are scattered from states below E_F to states above E_F, as a result of the above collisions. Hence the validity of Eq. (9.15). We can take into account the effect of electron–electron collisions on r^e by adding an imaginary component to the one-electron potential as described in Section 4.5. Needless to say that an evaluation of this quantity from first principles is extremely difficult (see, e.g., Inkson, 1971) and that in practice it is described by semiempirical formulas.

We assume that the real part of the one-electron potential is described by Eqs. (4.1) and (5.1). Although at the high temperatures used in thermionic emission experiments the lattice constant is larger (usually by 3% or so) than at room temperature, we may assume that the radius of the muffin-tin spheres and the spherical potential within each sphere are approximately the same with the corresponding quantities at zero temperature, which are known, from self-consistent calculations of the energy band structure, for many metals (Moruzzi *et al.*, 1978). We also assume that the real part of the potential for $z > 0$, given by

$$V_r(z) = V_s(z) - eFz \qquad (9.16)$$

is practically identical with that given by Eq. (9.1) for z in the region of the barrier maximum, so that Eqs. (1.28) and (1.29), which give the position and value of this maximum, remain valid. The imaginary component of the potential vanishes away ($z > 3$ Å) from the surface. We assume that the following semiempirical formula, which has been used in many calculations of very low energy electron diffraction at metal surfaces (see, e.g., McRae and Caldwell, 1976; Jennings, 1978; 1979; Schäfer, Schoppe, Holzl, and Feder, 1981), describes reasonably well the imaginary part of the potential

$$V_{im}(E,\mathbf{r}) = V_{im}(E) \qquad \text{for } z < 0$$
$$= V_{im}(E) \exp[-(z/\beta)^2] \qquad \text{for } z > 0 \qquad (9.17)$$
$$V_{im}(E) = -\alpha(|E - E_F|)^\gamma \qquad \alpha > 0 \qquad (9.18)$$

We emphasize that α and γ are bulk parameters and thus independent of the surface under consideration. They can be determined from optical data for the given metal (McRae, 1976). In practice, however, extracting the values of α and γ from such data is not free of ambiguities. The constant β in Eq. (9.17) determines the effective range of the imaginary component of the potential on the vacuum side of the interface and is, for practical purposes, an adjustable parameter. In Fig. 9.3 we show a one-dimensional (schematic) representation of the complex potential at the metal–vacuum interface when the real part of the surface barrier is approximated by Eq. (4.2). In this case z_0 and the distance $[d/2 + D_s]$ of the metal–vacuum interface (the plane $z = 0$) above the center of the top layer are additional adjustable parameters.

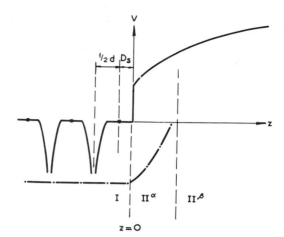

FIGURE 9.3

A one-dimensional (schematic) representation of the complex potential at a metal–vacuum interface. The wells represent atomic layers of muffin-tin atoms. The solid curve represents the real part and the dash–dot curve the imaginary part of the potential.

9.2. THE HERRING–NICHOLS FORMULA FOR THE ELASTIC REFLECTION COEFFICIENT

We separate space into the five different regions marked I, II^α, II^β, III, and IV in Fig. 9.1. Regions II^β and IV are nonreflecting regions. In these regions the wave function of an electron with $w \gtrsim 0$ [w as defined by Eq. (9.13); thermally emitted electrons satisfy this condition] can be adequately described by the WKB formulas (see, e.g., Landau and Lifshitz, 1958).

In region II^β, an electron with energy E and wave vector $\mathbf{k}_\parallel$, propagating in the negative z direction, is described by

$$\Psi^-_{E,\mathbf{k}_\parallel}(\mathbf{r}) = \exp(i\mathbf{k}_\parallel \cdot \mathbf{r}_\parallel)\psi_{tr}(z) \tag{9.19}$$

$$\psi_{tr}(z) = \frac{1}{[q(z)]^{1/2}} \exp\left[-i \int_{z_m}^z q(z)\,dz\right] \tag{9.20}$$

$$q(z) = \left(\frac{2m}{\hbar^2}\right)^{1/2}\left[E - \frac{\hbar^2 k_\parallel^2}{2m} - V(z)\right]^{1/2} \tag{9.21a}$$

$$V(z) = V_r(z) + iV_{im}(z) \tag{9.21b}$$

In region II^β the imaginary part of the potential is zero and $E - \hbar^2 k_\parallel^2/2m - V(z) > 0$, so that $q(z)$ is real and positive in region II^β. We have

$$\exp(i\mathbf{k}_\parallel \cdot \mathbf{r}_\parallel)\psi_{tr}(z) \qquad \text{(in region } II^\beta)$$
$$\Rightarrow \exp(i\mathbf{k}_\parallel \cdot \mathbf{r}_\parallel)[\psi_{in}(z) + \lambda_0\psi^*_{in}(z)] \qquad \text{(in region IV)} \tag{9.22}$$

where ψ_{in} represents a wave propagating in the negative z direction (incident on the barrier), and its complex conjugate ψ_{in}^* a wave propagating in the positive z direction (reflected wave). Obviously, $|\lambda_0|^2$ is the coefficient of reflection of an electron by the potential barrier in region III acting in isolation.

An electronic wave, in region II^β, propagating in the negative z direction can equally well be described as follows:

$$X_{E,\mathbf{k}_\parallel}^{(-)} = \exp(i\mathbf{k}_\parallel \cdot \mathbf{r}_\parallel)\chi_0^{(-)}(z) \tag{9.23}$$

$$\chi_0^{(\pm)}(z) = \frac{1}{[q(z)]^{1/2}} \exp\left[\pm i \int_0^z q(z)\,dz\right] \tag{9.24}$$

We note that our choice of lower limits in the integrals appearing in Eqs. (9.20) and (9.24) does not imply that the WKB approximation is valid outside region II^β. We note also that $q(z)$ is complex in region II^α because the potential is complex there. [It may appear to the reader that our choice of the lower limits in Eqs. (9.20) and (9.24) is not the most convenient one and that a more obvious choice would be to take as the lower limit in both cases a point in region II^β. Our present choice is convenient only when made in conjunction with additional approximations, namely, a parabolic approximation to the potential in region III and a WKB approximation to the wave function in region II^α (see Section 9.5).]

One can easily show that

$$\chi_0^{(+)}(z) = \exp\left[i \int_0^{z_m} q(z')\,dz'\right] \psi_{tr}^*(z) \tag{9.25}$$

$$\chi_0^{(-)}(z) = \exp\left[-i \int_0^{z_m} q(z')\,dz'\right] \psi_{tr}(z) \tag{9.26}$$

where z lies, of course, in region II^β.

Incident upon the potential field to the left of region II^β the wave described by Eq. (9.23) will be partly reflected by it and partly transmitted into it. The "reflected wave", denoted by $X_{E,\mathbf{k}_\parallel}^{(+)}$, consists (in the absence of thermal vibration of the atoms) of a specularly reflected beam and a number of "diffracted beams." At this stage we disregard the thermal vibration of the atoms. We have

$$X_{E,\mathbf{k}_\parallel}^{(+)} = \mu_0 \exp(i\mathbf{k}_\parallel \cdot \mathbf{r}_\parallel)\chi_0^{(+)}(z) + \sum_{\mathbf{g} \neq 0} B_\mathbf{g} \exp[i(\mathbf{k}_\parallel + \mathbf{g}) \cdot \mathbf{r}_\parallel]\chi_\mathbf{g}^{(+)}(z) \tag{9.27}$$

where $\mathbf{g}$ are the 2D reciprocal vectors corresponding to the crystallographic plane (surface) under consideration. For very low energy electrons (less than 10 eV or so above the vacuum level) the normal energy of the electron, $E - (\hbar^2/2m)(\mathbf{k}_\parallel$

+ g)^2, in any of the "diffracted" beams ($\mathbf{g} \neq 0$) lies well below the barrier maximum, so $\chi_{\mathbf{g}}^{(+)} \simeq 0$ in the outer part (large z) of region II$^\beta$. It follows that, for thermal electrons,

(transmitted wave in I)
$$\Rightarrow \exp(i\mathbf{k}_\parallel \cdot \mathbf{r}_\parallel)[\chi_0^{(-)}(z) + \mu_0\chi_0^{(+)}(z)] \qquad \text{(in outer II}^\beta) \qquad (9.28)$$

[It would seem from the above statement that the measured elastically reflected current, when the $\mathbf{g} \neq 0$ beams do not emerge from the metal, consists of the specular beam only. This is how the theoretical elastic reflection coefficient is calculated, but experimentally it is found that, because of the finite spread in energy and angular direction of the incident beam, electrons are scattered out of the specular beam (more so at lower energies) even when the surface is perfectly pure and smooth (Anderson and Kasemo, 1970; McRae and Caldwell, 1976.]

Using Eqs. (9.25) and (9.26) in Eq. (9.28) we obtain

(transmitted wave in I)
$$\Rightarrow \exp(i\mathbf{k}_\parallel \cdot \mathbf{r}_\parallel)[\psi_{\mathrm{Tr}}(z) + \mu_0\eta \exp(i\theta)\psi_{\mathrm{Tr}}^*(z)] \qquad \text{(in outer II}^\beta) \qquad (9.29)$$

where η and θ are, by definition, real numbers such that

$$\eta \exp(i\theta) = \exp\left[2i \int_0^{z_m} q(z)\, dz \right] \qquad (9.30)$$

Combining Eq. (9.29) with Eq. (9.22) we finally obtain

(transmitted wave in region I) $\Rightarrow \{[1 + \mu_0\eta \exp(i\theta)\lambda_0^*]\psi_{\mathrm{in}}(z)$
$$+ [\lambda_0 + \mu_0\eta \exp(i\theta)]\, \psi_{\mathrm{in}}^*(z)\} \exp(i\mathbf{k}_\parallel \cdot \mathbf{r}_\parallel) \qquad \text{(in region IV)} \qquad (9.31)$$

The first term on the right of the above equation represents a wave with its source at $z = +\infty$ propagating towards the metal (incident wave) and the second term represents the reflected wave propagating towards $z = +\infty$. It follows from Eq. (9.31) that the elastic reflection coefficient $r^e(\mathbf{E}, \mathbf{k}_\parallel)$ is given by

$$r^e(\mathbf{E}, \mathbf{k}_\parallel) = \left| \frac{\lambda + \mu}{1 + \mu\lambda^*} \right|^2 \qquad (9.32)$$
$$\lambda \equiv \lambda_0 \exp(-i\theta) \qquad (9.33a)$$
$$\mu = \eta\mu_0 \qquad (9.33b)$$

A formula of this type was first derived by Herring and Nichols (1949). It is worth noting that had we chosen the lower limit in the integrals of Eqs. (9.20) and (9.24) in the manner described in the parenthesis following Eq. (9.24) we would obtain the same formal result, Eqs. (9.32) and (9.33), with $\eta = 1$ and $\theta = 0$.

9.3. CALCULATION OF μ_0

Let an electron with energy E and wave vector $\mathbf{k}_\parallel$ in region II^β (see Fig. 9.3) be incident on the metal. This electron will be partly reflected by the potential field in regions I and II^α and partly transmitted into the metal. The complete wave function in region II (II^α and II^β) is given by

$$\Psi_{E,\mathbf{k}_\parallel} = B_0^- \chi_0^{(-)}(z) \exp(i\mathbf{k}_\parallel \cdot \mathbf{r}_\parallel) + B_0^+ \chi_0^{(+)}(z) \exp(i\mathbf{k}_\parallel \cdot \mathbf{r}_\parallel) \quad (9.34)$$
$$+ \sum_{\mathbf{g} \neq 0} B_\mathbf{g}^+ \chi_\mathbf{g}^{(+)}(z) \exp[i(\mathbf{k}_\parallel + \mathbf{g}) \cdot \mathbf{r}_\parallel]$$

The wave functions $\chi_\mathbf{g}^{(\pm)}$ are defined as follows. They are the solutions of the Schrödinger equation

$$\left\{ -\frac{\hbar^2}{2m}\frac{d^2}{dz^2} + V(z) - \left[E - \frac{\hbar^2}{2m}(\mathbf{k}_\parallel + \mathbf{g})^2 \right] \right\} \chi_\mathbf{g}^{(\pm)}(z) = 0 \quad (9.35)$$

in region II, which satisfy the following boundary conditions. $\chi_0^{(\pm)}(z)$ are given by Eqs. (9.24) in region II^β. We note that the same expression will describe $\chi_0^{(\pm)}$ in region II^α as well, if the WKB approximation is valid in that region, otherwise $\chi_0^{(\pm)}(z)$ in region II^α must be obtained by numerical integration of Eq. (9.35). For thermal electrons (total energy around the barrier maximum) $\chi_\mathbf{g}^{(+)}(z)$ ($\mathbf{g} \neq 0$) decrease and $\chi_\mathbf{g}^{(-)}(z)$ ($\mathbf{g} \neq 0$) increase exponentially in region II^β as z increases. The physical meaning of the various terms in Eq. (9.34) has already been explained in the previous section.

In the region of constant potential between the muffin-tin atoms of the surface layer and the interface at $z = 0$ (region I) the wave function has the following form:

$$\Psi_{E,\mathbf{k}_\parallel} = \sum_\mathbf{g} \{ A_\mathbf{g}^+ \exp(iK_\mathbf{g}^+ \cdot \mathbf{r}) + A_\mathbf{g}^- \exp(iK_g^- \cdot \mathbf{r}) \} \quad (9.36)$$

$$K_\mathbf{g}^\pm = \left(\mathbf{k}_\parallel + \mathbf{g}, \pm \left\{ \frac{2m}{\hbar^2}[E - iV_{\mathrm{im}}(E)] - (\mathbf{k}_\parallel + \mathbf{g})^2 \right\}^{1/2} \right) \quad (9.37)$$

where $V_{\mathrm{im}}(E)$ is the imaginary part of the potential inside the metal ($z < 0$). (The real part of the interstitial potential equals zero by definition.) We note that the same $\mathbf{g}$ vectors (n in number including $\mathbf{g} = 0$) appear in Eqs. (9.34) and (9.36). The coefficients $A_\mathbf{g}^\pm$ in Eq. (9.36) are obviously related to the $B_\mathbf{g}^\pm$ coefficients in Eq. (9.34) since the wave function and its derivative are continuous functions. These relations are conveniently expressed in terms of the reflection and transmission matrix elements of region II^α (this includes the potential step at $z = 0$ if such exists) as follows:

$$B_g^+ = T^{-+}(\mathbf{k}_{||} + \mathbf{g})A_g^+ + R^{++}(\mathbf{k}_{||} + \mathbf{g})B_g^- \tag{9.38}$$
$$A_g^- = R^{--}(\mathbf{k}_{||} + \mathbf{g})A_g^+ + T^{+-}(\mathbf{k}_{||} + \mathbf{g})B_g^- \tag{9.39}$$

Equation (9.38) tells us that a wave in region II^β $\{B_g^+\chi_g^+(z) \exp[i\mathbf{k}_{||} + \mathbf{g}) \cdot \mathbf{r}_{||}]\}$ which propagates or decays in the positive z direction derives from a wave in region I $[A_g^+ \exp(iK_g^+ \cdot \mathbf{r})]$ partly transmitted through region II^α, and from a wave in region II^β $\{B_g^-\chi_g^-(z) \exp[i(\mathbf{k}_{||} + \mathbf{g}) \cdot \mathbf{r}_{||}]\}$ partly reflected by the potential in region II^α. A similar interpretation applies to Eq. (9.39). $T^{-+}(\mathbf{k}_{||} + \mathbf{g})$ and the other matrix elements, which depend also on the energy although this has not been explicitly denoted in Eqs. (9.38) and (9.39) can be obtained easily, after $\chi_g^{(\pm)}(z)$ have been evaluated in region II^α, by matching an appropriate wave function on the left ($z = 0-$) or right ($z = 0+$) of the interface to its corresponding expression on the other side of the interface. In what follows it is convenient to use a matrix notation whereby $A^\pm$ and $B^\pm$ denote column vectors of n elements, $\{A_g^\pm\}$ and $\{B_g^\pm\}$, respectively, and T^{-+}, T^{+-}, etc. are $n \times n$ *diagonal* matrices with elements $[T^{-+}(\mathbf{k}_{||} + \mathbf{g})]$, $[T^{+-}(\mathbf{k}_{||} + \mathbf{g})]$ etc., respectively. In addition to Eqs. (9.38) and (9.39) we have

$$\mathbf{A}^+ = \mathbf{R}^c\mathbf{A}^- \tag{9.40}$$

where $\mathbf{R}^c$ is the reflection matrix defined by Eq. (5.65). We remember that $R_{gg'}^c$ is the amplitude of the diffracted beam described by $\exp(i\mathbf{K}_g^+ \cdot \mathbf{r})$ due to an incident beam described by $\exp(i\mathbf{K}_{g'}^- \cdot \mathbf{r})$. We note that only the potential field in region I enters the calculation of the matrix $\mathbf{R}^c$.

The combination of Eqs. (9.38)–(9.40) leads after some algebra to the following matrix equation:

$$\mathbf{B}^+ = \overline{\mathbf{M}}\mathbf{B}^- \tag{9.41}$$
$$\overline{\mathbf{M}} = \mathbf{T}^{-+}(\mathbf{I} - \mathbf{R}^c\mathbf{R}^{--})^{-1}\mathbf{R}^c\mathbf{T}^{+-} + \mathbf{R}^{++} \tag{9.42}$$

where $\mathbf{I}$ is the unit $n \times n$ matrix.

The reflection amplitude $\mu_0(\mathbf{E},\mathbf{k}_{||})$ is by definition [Eqs. (9.23)–(9.27)] given by

$$\mu_0(\mathbf{E},\mathbf{k}_{||}) = B_0^+/B_0^- \tag{9.43a}$$
$$= \overline{M}_{g=0 \ g'=0} \tag{9.43b}$$

The calculation of the diagonal matrices describing reflection by, and transmission through, region II$^\alpha$ is straightforward (see Section 9.5, therefore the problem of calculating μ_0 reduces to that of calculating R^c. We have already shown, in Section 5.3, how to calculate R^c, for the model potential described by Eq. (4.1), for a frozen lattice.

At this stage we must consider the thermal vibration of the atoms and its effect on the scattering of an electron by the metal.

9.4. EFFECT OF THERMAL VIBRATION OF THE ATOMS ON THE ELASTIC REFLECTION COEFFICIENT

When the atoms vibrate the perfect 2D periodicity of the metal parallel to the surface is destroyed. Therefore, when a beam of electrons with energy E and wave vector $\mathbf{k}_{\|}$ is incident on the metal surface, the backscattered electrons are not restricted to a discrete set of diffracted beams $\{E, \mathbf{k}_{\|} + \mathbf{g}\}$. We have seen that in thermionic emission only electrons with energies around the barrier maximum and $\mathbf{k}_{\|} \simeq 0$ matter and that, in the absence of thermal vibrations, the elastic reflection coefficient is entirely determined by the amplitude of the specularly reflected beam. Diffuse scattering, i.e., scattering into angles other than those of the diffracted beams, leads to a reduction in the flux of the specularly reflected beam (and of any other diffracted beam). If we assume that the electrons off the specular direction, or at least the vast majority of them, cannot escape from the metal because their normal energy lies below the barrier maximum, we can incorporate the effect of thermal vibration into our theory simply by calculating the effect that diffuse scattering has on the amplitude of the specularly reflected beam. [The reduction in the reflection coefficient calculated in the above manner may be slightly exaggerated; some electrons off the specular beam may escape from the metal.] It turns out that the effect of the lattice vibrations on the elastic reflection coefficient of thermal electrons is relatively small. This is not surprising. LEED experiments have shown that there is little diffuse scattering at low incident energies even at very high temperatures. This is clearly demonstrated in Fig. 9.4, taken from a paper by Tabor and Wilson (1970).

The problem of how to incorporate the thermal vibration of the atoms in the calculation of the amplitude of the (coherent) diffracted beams $\{\mathbf{k}_{\|} + \mathbf{g}\}$ has been solved (under certain assumptions) by LEED theorists (Duke and Laramore, 1970; Laramore and Duke, 1970; Holland, 1971; Pendry, 1974). The assumptions made by these authors and the essential results can be summarized as follows. Consider the scattering of an electron with energy E and wave vector $\mathbf{k}$ by a single muffin-tin atom. Scattering into an angular direction specified by

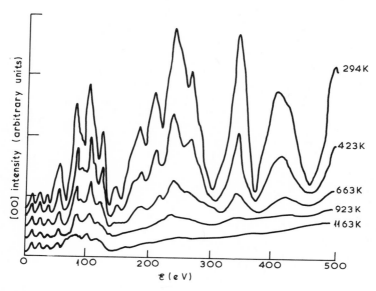

FIGURE 9.4

Intensity of specular reflection as a function of energy, taken at 8° from normal incidence on the (100) plane of niobium, at various temperatures. Curves for different temperatures have been displaced relative to one another for the sake of clarity. (From Tabor and Wilson, 1970.)

$\mathbf{k}'(|\mathbf{k}'| = |\mathbf{k}|)$ is proportional to $|t(\mathbf{k}',\mathbf{k})|^2$,

$$t(\mathbf{k}',\mathbf{k}) = 8\pi^2 \sum_{\substack{lm \\ l'm'}} t(l'm';lm)(-1)^{m'} Y_{l'-m'}(\Omega_{\mathbf{k}'}) Y_{lm}(\Omega_{\mathbf{k}}) \qquad (9.44)$$

$$t(l'm';lm) = -\delta_{ll'} \delta_{m'm} \left(\frac{2mE}{\hbar^2}\right)^{-1} \exp(i\delta_l) \sin\delta_l \qquad (9.45)$$

where δ_l are the scattering phase shifts (see Section 3.5.1) and δ_{ij} is the usual Kronecker delta. We note that the scattering amplitude given by Eq. (9.44) refers to an origin at the center of the atom. If we keep the origin fixed and displace the atom by $\Delta\mathbf{R}$, $t(\mathbf{k}',\mathbf{k})$ changes by a phase factor as follows:

$$t(\mathbf{k}',\mathbf{k}) \rightarrow t(\mathbf{k}',\mathbf{k}) \exp[i(\mathbf{k} - \mathbf{k}') \cdot \Delta\mathbf{R}] \qquad (9.46)$$

At high temperatures the atom vibrates around a mean position. However, because the atom moves slowly compared to the electron, it may be assumed stationary during the scattering process. In that case scattering of the electron by the vibrating atoms (at temperature T) is described by a scattering amplitude

$t(T;\mathbf{k}',\mathbf{k})$ which is the thermal average of the quantity on the right-hand side of Eq. (9.46). It can be shown that (Glauber, 1955)

$$\langle \exp[i(\mathbf{k} - \mathbf{k}') \cdot \Delta\mathbf{R}]\rangle_T = \exp\{-\tfrac{1}{2}\langle[(\mathbf{k} - \mathbf{k}') \cdot \Delta\mathbf{R}]^2\rangle_T\} \qquad (9.47)$$

Hence

$$t(T;\mathbf{k}',\mathbf{k}) = t(\mathbf{k}',\mathbf{k}) \exp(-M) \qquad (9.48)$$
$$M = \tfrac{1}{2}\langle[(\mathbf{k} - \mathbf{k}') \cdot \Delta\mathbf{R}]^2\rangle_T \qquad (9.49)$$

The factor $\exp(-2M)$ is known as the Debye–Waller factor. The above derivation ignores energy exchange (of the order of k_BT) between the electron and the vibrating atoms so that $|\mathbf{k}| = |\mathbf{k}'| = (2mE/\hbar^2)^{1/2}$ in the above formulas (for thermal electrons $E - E_F \simeq \phi \simeq 4$ eV $\gg k_BT$). For a simple crystal (one atom per unit cell) at high temperatures ($T \gg \Theta_D$) one obtains the following approximate formula for M (see, e.g., Pendry, 1974)

$$M = \xi|\mathbf{k} - \mathbf{k}'|^2 \qquad (9.50a)$$
$$\xi = 3\hbar^2T/2m_Ak_B\Theta_D^2 \qquad (9.50b)$$

where m_A is the mass of the atom and Θ_D is the Debye temperature of the solid which constitutes a measure of the compressibility of the solid. Equations (9.50) describe, at best, the vibration of an atom in the bulk of the metal. We expect the vibration of a surface atom to have on the average a larger amplitude than that of an atom in the interior of the crystal. Since the mean amplitude of the vibration is inversely proportional to the Debye temperature, the above implies that the effective of value Θ_D for a surface atom is in general smaller than the bulk value of Θ_D.

It can be shown (Pendry, 1974) that when Eq. (9.50a) is substituted in Eq. (9.48) the resulting formula for $t(T,\mathbf{k}',\mathbf{k})$ can be transformed into an expression which is formally identical with that given by Eqs. (9.44) and (9.45), except that the phase shifts now depend on the temperature as follows:

$$\exp[i\delta_l(T)] \sin[\delta_l(T)] = \sum_{l',l''} i^{l''} \exp(-2\xi k^2)j_{l''}(-2i\xi k^2)$$
$$\times \exp(i\delta_{l'}) \sin \delta_{l'} \left[\frac{4\pi(2l' + 1)(2l'' + 1)}{(2l + 1)}\right]^{1/2} B_{l''0}(l'0,l0) \quad (9.51)$$

where $k^2 = 2mE/\hbar^2$, $j_l(z)$ is the spherical Bessel function and $B_{l''0}$ is the integral of three spherical harmonics defined by Eq. (3.83). A program which evaluates

$\delta_l(T)$ given the zero temperature phase shifts δ_l, can be found in the book of Pendry (1974).

We note that $\delta_l(T)$ is a complex quantity with a positive imaginary part, which implies that the outgoing flux is less than the incident flux on the atoms. This is due to the fact that all flux scattered incoherently has been neglected in the above averaging procedure.

It is obvious that, having an expression which describes the scattering of an electron by the individual (vibrating) atom in terms of temperature-dependent phase shifts, which is formally identical with the corresponding expression at zero temperature (atom fixed at its mean position), allows us to use Eq. (3.98) with δ_l replaced by $\delta_l(T)$, to calculate the scattering matrix for one layer at any given temperature. Knowing this matrix we can calculate the reflection matrix $\mathbf{R}^c$ for the semi-infinite crystal exactly as for the frozen lattice (see Section 5.3). We must remember, however, that the lattice constant (the lattice at the higher temperatures gives the mean position of the atoms) is larger at the higher temperatures.

9.5. CALCULATION OF THE SURFACE BARRIER SCATTERING MATRIX ELEMENTS

Scattering by the surface potential barrier is described by the scattering matrix elements associated with region II^α ($T^{-+}, T^{+-}, R^{++}, R^{--}$) and also by the parameters λ_0, η, and θ, which appear in Eq. (9.32). For an accurate quantitative analysis of a given set of data it may be necessary to evaluate these quantities numerically. On the other hand for certain barriers these quantities can be calculated to a good degree of approximation using, at least partly, analytic formulas. The evaluation of the transmission (T^{-+} and T^{+-}) and reflection amplitudes (R^{--} and R^{++}) defined by Eqs. (9.38) and (9.39) is greatly simplified if we assume that the WKB approximation remains valid in region II^α (for $z > 0$) as well as region II^β. The WKB formulas for T^{-+}, R^{--}, etc. are straightforward and need not be given here. We should point out, however, that the detailed shape of the potential variation in region II^α is not known and that most of the early work on the periodic deviations from the Schottky line has primarily been directed at understanding the nature of this region by comparing theoretical models of this barrier with experiment (Juenker, Colladay, and Coomes, 1953; Juenker, 1955; Cutler and Gibbons, 1958; Belford, Kupperman, and Phipps, 1962; Yang and Yang, 1970). In these calculations, however, the energy-band structure, inelastic electron collisions and the thermal vibration of the atoms were neglected.

For the purpose of evaluating λ_0, as defined by Eq. (9.22) we assume (Guth and Mullin, 1941) that the potential barrier in region III (see Fig. 9.1) is adequately described by the first two nonzero terms of its Taylor expansion around z_m (the position of the barrier maximum). We have

$$V(z) = V_{\max} - 2(eF^3)^{1/2}(z - z_m)^2 \qquad (z \text{ in region III}) \qquad (9.52)$$

The above, known as the parabolic approximation to the potential is shown schematically by the broken curve in Fig. 9.1. Substituting a wave function of the following general form (for an electron with energy E and wave vector $\mathbf{k}_{\parallel}$)

$$\psi_{E,\mathbf{k}_{\parallel}}(\mathbf{r}) = \exp(i\mathbf{k}_{\parallel} \cdot \mathbf{r}_{\parallel})\psi(z) \qquad (9.53)$$

into the Schrödinger equation

$$\left[-\frac{\hbar^2}{2m}\nabla^2 + V(z) - E \right]\psi_{E,\mathbf{k}_{\parallel}} = 0 \qquad (9.54)$$

where $V(z)$ is given by Eq. (9.52), we obtain the following equation for ψ:

$$\frac{d^2\psi}{dt^2} + \left(\delta + \frac{1}{4}t^2 \right)\psi(t) = 0 \qquad (9.55)$$

where

$$\delta = \frac{1}{\sqrt{8}}\left(\frac{2m}{\hbar^2} \right)^{1/2}(eF^3)^{-1/4}w \qquad (9.56)$$

w is the energy defined by Eq. (9.13) and

$$t = 8^{1/4}\left(\frac{2m}{\hbar^2} \right)^{1/4}(eF^3)^{1/8}(z - z_m) \equiv \alpha_0(z - z_m) \qquad (9.57)$$

Equation (9.55) is known as Weber's equation and its solutions, corresponding to different asymptotic behavior as $t \to \pm\infty$, are readily available in the literature (see, e.g., Whittaker and Watson, 1952). A solution exists which has the following asymptotic behavior:

$$\psi \simeq \psi_{\text{tr}}(t) \qquad \text{for } t \ll 0 \qquad (9.58a)$$
$$\simeq \psi_{\text{in}}(t) + \lambda_0\psi_{\text{in}}^*(t) \qquad \text{for } t \gg 0 \qquad (9.58b)$$

where

$$\psi_{\text{tr}} = \left(\frac{2}{\alpha_0}\right)^{1/2} \delta^{-i\delta/2} \left[\frac{1}{(|t|)^{1/2}}\right] \exp\left[i\left(\frac{t^2}{4} + \delta \ln |t|\right)\right], \qquad t \ll 0$$

(9.59a)

$$= \frac{1}{[q(z)]^{1/2}} \exp\left[-i\int_{z_m}^{z} q(z)\,dz\right], \qquad z \ll z_m$$

(9.59b)

$$\psi_{\text{in}}(t) = \left(\frac{4\pi}{\alpha_0}\right)^{1/2} \frac{\delta^{-i\delta/2} e^{-\pi\delta/2}}{\Gamma(1/2 - i\delta)} \left(\frac{1}{\sqrt{t}}\right) \exp\left[-i\left(\frac{t^2}{4} + \delta \ln t\right)\right], \qquad t \gg 0$$

(9.60)

$$\lambda_0 = \Gamma^*(1/2 - i\delta)(2\pi)^{-1/2} \exp\left(-\frac{\pi\delta}{2} - i\delta \ln \delta - \frac{\pi i}{2}\right)$$

(9.61)

$\Gamma(x)$ is the well-known gamma function and the asterisk denotes, as usual, complex conjugation. We note that in deriving the above expressions we have implicitly assumed that there exists a region of z, on either side of the barrier maximum, where the parabolic approximation [Eq. (9.52)] remains valid while at the same time $(z - z_m)$ is sufficiently large for the asymptotic expressions of the wave function [Eqs. (9.58)–(9.60)] to be valid. That this is the case to a reasonably good approximation for values of the parameters appropriate to a thermionic emission experiment has been confirmed by explicit calculations (see, e.g., Cutler and Gibbons, 1958).

The first term in Eq. (9.58b), given explicitly by Eq. (9.60), represents an electronic wave incident on the barrier from the right. Since the potential in region III joins smoothly with the nonreflecting potential in region IV, we may assume that the source of ψ_{in} lies at $z = +\infty$. This wave is partly reflected by the potential barrier in region III and partly transmitted through it. The reflected wave, represented by the second term in Eq. (9.58b), propagates towards $z = +\infty$, and the transmitted wave, given by ψ_{tr}, propagates away from the barrier maximum in the negative z direction. The potential in region III, where the parabolic approximation is valid, joins smoothly with the potential in the nonreflecting region II$^\beta$, therefore the explicit WKB expression for ψ_{tr} given by Eq. (9.59b) is valid in region II$^\beta$ as well. We note that in region II$^\beta$ $\psi_{\text{tr}}(z)$ as given by the above equation is identical with the earlier definition of this quantity [Eq. (9.20)], so that the λ_0 in Eq. (9.58b) is identical with the λ_0 defined earlier by Eq. (9.22).

One can easily show, starting from Eq. (9.61) and using standard properties of the gamma function that

$$|\lambda_0| = (1 + e^{2\pi\delta})^{-1/2}$$

(9.62)

We need, also, an expression for the argument of λ_0. We note that according to Eq. (9.62), $|\lambda_0| \simeq 1$ for $\delta < -1$ [normal energy of the electron a bit below the barrier maximum according to Eq. (9.56)] in which case the reflection coefficient r^e is also equal to unity. Similarly $|\lambda_0| \simeq 0$ for $\delta > 1$ (energy a bit above the barrier maximum). We therefore need an expression for $\arg\{\lambda_0\}$ only when $|\delta| \ll 1$. We have (see, e.g., Jahnke and Emde, 1945)

$$\arg\{\Gamma(1/2 - i\delta)\} \simeq C\delta \qquad \text{for } |\delta| \ll 1 \qquad (9.63)$$
$$C \equiv 1.9635 \qquad (9.64)$$

Substituting Eq. (9.63) into Eq. (9.61) we obtain

$$\arg\{\lambda_0\} \simeq -(\delta \ln |\delta| + C\delta + \pi/2) \qquad \text{for } |\delta| \ll 1 \qquad (9.65)$$

The coefficients η and θ, defined by Eq. (9.30), can be calculated as follows. We note that $V_{\text{im}}(E,z)$ is different from zero only in the immediate vicinity of the interface [see Eq. (9.17)] and that there

$$|V_{\text{im}}(E,z)| \ll W - V_r(z) \qquad (9.66)$$

where W is the normal energy of the electrons, defined by

$$W = E - \hbar^2 k_\parallel^2/2m \qquad (9.67a)$$
$$= V_{\text{max}} + w \qquad (9.67b)$$

It follows from Eq. (9.66) that $q(z)$ is to very good approximation given by

$$q(z) \simeq \left\{ \frac{2m}{\hbar^2}[W - V_r(z)] \right\}^{1/2} - \frac{i}{2}\left(\frac{2m}{\hbar^2}\right)\frac{V_{\text{im}}(E,z)}{[W - V_r(z)]^{1/2}} \qquad (9.68)$$

Substituting Eq. (9.68) into Eq. (9.30) we obtain

$$\eta \simeq \exp\left\{ \left(\frac{2m}{\hbar^2}\right)^{1/2} \int_0^{z_m} \frac{V_{\text{im}}(E,z)\, dz}{[W - V_r(z)]^{1/2}} \right\} \qquad (9.69)$$

$$\theta \simeq 2 \int_0^{z_m} \left\{ \frac{2m}{\hbar^2}[W - V_r(z)] \right\}^{1/2} dz \qquad (9.70)$$

The numerical evaluation of the integral in Eq. (9.69) requires very little computing time because of the very short range of $V_{\text{im}}(E,z)$ and for that reason no additional approximation need be made in evaluating this quantity. In order to evaluate θ we proceed as follows (Juenker *et al.*, 1953; Juenker, 1955). We write

$$\theta = 2 \int_{\zeta}^{z_m} \left\{ \frac{2m}{\hbar^2} [W - V_r(z)] \right\}^{1/2} dz + 2 \int_0^{\zeta} \left\{ \frac{2m}{\hbar^2} [W - V_r(z)] \right\}^{1/2} dz$$

(9.71)

where ζ lies sufficiently near to z_m for the parabolic approximation to be valid, and at the same time sufficiently away from it for the asymptotic expression given by Eq. (9.59) to be valid. In that case the first integral in Eq. (9.71) is obtained by comparing Eq. (9.59b) with Eq. (9.59a). The second integral in Eq. (9.71) can be simplified as follows. We note that an accurate evaluation of θ is required only for small values of w, hence we may replace W in the integrand by its equal according to Eq. (9.67b), expand the integrand in Taylor series around $w = 0$, and keep only the first two terms of this series. The resulting integral can be performed analytically at least for certain barriers. When the potential barrier is given by [see Eqs. (4.2) and (9.16)]

$$V_r(z) = E_F + \phi - \frac{e^2}{4(z + z_0)} - eFz$$

(9.72)

one obtains, after some calculation, the following formula for θ:

$$\theta \simeq - \delta \ln |\delta| - (g + C) \delta + \sigma - (8me^2 z_0)^{1/2}/\hbar$$

(9.73)

$$g \equiv 4 - C - \ln 12 - \ln \sigma$$

(9.74)

$$\sigma \equiv \frac{4}{3} \left(\frac{me^2}{\hbar^2} \right)^{1/2} \left(\frac{e}{F} \right)^{1/4}$$

(9.75)

where δ is given by Eq. (9.56), C by Eq. (9.64) and z_0 is the adjustable parameter in Eq. (9.72) for the potential barrier.

9.6. DEVIATIONS FROM THE SCHOTTKY LINE

If $\bar{t}$ in Eq. (9.10) were entirely independent of the applied field, a plot of $\ln(J)$ versus $\sqrt{F}$ would be a straight line (Schottky line) with slope

$$m_S = e^{3/2}/k_B T$$

(9.76)

We have already pointed out (in Chapter 1) that in practice one finds small oscillations about this line which increase in amplitude and period as F increases (see Fig. 1.6). The physical origin of these oscillations has been described, qualitatively, in Chapter 1. A quantitative analysis of the deviations from the Schottky line must be based on an accurate evaluation $\bar{t}$ as defined by Eq. (9.11). The following simplifications are possible. The dependence of $r(\mathbf{E},\mathbf{k}_{\parallel})$, given by Eqs. (9.15) and (9.32), on the field comes exclusively from λ, as defined by Eq. (9.33a). From Eqs. (9.62) and (9.56) we see that $|\lambda| = 1$, and therefore $r = 1$, for $w < 0$, except in the immediate vicinity of the barrier maximum. On the other hand, for $w > 0$, $|\lambda| = 0$ except again in the immediate vicinity of the barrier maximum. The region around the barrier maximum where $|\lambda|$ is neither zero nor unity increases with the applied field, but even at the higher applied fields used in a typical experiment this region is only a small fraction of $k_B T$ (about one tenth or so) for $T \simeq 1000$ K. Hence only a small fraction of the emitted current comes from this energy region. It is for this reason that the deviations from the Schottky line are generally small and, in fact, negligible in the low-field region. In this region, provided space charge limitation does not occur and that patchiness on the surface can be avoided (see e.g., Herring and Nichols, 1949; Nottingham, 1956), the experimental Schottky plot will indeed be a straight line with slope m_S. By extrapolating this line to zero field we obtain an "extrapolated to zero field" current density given by the Richardson equation

$$J(0) = A\bar{t}(T,0)T^2 \exp(-\phi/k_B T) \tag{9.77}$$

In the limit of zero field, Eq. (9.62) gives

$$\begin{aligned} |\lambda_0| &= 1 && \text{for } w < 0 \\ &= 0 && \text{for } w > 0 \end{aligned} \tag{9.78}$$

and, therefore, according to Eqs. (9.15) and (9.32) we obtain

$$\begin{aligned} r(\mathbf{E},\mathbf{k}_{\parallel}) &= |\mu(\mathbf{E},\mathbf{k}_{\parallel})|^2 && \text{for } w > 0 \\ &= 1 && \text{for } w < 0 \end{aligned} \tag{9.79}$$

In order to obtain $\bar{t}(T,0)$, we substitute Eq. (9.79) into Eq. (9.11) and noting that most of the contribution to the integral in Eq. (9.11) comes from a very narrow region of energy $\mathbf{E} \lesssim k_B T$ and $\mathbf{k}_{\parallel}$ space $\mathbf{k}_{\parallel} < (2mk_B T/\hbar^2)^{1/2}$, we replace $\mu(\mathbf{E},\mathbf{k}_{\parallel})$ over that region by $\mu(0,0)$. Hence we obtain

$$\bar{t}(T,0) = 1 - |\mu(0,0)|^2 \tag{9.80}$$

which shows that the preexponential term in Richardson's equation is determined entirely by the potential field inside the metal and the immediate vicinity of the metal–vacuum interface ($z < \beta$). It is worth noting that Eq. (9.78) has been obtained on the basis of the parabolic approximation to the potential in the region of the barrier maximum. This approximation is valid in the range of applied field used in a typical experiment and it is therefore legitimate to use the above formulas to obtain the "extrapolated to zero field" current density. This corresponds to the procedure by which this quantity is obtained in practice. However, we must emphasize that the parabolic approximation breaks down in the limit of zero field, and therefore Eq. (9.78) does not give the reflection coefficient correctly just above the barrier (the vacuum level for zero field). Obviously the reflection coefficient equals unity at $w = 0+$ diminishing rapidly to zero as w increases.

The deviation from the Schottky line is defined by

$$\Delta J(F) = \left\{ \ln \left[\frac{J(F)}{J(0)} \right] - m_S F^{1/2} \right\} \Big/ \ln(10) \qquad (9.81)$$

where m_S is the theoretical slope of the Schottky line [Eq. (9.76)] and $J(0)$ is the value of the "extrapolated to zero field" emitted current density. Substituting Eqs. (9.10) and (9.77) into Eq. (9.81), we obtain

$$\Delta J(F) = \ln \left[\frac{\bar{i}(T,F)}{\bar{i}(T,0)} \right] \Big/ \ln(10) \qquad (9.82)$$

$\bar{i}(T,0)$ is given by Eq. (9.80). We obtain $\bar{i}(T,F)$ from Eq. (9.11) by putting

$$1 - r = 1 - \left| \frac{\lambda(w) + \mu(0,0)}{1 + \mu(0,0)\lambda^*(w)} \right|^2 \qquad (9.83)$$

We have replaced $\mu(E, k_\parallel)$ in Eq. (9.32) by its value at $E = 0$, $k_\parallel = 0$, as we have done in deriving Eq. (9.80) and for exactly the same reason. We obtain

$$\bar{i}(T,F) \frac{1}{k_B T} \int_{-\infty}^{+\infty} \left[1 - \left| \frac{\lambda(w) + \mu(0,0)}{1 + \mu(0,0)\lambda^*(w)} \right|^2 \right] \exp\left(-\frac{w}{k_B T} \right) dw \qquad (9.84)$$

We have already noted that $|\lambda| = 1$, in which case the integrand in Eq. (9.84) vanishes, for $w < 0$ except in the immediate neighborhood of $w = 0$, and that $\lambda = 0$ for $w > 0$ except again in the immediate neighborhood of $w = 0$; the region where $|\lambda|$ is neither unity nor zero is very small (a fraction of $k_B T$). The contribution to the integral from this region must in general be calculated numerically. The contribution to the integral from outside this region (where $\lambda \simeq 0$) can be obtained analytically.

We note that in the earlier theories of the periodic deviations from the Schottky line (see, e.g., Juenker, 1955; Miller and Good, 1953b; Cutler and Gibbons, 1958) it is assumed that $|\mu| \ll 1$. This condition is satisfied for the free-electron model used in these theories, but it may not be true for a real metal surface (see Section 9.7). When $|\mu| \ll 1$, the integral in Eq. (9.84) can be performed analytically to a very good degree of approximation. The way to do this is described in detail by Juenker et al. (1953) and Juenker (1955). They find

$$\Delta J(F) = f_1 + f_2 \tag{9.85a}$$

$$f_1 \simeq 2.0 \times 10^{10} T^{-2} y_F^{-6} \tag{9.85b}$$

$$f_2 \equiv 4.9 \times 10^5 |\mu| T^{-1} y_F^{-3.3} \cos(y_F + \pi/2 - \delta' + 0.6) \tag{9.85c}$$

$$y_F \equiv 357.1 F^{-1/4} \tag{9.85d}$$

where F is expressed in V/cm. The phase factor δ' is determined by the argument of μ and by the shape of the barrier in region II^α (see Fig. 9.1). Very similar formulas have been obtained by Miller and Good (1953), by Cutler and Gibbons (1958), and by Yang and Yang (1970; 1971). The deviations from the Schottky line were calculated by Juenker and his co-workers (Juenker et al., 1953; Juenker, 1955) for the model potential defined by Eq. (1.27). Practically identical results have been obtained (for the same potential) by Miller and Good (1953b). The solid line in Fig. 9.5, shows the periodic term f_2 [Eq. (9.85c)] as calculated by these authors. The nonperiodic term f_1 is very small when $|\mu| \ll 1$. [This is not necessarily true for larger values of $|\mu|$ as can be seen from Figs. 9.6 and 9.7;

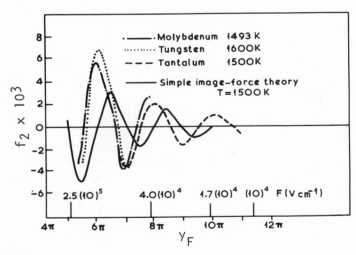

FIGURE 9.5

The periodic deviation from the Schottky line for W, Ta, and Mo polycrystalline emitters. The solid line was calculated for a free-electron-metal and an image-potential surface barrier. (From Cutler and Gibbons, 1958.)

see, also, Modinos (1982).] The experimental results shown in this figure [taken from the paper of Cutler and Gibbons (1958)] were obtained by a number of authors (Houde, 1952; Munick, LaBerge, and Coomes, 1950; Haas and Coomes, 1955) from *polycrystalline* emitters. It is not surprising that the theoretical result (valid for a perfectly uniform surface of a free-electron-like metal) does not agree very well with these experimental results. What is remarkable is the fact that the observed deviations from the Schottky line, for three different emitters, appear to be practically the same. This is probably due to the fact that tungsten, molybdenum, and tantalum have similar electronic properties and that in all three cases the emitted current is a weighted average with the low-work-function regions of the emitter contributing most. In the calculation of Cutler and Gibbons (1958) a small dip was introduced in the potential to simulate the oscillations (see Fig. 13) at the surface of the free-electron metal. Although the results reported by these authors were in better agreement with the experimental data shown in Fig. 9.5, subsequent, entirely numerical, calculations by Belford, Kupperman, and Phipps (1962) gave the same deviation from the Schottky line for the image potential barrier [Eq. (1.27)] and for the barrier with a dip at the surface of Cutler and Gibbons. (The numerical results of Belford *et al.* for the image barrier were in complete agreement with the analytic results shown by the solid line in Fig. 9.5.)

It would seem to us that there is no reason for the periodic deviation from the Schottky line (especially the amplitude, but also the phase at the higher applied fields) to be the same for different emitting planes. Obviously, in order to establish whether this is the case or not, one requires measurements from a number of single crystal planes, and such measurements have not been performed to this date except for tungsten (111) (see Fig. 1.6).

9.7. THERMIONIC EMISSION FROM Cu(100): THEORETICAL RESULTS

In a recent publication (Modinos, 1982) a calculation was presented of the energy distribution of the thermally emitted electrons and of the deviation from the Schottky line for Cu(100). This is the first thermionic emission calculation which takes into account (in the manner described in the preceding sections of this chapter) the energy band structure of the metal, inelastic electron–electron collisions and the thermal vibration of the atoms. Admittedly, copper is not an ideal thermionic emitter because of its relatively low melting point (1356 K) which limits the working temperature region to 800–1100 K or so. [At present thermionic emission data for copper exist only for polycrystalline emitters (Wilson, 1966).] We have chosen to study Cu(100) for two reasons. More is known from independent experiments about this surface (McRae and Caldwell, 1976; Haas and

Thomas, 1977) than any other. Secondly, it is a very interesting plane to study. This is so because of the energy gap which exists around the vacuum level along the ΓX symmetry line in the 3D Brillouin zone [normal to the (100) plane] which means that thermal electrons do not penetrate into the metal and for that reason the energy distribution and the current density which one measures in a thermionic emission experiment reflect strongly the properties of the electron potential in the *surface* region. The same may be true of course of other surfaces; e.g., Ni(100) has an energy gap around the vacuum level along the normal to the plane, similar to that of Cu(100), and nickel is a better thermionic emitter than copper because of its higher melting point (1726 K).

In Fig. 9.6 we show the calculated energy band structure of copper along the ΓX symmetry direction for three different values of the lattice constant α. The solid line ($a = 6.831$ a.u.) corresponds to $T \simeq 0$ K, the broken line ($a = 6.995$ a.u.) corresponds to $T \simeq 800$ K, and the dash–dot line ($a = 7.054$ a.u.) corresponds to $T \simeq 1060$ K. The work function $\phi \simeq 4.5$ eV in the temperature region 800–1100 K (Haas and Thomas, 1977) so that the vacuum level, at 12.05 eV above the muffin-tin zero, lies near the middle of the energy gap shown in Fig. 9.6. It is clear from Fig. 9.6 that electrons with $\mathbf{E} \simeq 0$ cannot penetrate into the metal and that a thermionic current at these energies exists because of finite absorption (nonzero imaginary component of the potential) in the top few layers of the metal and in the region of the metal–vacuum interface.

Our calculations of the thermally emitted current from Cu(100) were based on an one-electron (complex) potential of the form described in Section 9.1, with $V_r(z)$ given by Eq. (9.72). This potential is shown (schematically) in Fig. 9.3. The temperature dependent phase shifts (see Section 9.4) were calculated on the basis of Eq. (9.51) using the parametrized zero temperature phase shifts of Cooper, Kreiger, and Segall (1971), and an effective Debye temperature $\theta_D = 245$ K, which is roughly two-thirds of the Debye temperature in the bulk of the crystal [see comments following Eq. (9.50b)].

The parameters α and γ which determine the imaginary part of the potential inside the metal [Eq. (9.18)] have been estimated by McRae (1976) from optical data. His estimates are

$$\alpha = 0.02\,[V_{im}(E)\text{ in eV}] \tag{9.86a}$$
$$\gamma = 1.7 \tag{9.86b}$$

With the above choice of values $V_{im}(\mathbf{E} \simeq 0) \simeq -0.253$ eV.

McRae and Caldwell (1976) obtained a value for β, the parameter in Eq. (9.17) which determines the range of the imaginary part on the vacuum side of the interface, by comparing the experimental low-energy elastic reflection coefficient at a Cu(100) surface, shown in Fig. 9.2, with a theoretical curve for the same quantity. The real part of the surface barrier, in their theory, is given by Eq.

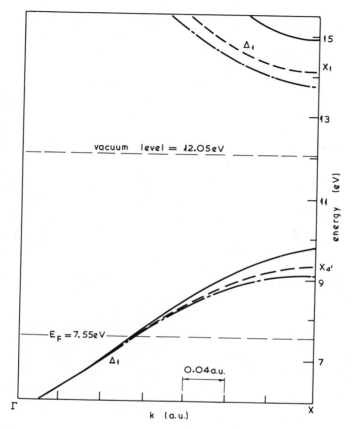

FIGURE 9.6

The energy band structure of copper along the ΓX direction for three different values of the lattice constant a. Solid curve: $a = 6.831$ a.u.; broken curve: $a = 6.995$ a.u.; dash–dot curve: $a = 7.054$ a.u. The energy is measured from the constant potential between the muffin-tin spheres.

(9.72) with $F = 0$, $D_s = 0$, and $z_0 = 0.5645$ a.u. [with this value of z_0, $V(0+)$ $= 0$]. With the imaginary part of the potential given by Eqs. (9.17) and (9.86) they were able to fit the data by putting $\beta \simeq 2$ a.u. Using these parameter values we found that the "extrapolated to zero field" elastic reflection coefficient, given by $[1 - \bar{t}(T,0)]$, equals approximately 0.69 which is larger than their value (0.57) by approximately 18%. This difference may be, partly, due to the fact that in their calculation the field is zero, whereas in our case it is not [see comments following Eq. (9.80)]. We note, however, that in the analysis of McRae and Caldwell the diffraction by the metal was treated approximately (two-beam approximation) and that may have introduced some error into their calculation. On the other hand, in our analysis the wave function of the electron in region II$^\alpha$ of the surface barrier (Fig. 9.3) was evaluated in the WKB approximation and this may have introduced some error in our calculation of the reflection coefficient.

Table 9.1 shows how $\mu(0,0)$ and $\bar{t}(T,0)$ depend on the parameters of the potential and on the temperature. The set of parameter values denoted by I are those of McRae and Caldwell (1976). The corresponding value of the transmission coefficient is $\bar{t}(T,0) \simeq 0.308$. Choosing a larger value for β (3.5 instead of 2.0 a.u.) leads to values of $\bar{t}(T,0)$ in the region 0.39–0.42 depending on the values of z_0 and D_s (sets: III, IV, V). It is clear that the latter two parameters do not affect the magnitude of $\bar{t}(T,0)$ in any significant way. We note (see column VI) that the larger transmission coefficient can also be obtained by keeping $\beta = 2$ a.u. and increasing instead the value of $V_{im}(E = 0)$, i.e., by increasing the value of α in Eq. (9.18). (We note, however, that ideally α being a bulk parameter must be determined by independent experiments.) The result under III was obtained with the same parameter as in II, but with the atoms fixed at their mean positions (thermal vibrations neglected). We see that the thermal vibration of the atoms reduces the reflection coefficient $(1 - \bar{t})$ by less than 10%. Since our calculation overestimates this reduction (see Section 9.4) we may conclude that the effect of thermal vibration on electron reflection at the surface in the limit of zero energy $(E = 0)$ is relatively small compared to that of electron–electron collisions (represented by the imaginary part of the potential) which is independent of the temperature. For this reason, $\bar{t}(T,0)$ at $T = 1060$ K (result VII) is only 7% larger than the corresponding quantity at $T = 800$ K (result II).

Figure 9.7 shows the results of our calculation for the total energy distribution of the emitted electrons for the sets of parameters denoted by II, III, and VII

TABLE 9.1

Dependence of $\mu(0,0)$ and of $\bar{t}(T,0)$ on the Parameters of the Optical Potential and on the Temperature

	I	II	III	IV	V	VI	VII
$|\mu(0,0)|$	0.832	0.781	0.816	0.762	0.758	0.750	0.762
$\arg[\mu(0,0)]$	2.033	2.033	2.017	2.200	3.055	2.198	2.104
$\bar{t}(T,0)$	0.308	0.390	0.333	0.419	0.425	0.437	0.419

I: $T = 800$ K; $z_0 = 0.5645$ a.u.; $D_s = 0.0$; $V_{im}(E = 0) = -0.253$ eV; $\beta = 2.0$ a.u.
II: $T = 800$ K; $z_0 = 0.5645$ a.u.; $D_s = 0.0$; $V_{im}(E = 0) = -0.253$ eV; $\beta = 3.5$ a.u.
III: $T = 800$ K; $z_0 = 0.5645$ a.u.; $D_s = 0.0$; $V_{im}(E = 0) = -0.253$ eV; $\beta = 3.5$
    a.u.; the thermal vibration of the atoms has not been taken into account.
IV: $T = 800$ K; $z_0 = 1.129$ a.u.; $D_s = 0.0$; $V_{im}(E = 0) = -0.253$ eV; $\beta = 3.5$ a.u.
V: $T = 800$ K; $z_0 = 1.129$ a.u.; $D_s = 0.7$ a.u.; $V_{im}(E = 0) = -0.253$ eV; $\beta = 3.5$
    a.u.
VI: $T = 800$ K; $z_0 = 1.129$ a.u.; $D_s = 0.0$; $V_{im}(E = 0) = -0.4$ eV; $\beta = 2.0$ a.u.
VII: $T = 1060$ K; $z_0 = 0.5645$ a.u.; $D_s = 0.0$; $V_{im}(E = 0)$; $= -0.253$ eV; $\beta = 3.5$
    a.u.

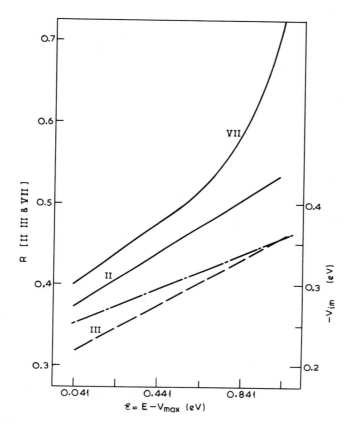

FIGURE 9.7

The enhancement factor $R(E)$, as defined by Eq. (9.87), for copper at two temperatures: $T = 800$ K (curve II); $T = 1060$ K (curve VII). Curve III was obtained with the thermal vibration of the atoms neglected (other parameters as for II). The dash–dot curve gives the variation of the imaginary component of the potential with the energy.

in Table 9.1. (All calculations were done for an applied field $F = 0.576 \times 10^{-3}$ V/Å.) For obvious reasons we have chosen to plot the enhancement factor $R(E)$, defined by

$$R(E) = j(E)/j_0(E) \qquad (9.87)$$

where $j_0(E)$ is the "free-electron" distribution defined by Eq. (9.14), rather than $j(E)$. Curve III, obtained with the same parameter values as curve II except that the thermal vibration has been disregarded, shows that the latter does not affect the shape of $R(E)$ in any significant way. The dash–dot curve on the same figure shows the variation of the imaginary part of the potential with the energy [Eqs. (9.18) and (9.86)]. It is clear that in this instance the variation of the $R(E)$ curve derives to a large extent from the corresponding variation of V_{im} with the energy. Note, however that $R(E)$ increases faster than V_{im} presumably because as the energy moves away from the center of the gap (see Fig. 9.6) the "incident" electron penetrates further into the metal and therefore the probability of it being scattered inelastically (absorbed) increases. At energies approaching the edges of the gap the penetration of the electron into the metal, and therefore absorption, may increase considerably. This is demonstrated quite clearly by curve VII obtained at the higher temperature. The (assumed) lattice constant for this temperature leads to the band structure shown by the dash-dot curve in Fig. 9.6. We see that the upper edge of the gap has been lowered so that electrons with $E \simeq 1$ eV ($E \simeq 13.0$ eV)

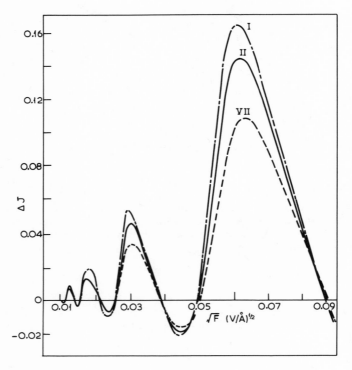

FIGURE 9.8

The deviation from the Schottky line. Its dependence on temperature and on the range of the imaginary component of the potential outside the metal. The three curves correspond to different sets of parameter values, as described in Table 9.1.

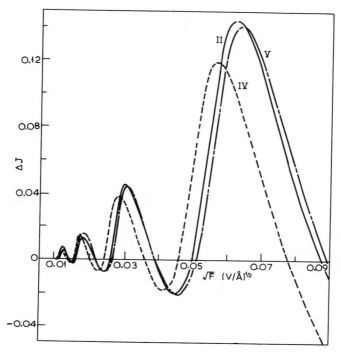

FIGURE 9.9

The deviation from the Schottky line. Its dependence on the surface barrier. The three curves correspond to different sets of parameter values, as described in Table 9.1.

are now nearer to the edge of the gap, they penetrate more into the metal, absorption increases considerably with a consequent decrease in the reflection coefficient which leads to a correspondingly large increase in $R(E)$.

The theoretical results on the deviation from the Schottky line obtained from Eqs. (9.82) and (9.84) are summarized in Figs. 9.8 and 9.9. The three curves in Fig. 9.8 correspond to the set of parameter values denoted by I, II, and VII in Table 9.1. Comparing curves I and II we see that increasing the range of the imaginary component of the potential outside the crystal leads to a decrease in the amplitude but has no effect on the "period" or the phase of the deviations from the Schottky line. The same holds true as to the dependence of these deviations on the magnitude of the imaginary component. An increase in the latter leads to a decrease in the amplitude of the deviations but has no effect on the period or the phase of these deviations. Comparing curve II and VII we see that an increase in the temperature has no effect on the period and phase of these deviations but leads again to a decrease in the amplitude of these deviations. This result is in agreement with observations on polycrystalline emitters (Haas and Thomas, 1972; Nottingham, 1956). Figure 9.9 demonstrates the dependence of the deviation from the Schottky line on the parameters (z_0 and D_s) which determine the real part of the potential at the interface. The three curves in this figure correspond to the set of parameter values denoted by II, IV, and V in Table 9.1. It is seen that a change in z_0, with D_s kept constant, leads to a noticeable change in both the phase and the amplitude of the deviation especially at the higher applied fields (compare curves II and IV) but when both D_s and z_0 are changed the corresponding changes in the deviation are much less (compare curves II and V). It would seem to us, at this stage, that if the bulk parameters are known from independent experiments the range of the imaginary component of the potential can be determined from measurements of the energy distribution and of the current density [the magnitude of $\bar{\iota}(T,0)$ and the amplitude of the deviations from the Schottky line suffice to determine β] but, whereas we expect to be able to reproduce the experimentally observed phase and "period" of the deviations by an appropriate choice of z_0 and D_s, these parameters cannot be determined uniquely (as is clear from Fig. 9.9) from such data. Only if additional information on these parameters is available from independent experiments, e.g., LEED, photoemission, or field emission experiments, can z_0 and D_s be uniquely determined. We must also emphasize that representing the potential in the region of the metal–vacuum interface by Eqs. (9.17) and (9.72) is in itself an approximation. Obviously, if we choose $D_s > 4$ a.u., there is no doubt that the potential for $z > 0$ (see Fig 9.3) will be adequately described by Eq. (9.72), but we cannot, of course, assume that the actual potential between the muffin-tin spheres of the top layer and the plane $z = 0$ is constant or, for that matter, that it is a function of z only. In principle the potential in this region, at least its real part, must be calculated self-consistently. The same applies to the muffin-tin potential of the top one or two layers which may be different from that in the bulk of the crystal. Were the potential in the interface region (between the muffin-tin spheres of the top layer and the plane $z = 0$) known from such calculations, we could, at least in principle, take its effect into account in our evaluation of μ_0, by treating this region as an additional "scattering layer." In the meantime the best one can hope for, from an effective potential, such as the one described by Eqs. (9.17) and (9.72) is that for particular values of the parameters this potential scatters the electron in the interface region in the same manner as the more realistic self-consistent potential does.

At the present time no data exist on the deviation from the Schottky line from single-crystal planes of copper or any other metal except for the data of Stafford and Weber (1963) on W(111) which we have already mentioned. Reliable measurements of $\bar{\iota}(T,0)$, i.e., of the preexponential term in Richardson's equation [Eq. (9.77)], and of the energy distribution of the thermally emitted electrons do not exist either (see Chapter 2). It has been the aim of this chapter to demonstrate that such measurements can provide us with useful information on the electronic properties of metal surfaces. Hopefully the knowledge that such measurements can be analyzed in a systematic way will provide additional stimulus for performing these experiments.

Thermionic emission from semiconductors can be treated along the same lines as for metals. We know, however, that simple models of the surface (e.g., terminating abruptly the bulk potential at the metal–vacuum interface) which are reasonable approximations for metals are entirely unsuitable for semiconductor surfaces which practically always reconstruct. Band bending at the surface due to doping or field penetration could complicate the analysis further, as we have seen in our discussion of field emission from semiconductors (Chapter 8).

SECONDARY ELECTRON EMISSION SPECTROSCOPY

10.1. INTRODUCTION

When an electron with energy E_p (in practice $10 < E_p < 500$ eV) is incident on a solid surface it may be scattered back into the vacuum either elastically or inelastically. Part of its energy, however, passes through a series of collisions (cascade) to other (secondary) electrons, and some of these secondary electrons are emitted from the solid. Obviously, in practice one cannot distinguish between a primary backscattered electron and an emitted secondary electron, and, therefore, the distinction between the two is made for convenience, on a rather artificial basis. This is better explained by reference to Fig. 10.1, which shows a typical energy distribution of the electrons emitted from a metal surface. The electrons in region III correspond to elastically scattered electrons and electrons which have collided with phonons losing energy of the order of a few hundredths of an eV. Electrons which lost more energy appear in region II. This region is characterized by certain peaks corresponding to inelastic collisions of the incident (primary) electron with plasmons (collective oscillations of the electrons) and with other electrons. The energy of these peaks is fixed relative to the primary energy (they lie between 2 and 50 eV below it) and is determined by the excitation energy of plasmons and of interband electronic transitions. Region I consists of the true secondary electrons which are generated by the cascade process. In this chapter we are exclusively concerned with the electrons in this region, referring to them, simply, as secondary electrons. Because of the series of (random) collisions that (presumably) occur between the initial one (involving the incident electron) and the final event, i.e., the escape of the secondary electron from the metal, the latter should retain very little information about the incident electron. Consequently the *shape* of the energy and angular distribution of the emitted secondary electrons should be practically independent of the energy E_p and of the direction of the incident electron beam, provided E_p is sufficiently large. This is certainly true for metals of high symmetry, as can be seen, for example, in Fig. 10.2 (Schäfer, Schoppe, Holzl, and Feder, 1981), which shows the energy distribution of the secondary electrons normal to tungsten (100) for different primary energies. We see that the overall shape

327

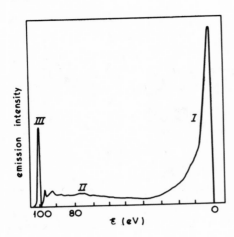

FIGURE 10.1

Secondary electron energy spectrum emitted from a metal surface under bombardment by 100-eV electrons. E denotes the kinetic energy of the emitted electron. Electrons in region I are true secondary electrons, those in II and III are, respectively, inelastically and elastically scattered primary electrons.

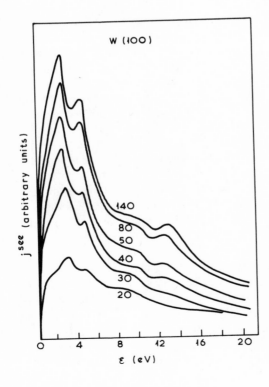

FIGURE 10.2

SEE energy distributions normal to a W(100) surface. The energy is measured from the vacuum level. The primary electron beam (E_p = 20–140 eV) is incident at an angle of 45° to the surface normal. (From Schafer *et al.*, 1981.)

of the distribution and the positions of the fine structure are independent of the primary energy above $E_p \simeq 40$ eV. It has also been established (Willis and Christensen, 1978) that the SEE (secondary electron emission) spectra of tungsten for these energies are independent of the direction of the incident beam. We note that the peaks in the SEE spectra in Fig. 10.2 are less pronounced and slightly shifted to higher energies when $E_p < 40$ eV, which suggests that at these primary energies, the electron excitation process, induced by the primary electron beam, affects the distribution of the emitted electrons. A more pronounced dependence on the energy and a dependence on the direction of the incident beam has been observed in SEE energy distributions from graphite, a strongly anisotropic material, by Willis, Feuerbacher, and Fitton (1971a; 1971b) and in SEE angle-resolved energy distributions from a silicon (111) 7 × 7 surface by Best (1975; 1976; 1979). The graphite results are demonstrated in Fig. 10.3, which shows the (angle-integrated) energy distribution of the emitted secondary electrons for different primary energies (at the same angle of incidence), and for two different angles of incidence at the same primary energy $E_p = 29$ eV. Obviously, the dependence of the SEE spectra on the energy and direction of the incident beam cannot be understood without a close examination of the excitation processes (involving both electron interband transitions and plasmon excitations) induced by the incident beam.

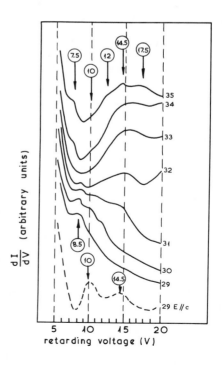

FIGURE 10.3

Energy distribution of secondary electrons emitted from graphite at primary-beam energies of 29–35 eV. The solid curves were obtained for an angle of incidence $\theta_i = 40°$, and the dashed curve for normal incidence ($\theta_i = 0°$) to the basal plane. The peaks appearing at 10 and 14.5 eV are associated with $E \| c$ (electric field parallel to the c axis) interband transitions. The curves corresponding to different primary energies have been shifted vertically for the sake of clarity. The relative intensities of the peaks in the SEE spectrum are independent of the primary beam energy E_p for $E_p > 50$ eV. (From Willis *et al.*, 1971b.)

There are a number of theories on SEE from metals, most of them based on the Sommerfeld model (see Section 1.1) of metals. These theories (Van der Ziel, 1953; Wolff, 1954; Streitwolf, 1959; Stolz, 1959; Amelio, 1970; Chung and Everhart, 1974; Schou, 1980) differ from each other in the way they approximate the cascade process. Most of them give a formula for the energy distribution of the emitted secondary electrons which has (approximately) the following form (see, e.g., Seah, 1969):

$$j^{SEE}(E) \propto (E + E_0)^{-x} P(E) \qquad (10.1)$$

where E denotes the kinetic energy of the emitted electron (it is equal to its total energy measured from the vacuum level). The first factor in the above equation is proportional to the number of secondary electrons with energy E on the metal side of the metal–vacuum interface; the values of E_0 and x are different in the different theories. In Wolff's theory, for example, $x \simeq 2$ and $E_0 = E_i$ (E_i is the difference between the vacuum level and the bottom of the conduction band). The second factor in Eq. (10.1) is a transmission coefficient which determines what fraction of the secondary electrons inside the metal go over the barrier (are emitted) into the vacuum. In Wolff's theory, the momentum distribution inside the metal is spherically symmetric and, therefore,

$$P(E) \simeq 1 - \left(\frac{E_i}{E + E_i}\right)^{1/2} \qquad (10.2)$$

The above equation applies when the energy analyzer accepts every emitted electron with energy E, whatever the direction of its velocity. When the analyzer accepts only the electrons emitted within a small solid angle around a particular direction (as in the experimental setup shown in Fig. 10.4) the expression for the transmission coefficient must be modified accordingly. If, for example, only the electrons emitted in the direction normal to the surface are collected, $P \simeq E/(E_i + E)$. In practice E_i is replaced by an adjustable E_n. Seah (1969) was able to fit the energy distribution of the secondary electrons emitted in the normal direction from copper and silver surfaces using the expression

$$j^{SEE}(E, \text{normal to the surface}) \propto [(E + E_0)^x (E + E_n)]^{-1} E \qquad (10.3)$$

and the following values for the parameters, Cu: $E_0 = 4.5$ eV, $x = 1.6$, $E_n = 0.35$ eV; Ag: $E_0 = 4.5$ eV, $x = 2.0$, $E_n = 0.35$ eV. The corresponding energy distribution curves have a maximum at $E = 0.6$ eV for Cu and at $E = 0.8$ eV for Ag, the full width at half-maximum being 5.5 and 3.4 eV, respectively. Seah observed [in the case of Ag(111)] two peaks at 12.8 eV and 17.3 eV superimposed on the cascade energy distribution curve [described by Eq. (10.3)], and suggested

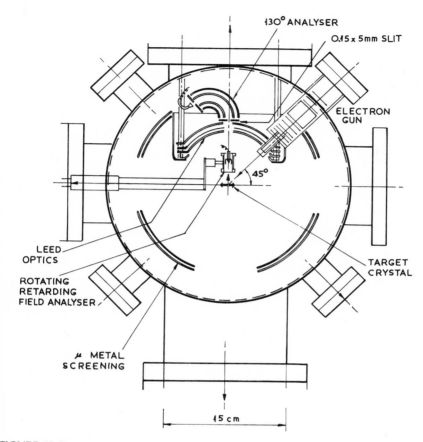

FIGURE 10.4

Apparatus used for studying angle resolved energy distributions of emitted secondary electrons. The primary electron beam (E_p = 100 eV) is incident at 45° to the target crystal normal. In the diagram are shown the LEED–Auger optics, the angular selective slits, the 130°-sector cylindrical electrostatic analyzer, and a small retarding field analyzer which could be rotated out of the plane of the figure so as to provide a cross check on the gross features of the SEE energy distribution. (From Willis and Christensen, 1978.)

that these were related to bulk energy-band structure effects which are not taken into account by the (free-electron) theories of SEE leading to Eq. (10.3). Subsequent experiments (Willis, 1975; Willis and Christensen, 1978; Schäfer, Schoppe, Holzl, and Feder, 1981) using a small-angle-selective analyzer (the apparatus used by Willis and Christensen is shown schematically in Fig. 10.4) have shown that the SEE energy distribution from the (100), (111), and (110) planes of tungsten, measured along a particular direction (not only the normal to the surface), exhibits considerable fine structure. This is clearly demonstrated in Fig. 10.5a, which shows the SEE energy distribution normal to a W(100) surface as measured by Willis and Christensen (1978) (broken line), and as measured by Shaffer *et al.* (1981) (solid line). The broken–solid line in the same figure is a theoretical background (cascade) curve obtained by Schaffer *et al.* from Eq. (10.3) with $x = 1.88$, $E_0 = 2.7$ eV and $E_n = 4.55$ eV. We note that although the background curves in the Willis and Schafer *et al.* experiments are different, the positions of fine structure in the energy distribution curves are (overall) in good agreement with each other. Peaks A and B in Willis's curve coincide (respectively) with peaks M_1 and M_2 in the Schafer *et al.* curve to within (approximately) 0.3 eV. The less pronounced peaks M_3 and M_4 in the Schafer *et al.* curve correspond, but not as closely, with (respectively) peaks C (more of a shoulder than a peak) and D in Willis's curve. The very broad peak E in Willis's curve does not appear in the Schafer *et al.* experiment. Figure 10.5b shows the energy distribution of the electrons emitted from W(100) in the normal direction, when a beam of photons of energy $h\nu = 22$ eV (solid line) and $h\nu = 35$ eV (broken line) in incident on the surface. The remarkable similarity between the Schafer *et al.* SEE curve and the Weng *et al.* photoemission curve which is expected on theoretical grounds suggests that of the two experimental SEE spectra shown in Fig. 10.5a the one obtained by Schafer *et al.* is the more accurate, and that the broad peak E in Willis's curve is probably an experimental artifact.

In the present chapter we shall be concerned with the interpretation of the fine structure in the energy and angle-resolved spectra of the emitted secondary electrons, but not with the shape and magnitude of the corresponding background distribution curve. We note that while the shape of this curve may not depend on the energy and direction of the incident electron, its magnitude does. A calculation of the background (cascade) distribution curve from first principles for realistic models of the target surface and of the cascade process appears to be very difficult (see, e.g., Wolff, 1954; Hachenberg and Brauer, 1959; Ganachaud and Cailler, 1979; Schou, 1980; Salehi and Flinn, 1980).

10.2. BULK DENSITY OF STATES EFFECTS IN SECONDARY ELECTRON EMISSION SPECTRA

Willis and Christensen (1978) and Christensen and Willis (1979) have shown that much of the fine structure in the energy distribution of the electrons emitted along a given direction from a crystal surface can be related to the bulk energy band structure of the crystal under consideration. [Peaks in the bulk density of states (certainly peaks in the surface density of states) can lead to fine structure in the angle-integrated energy distribution of the emitted secondary electrons

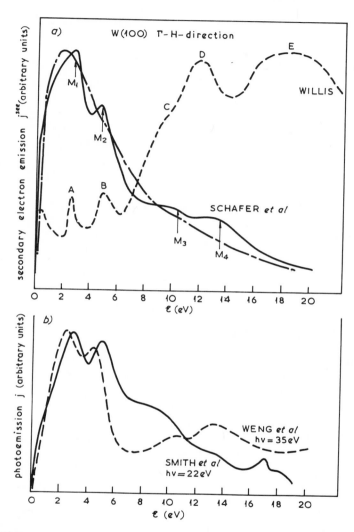

FIGURE 10.5

(a) The experimental SEE spectrum of electrons emitted in the normal direction from W(100) obtained by Willis and Christensen (broken line), and by Schafer et al. (solid line). The broken–solid line is a theoretical cascade curve [Eq. (10.3) with $x = 1.88$, $E_0 = 2.7$ eV, and $E_n = 4.55$ eV]. (b) Experimental photoemission spectrum of electrons emitted in the normal direction from W(100) obtained by Weng et al. (1978) (broken line) and by Smith, Anderson, Hermanson, and Lapeyere (1976) (solid line). (From Schäfer et al., 1981.)

as well. Such structure has been observed, for example, in SEE energy distributions from graphite (see Fig. 10.3) by Willis *et al.* (1971b) and by Willis, Fitton, and Painter (1974), but, obviously, fine structure of this kind is bound to be more pronounced in angle-resolved spectra. Theoretically, the angle-integrated distribution is obtained, simply, by integrating the calculated $j^{SEE}(E;\theta,\phi)$ $d\Omega$, as defined below, over the appropriate angle (determined by the angular acceptance of the analyzer)]. In the case of metals it may be assumed that, as a result of the cascade process an electron is to be found in a state Ψ^B, as defined by Eq. (4.37), with a probability $P_c(E^B)$ which is a smooth slowly varying function of the energy. It could be, for example, that $P_c(E) \propto (E + E_0)^{-x}$ as in Eq. (10.1). An electron described by Eq. (4.37) will contribute to $j^{SEE}(E,\theta,\phi)$ $d\Omega$ (the number of electrons, with energe E, emitted per unit area per unit time within a solid angle $d\Omega = \sin \theta \, d\theta \, d\phi$ in the direction specified by the polar angle θ, measured from the surface normal, and the azimuthal angle ϕ) the amount

$$j_B \, d\Omega = P_c(E) \, \delta(E - E^B)(2V)^{-1} \sum_g |D_g^B|^2 (2/m)^{1/2}$$
$$\times [E^B - (\hbar^2/2m)(\mathbf{k}_\parallel + \mathbf{g})^2]^{1/2}$$
$$\times \Theta(E^B - (\hbar^2/2m)(\mathbf{k}_\parallel + \mathbf{g})^2) \, \delta(\mathbf{q}_\parallel - \mathbf{k}_\parallel - \mathbf{g}) \, d^2q_\parallel \quad (10.4)$$

where Θ is the step function [defined by Eq. (5.58)] and $\mathbf{q} = (\mathbf{q}_\parallel, \mathbf{q}_z)$ is defined by the following relations:

$$q^2 = 2mE/\hbar^2 \qquad\qquad (10.5a)$$
$$q_\parallel = q \sin \theta \qquad\qquad (10.5b)$$
$$(q_x, q_y, q_z) = (q_\parallel \cos \phi, q_\parallel \sin \phi, q \cos \theta) \qquad (10.5c)$$
$$d^2q_\parallel = q^2 \cos \theta \, d\Omega \qquad\qquad (10.5d)$$

The contribution to $j^{SEE}(E,\theta,\phi)$ $d\Omega$ from all one-electron states which have their source deep inside the metal (these are described in Section 4.4.1) is given by

$$j_b^{SEE}(E,\theta,\phi) \, d\Omega = \sum_B j_B \, d\Omega \qquad\qquad (10.6)$$

Substituting Eqs. (10.4) and (10.5) into Eq. (10.6) and transforming the sum over $B \equiv (E^B, \mathbf{k}_\parallel, \alpha)$ into an integral [using Eq. (4.19)] we obtain

$$j_b^{SEE}(E,\theta,\phi) = \frac{\cos^2 \theta}{4\pi^3} \left(\frac{\hbar q^3}{m} \right) P_c(E) \iint_{SBZ} d^2k_\parallel$$
$$\times \left[\sum_\alpha \left(\frac{\partial k_{\alpha z}}{\partial E} \right) \sum_g |D_g^B|^2 \, \delta(\mathbf{k}_\parallel + \mathbf{g} - \mathbf{q}_\parallel) \right]_{E^B = E} \quad (10.7)$$

We note [see, e.g., comments following Eq. (4.19)] that the above formula is valid when the potential does not depend on the spin. Spin–orbit interaction and the consequent splitting of the energy bands in tungsten (and other heavy metals) does not seem to affect the (angle-integrated) energy distribution of the field-emitted electrons (studied in Chapter 5), and it is not likely to affect the angle-integrated SEE energy distribution, but it may have a significant effect on angle-resolved SEE spectra. The more general form of Eq. (10.7) which takes into account spin–orbit coupling is obtained as follows. We remember that when the spin–orbit interaction is neglected the complete wave function is given by a spatial component [say that described by Eq. (4.37)] times a spin function corresponding to spin up or spin down. We have not denoted explicitly this spin function because (in the absence of magnetic fields) it does not enter into the calculation of the local density of states, the emitted current, etc. except for the familiar spin-degeneracy factor of 2 in the formulas which give the total (spin up plus spin down) density of states etc. When the spin–orbit interaction is taken into account a scattering state with its source at $z = -\infty$ (similar arguments apply to any other one-electron state) will be given [instead of by Eq. (4.37)] by

$$\Psi^B = \frac{1}{(2V)^{1/2}} \sum_{\mathbf{g}} \left[D_{\mathbf{g}}^B(\uparrow)\begin{pmatrix}1\\0\end{pmatrix} + D_{\mathbf{g}}^B(\downarrow)\begin{pmatrix}0\\1\end{pmatrix} \right]$$
$$\times \exp(i[\mathbf{k}_{\parallel} + \mathbf{g}] \cdot \mathbf{r}_{\parallel})\phi_{\mathbf{k}_{\parallel}+\mathbf{g}}^+(z), \qquad z > 0 \quad (10.8)$$

$B \equiv (\mathbf{E}^B, \mathbf{k}_{\parallel}, \alpha)$ where $\mathbf{E}^B$ is the energy of the Bth state measured from the vacuum level, $\mathbf{k}_{\parallel}$ is a wave vector in the SBZ, and α is a band index. [A one-to-one correspondence between Ψ^B (the electron states of the semi-infinite crystal) and ψ^B (the Bloch waves of the infinite crystal propagating in the positive z direction) can be established in exactly the same way as before (see Sections 4.2.1 and 4.4.1); Ψ^B can be constructed (in essentially the same way as before) from a knowledge of the Bloch waves of the infinite crystal. The latter can be calculated using a generalized version of the method described in Chapter 3 which takes into account relativistic effects including spin–orbit coupling (Feder, 1972; 1974; 1976). A new set of α-bands is, of course, obtained when the spin–orbit coupling is taken into account. Compare, for example, the energy-band structure of tungsten with spin–orbit interaction taken into account (shown in Fig. 5.7c) with that which is obtained by neglecting spin–orbit coupling (shown in Fig. 5.7b).] Because spin is no longer a good quantum number, the amplitude of the wave function described by Eq. (10.8) has a spin-up component $D_{\mathbf{g}}^B(\uparrow)$ and a spin-down component $D_{\mathbf{g}}^B(\downarrow)$ for each reciprocal vector $\mathbf{g}$ that contributes to the wave function. [The spin part of the wave function is described in terms of the well-known eigenvectors $\begin{pmatrix}1\\0\end{pmatrix}$ and $\begin{pmatrix}0\\1\end{pmatrix}$ corresponding to spin up and spin down, respectively.] The number of electrons emitted along a given direction (θ, ϕ) with energy $\mathbf{E}$ is now given [instead of by Eq. (10.7)] by

$$j_b^{\text{SEE}}(\mathbf{E},\theta,\phi) = j_b^{\text{SEE}\uparrow}(\mathbf{E},\theta,\phi) + j_b^{\text{SEE}\downarrow}(\mathbf{E},\theta,\phi) \qquad (10.9)$$

where $j_b^{\text{SEE}\uparrow(\downarrow)}$ represents the emitted electrons with spin up (down). One can easily see that

$$j_b^{\text{SEE}\uparrow}(\mathbf{E},\theta,\phi) = \frac{\cos^2\theta}{8\pi^3}\left(\frac{\hbar q^3}{m}\right) P_c(\mathbf{E}) \iint_{\text{SBZ}} d^2 k_{\parallel} \qquad (10.10)$$

$$\times\left[\sum_{\alpha}\left(\frac{\partial k_{\alpha z}}{\partial \mathbf{E}}\right)\sum_{\mathbf{g}}|D_{\mathbf{g}}^B(\uparrow)|^2\,\delta(\mathbf{k}_{\parallel}+\mathbf{g}-\mathbf{q}_{\parallel})\right]_{\mathbf{E}^B\,=\,\mathbf{E}} \qquad (10.10)$$

$j_b^{\text{SEE}\downarrow}(\mathbf{E},\theta,\phi)$ is given by Eq. (10.10) with $D_{\mathbf{g}}^B(\uparrow)$ replaced by $D_{\mathbf{g}}^B(\downarrow)$. It is clear that $j_b^{\text{SEE}\uparrow}$ may be different from $j_b^{\text{SEE}\downarrow}$, i.e., the SEE current along a particular direction may be spin-polarized.

Let us (for the moment) assume that there is no other contribution to $j^{\text{SEE}}(\mathbf{E},\theta,\phi)$ apart from j_b^{SEE}. The calculation of this quantity on the basis of Eqs. (10.9) and (10.10) is straightforward but laborious.

A great simplification of Eq. (10.10) occurs, if the $D_{\mathbf{g}}^B$ coefficients in this equation are put equal to a constant, as suggested by Christensen and Willis (1979). Putting

$$|D_{\mathbf{g}}^B(\uparrow)|^2 + |D_{\mathbf{g}}^B(\downarrow)|^2 \simeq \overline{D} \qquad (10.11)$$

where $\overline{D}$ is a positive number, one obtains

$$j_b^{\text{SEE}}(\mathbf{E},\theta,\phi) = \frac{\cos^2\theta}{8\pi^3}\left(\frac{\hbar q^3}{m}\right) P_c(\mathbf{E})\overline{D}\rho(\mathbf{E};\theta,\phi) \qquad (10.12)$$

where ρ is the effective [for the direction (θ,ϕ)] density of states given by

$$\rho(\mathbf{E};\theta,\phi) = \iint_{\text{SBZ}} d^2 k_{\parallel} \sum_{\alpha}\left(\frac{\partial k_{\alpha z}}{\partial \mathbf{E}}\right)\sum_{\mathbf{g}}\delta(\mathbf{k}_{\parallel}+\mathbf{g}-\mathbf{q}_{\parallel}) \qquad (10.13)$$

We remember [see Eq. (4.4)] that $k_{\alpha z} = k_{\alpha z}(\mathbf{E},\mathbf{k}_{\parallel})$ and that $\mathbf{q}_{\parallel}$ is determined by (θ,ϕ) according to Eqs. (10.5). The important thing about Eq. (10.13) is that it is completely determined from a knowledge of the bulk energy band structure. We note that $P_c(\mathbf{E})$ in Eq. (10.12) is a slowly varying function of the energy, and, therefore, any structure in the SEE spectra arises, in this approximation, from $\rho(\mathbf{E};\theta,\phi)$. Because of the approximation described by Eq. (10.11), and, also, because inelastic collisions (which are responsible for the broadening of the electron energy levels) have not been taken into account, only spectral positions of structure can be analyzed on the basis of Eq. (10.12), not line shapes or intensities. A lot of spectra, similar to those shown in Fig. 10.6, from the (110), (111), and (100) planes of tungsten, have been obtained and analyzed in this way by Willis and Christensen (1978) and by Christensen and Willis (1979).

Column (b) in Fig. 10.6 shows the experimental SEE spectra from W(100) for different polar angles $0° \leq \theta \leq 70°$ in the azimuthal plane $\phi = 0$. This plane is normal to the (100) plane and crosses it along the $\overline{\Gamma X}$ direction (see Fig. 4.4). The cross-hatched (in column b) peak is an experimental artifact which appears in all the spectra taken by Willis and Christensen. The overall decrease in the spectral intensity with increasing θ, seen in column (b), derives from the $\cos^2 \theta$ factor in Eqs. (10.10) and (10.12). The comparison with $\rho(E;\theta,\phi)$ is made easier if one obtains an enhanced experimental spectrum by isolating to some degree the fine structure of the spectrum from the background cascade distribution. Willis and Christensen have chosen to do this by subtracting from a given experimental SEE spectrum an arbitrary linear background curve (straight lines

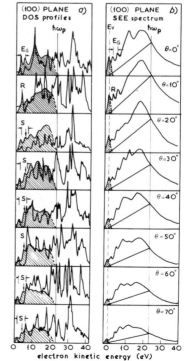

FIGURE 10.6

Column (b) shows the experimental SEE spectrum from W(100) for different polar angles $0° \lesssim \theta \lesssim 70°$ in the azimuthal plane $\phi = 0$. (This plane is normal to the (100) plane and crosses it along the $\overline{\Gamma X}$ direction of the SBZ.] In column a the enhanced experimental spectrum shown shaded, is compared with the corresponding one-dimensional density of states (DOS) given by $\rho(E,\theta,\phi = 0)$. (From Willis and Christensen, 1978.)

in column b) which extends from $E = 0$ to approximately $E \simeq 24$ eV. [We note that for energies above this limit, which corresponds approximately to the mean value of the main surface and bulk plasmon losses in tungsten (Scheibner and Tharp, 1967; Raether, 1965), a large imaginary component of the potential eliminates the fine structure from the SEE spectra.] The resulting enhanced SEE spectrum is shown (shaded) in column (a) of Fig. 10.6. The solid lines in the same column show $\rho(E;\theta,\phi = 0)$ calculated by Christensen and Willis by extending the relativistic augmented-plane-wave calculation of the energy band structure of tungsten by Christensen and Feuerbacher (1974) to higher energies (up to 50 eV above the Fermi level).

The relation between the energy-band-structure, the density of states $\rho(E;\theta,\phi)$ and the corresponding SEE spectrum is best illustrated in Fig. 10.7, which shows the energy bands along the ΓH symmetry line of the 3D Brillonin zone [normal to the (100) plane] besides the corresponding $\rho(E;\theta = 0,\phi = 0)$ curve. In this case $\mathbf{q}_{\parallel} = 0$ and therefore Eq. (10.13) becomes

$$\rho(E;\theta = 0;\phi = 0) = \sum_{\alpha} \left(\frac{\partial k_{\alpha z}}{\partial E} \right) = \sum_{A} \left(\frac{\partial E_A}{\partial k_z} \right)^{-1} \qquad (10.14)$$

The second step in the above equation follows from Eq. (4.5). We shall refer to the bands $E_A(\mathbf{k}_{\parallel} = 0, k_z)$, $A = 1,2, \ldots$ in Fig. 10.7 as first, second, etc. in order of increasing energy. We see that the edges of the band gap E_G between bands 6 and 7 are clearly identified (peaks A and B) in the (enhanced) experimental SEE spectrum (shown by the broken line in Fig. 10.7). It is worth noting that the emission intensity is small but not zero for energies within the gap. The peak (or

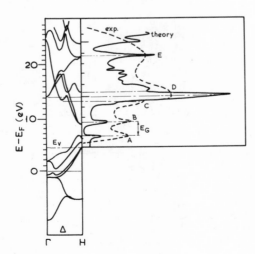

FIGURE 10.7

The energy band structure of tungsten along the ΓH direction [normal to the (100) plane], the corresponding one-dimensional density of states $\rho(E,\theta = 0,\phi = 0)$ (theory), and the experimental SEE spectrum (exp) of electrons emitted in the normal direction from the (100) plane. (From Christensen and Willis, 1979.)

shoulder), denoted by C, at $E \simeq 9.3$ eV has been attributed by Christensen and Willis to emission from the (local) minimum in band 7 at a value of k_z almost halfway between Γ and H. Although this interpretation is plausible in terms of the band structure shown in Fig. 10.7, there is now good reason to believe (see next section) that this peak is due to a surface resonance. Although there is some doubt (see Fig. 10.5) as to the exact position of the fourth peak in the SEE spectrum (peak D), it is evident from Fig. 10.7 that it derives from a corresponding peak in $\rho(E,\theta = 0,\phi = 0)$ associated with the almost triply degenerate level at Γ at 14.4 eV above the Fermi level. The very broad peak (E) observed by Willis and Christensen was attributed by these authors to a peak in ρ due to a critical point at H at 22.3 eV above E_F (bands 10 and 11). We have already pointed out that this peak was not observed in the experiments of Shafer *et al.* (1981).

Let us now examine the SEE spectra for $10° < \theta < 70°$ shown in Fig. 10.6. We note that the gap E_G between bands 6 and 7 exists practically throughout the Brillouin zone, although its magnitude varies through the zone (Willis and Christensen, 1978). The edges of this gap are marked by vertical lines in column (a) of Fig. 10.6. A remarkable feature of the spectra corresponding to $20° < \theta < 70°$ is the peak S at an energy within the gap. It is not difficult to guess the physical origin of this peak. It must be associated with a surface resonance of the kind described in Section 4.4.3 and illustrated (schematically) in Fig. 4.9b. In Fig. 10.8 the measured values of the position of this surface resonance peak and of the band-gap edges are plotted against $\theta = \sin^{-1}[\hbar k_{\|}/(2mE)^{1⁄2}]$; we note that $k_{\|}$ runs along the $\overline{\Gamma X}$ direction of the SBZ (shown in Fig. 4.4). It is seen that the agreement between theory (solid lines) and experiment (points) is very good for the band-gap edges. It is also worth noting that the dispersion curve of the surface resonance is symmetric about $k_{\|} = \pi/a \simeq 1 \text{Å}^{-1}$ (corresponding to $\theta = 70°$; a is the lattice constant of tungsten) as expected from symmetry arguments. Another remarkable feature of the experimental results shown in column (a) of Fig. 10.6 is the sharp peak R which occurs in the spectrum for $\theta = 10°$. It constitutes an example of what Willis and Christensen call a "lower-band-edge-resonance." [Similar peaks have been observed by these authors in the emission spectra from W(111).] This peak implies that a resonance of some sort (similar to that shown schematically in Fig. 4.9C) may exist at the lower edge of the band gap, leading to an enhancement of the density of states peak at this energy.

The remaining peaks (other than S and R) in the off-normal spectra shown in Fig. 10.6 appear to correspond to peaks in the density of states $\rho(E;\theta,\phi = 0)$, but this correspondence is more obvious for some directions (e.g., $\theta = 30°$) and less so for other directions (e.g., $\theta = 60°$). A more detailed analysis of these peaks in terms of the bulk energy-band structure can be found in the paper of Christensen and Willis (1979). In the same paper the reader will find a detailed exposition of SEE spectra (analogous to those shown in Fig. 10.6) from the (110) and (111) planes of tungsten and a comparison between these spectra and the corresponding

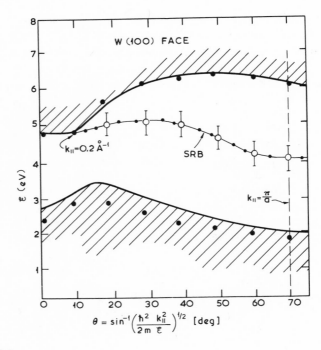

FIGURE 10.8

Angular dependence of the W(100) bulk band gap E_G and the SEE spectral peak S (Fig. 10.6). The experimental results are denoted by points. The solid lines show the gap in the calculated energy band structure. (From Willis and Christensen, 1978.)

(one-directional) density of states defined by Eq. (10.13). It follows from their analysis that a majority of the peaks in the SEE spectra do correspond to peaks in the above-mentioned density of states. They refer to structure in the SEE spectra which cannot be attributed to bulk density of states effects as "anomalies." Anomalous structure includes the following:

(i) emission from an energy-gap region where $\rho(E;\theta,\phi)$, as defined by Eq. (10.13), equals zero. It is evident, for example, from the SEE spectra from W(100) shown in Fig. 10.6, that although emission is small for energies within the band-gap (it goes through a minimum at the middle of the gap), it is not zero. This can be understood as follows. The electronic wave functions in the gap region have the form shown (schematically) in Fig. 4.9a. It is clear that an electron incident on the metal from outside can penetrate a short distance into the metal even when its energy lies within a band gap. That being the case, it can scatter inelastically into a bulk state (of higher or lower energy) and be carried away into the interior of the metal. It follows from time-reversal symmetry that the opposite process is equally possible and that, therefore, an electron can "jump," via an inelastic collision, from a state whose source lies deep inside the metal into a scattering state (such as the one shown in Fig. 4.9a) whose energy lies in the band-gap region and be carried way (emitted) into the vacuum. [The significance of this process has already been pointed out in relation to thermionic emission (Chapter 9).]

(ii) Emission from surface states (resonances) within an energy band-gap. Peak S in the off-normal spectra in Fig. 10.6 is an example of such structure.

(iii) A particularly enhanced emission from *lower* edges of band gaps. Peak R in the SEE spectrum ($\theta = 10°$) shown in Fig. 10.6 provides an example of this kind of structure.

(iv) Enhanced emission due to a "vacuum-bulk resonance" (VBR). This is a term used by Christensen and Willis to describe peaks in the SEE spectra, which seem to occur at particular values of E and $\mathbf{k}_{\parallel}$ such that, for some band $E_A(\mathbf{k}_{\parallel}; k_z)$

$$\frac{1}{\hbar} \frac{\partial E_A}{\partial k_z} (\mathbf{k}_{\parallel}; k_z) \Bigg|_{E_A = E} = \frac{[2mE - (\hbar k_{\parallel})^2]^{1/2}}{m} \tag{10.15}$$

i.e., when the normal component of the group velocity of the electron (for the given band) inside the metal equals the normal component of its velocity outside the metal. Such peaks are not observed in the SEE spectra from W(100), but a number of peaks in the spectra from W(111) have been identified, by Christensen and Willis, as VBR peaks.

Most probably, the peaks described by (iii) and (iv) are associated with surface resonances of the type described schematically in Fig. 4.9c, and probably for some (not clearly understood) reason such resonances are more likely to occur when the conditions specified in (iii) and (iv) above are satisfied.

It is obvious that while a classification of the structure observed in SEE spectra along the lines suggested by Christensen and Willis may be very useful, it cannot, by itself, provide the basis for a quantitative analysis of these spectra.

10.3. QUANTITATIVE ANALYSIS OF THE FINE STRUCTURE IN SECONDARY ELECTRON EMISSION SPECTRA

Feder and Pendry (1978) proposed a theoretical analysis of the fine structure of SEE spectra based on the following proposition:

$$j^{SEE}(E, \mathbf{q}_{\parallel}) = j(E, \mathbf{q}_{\parallel}; T') \tag{10.16}$$

where j^{SEE} is the SEE distribution [$\mathbf{q}_{\parallel}$ is defined by Eqs. (10.5)], and $j(T')$ is the distribution of electrons emitted from a very hot solid. $T' = T'(E)$ is an effective temperature varying slowly with the energy. It is assumed, in other words, that states close together in energy are in thermodynamic equilibrium corresponding to an effective temperature $T'(E)$. The above assumption is justified by the observation that the shape of the energy distribution of the emitted electrons is, at least for some materials, practically independent of the energy and direction of the incident electron beam, which implies that the distribution of the electrons over states is essentially determined by statistics. The calculation of $j(E,\mathbf{q}_{\parallel},T')$ can be transformed into a LEED calculation as follows. We assume, as in the case of thermionic emission (Chapter 9), that the metal is in thermal equilibrium [the effective temperature is given by $T'(E)$] with a free-electron gas occupying the half space from the metal surface to $z = +\infty$. Following Feder and Pendry we denote the number of electrons with energy E and wave vector $\mathbf{q}_{\parallel}$ arriving at the surface (per unit area per unit time) from the right by

$$j^0(E,\mathbf{q}_{\parallel};T') = C_0(E) \tag{10.17}$$

j^0 is calculated for a free-electron gas at equilibrium and is, therefore, independent of $\mathbf{q}_{\parallel}$. We note that $C_0(E)$ is a smooth function of the energy but it is otherwise unknown because $T'(E)$ is unknown. Obviously, since we assume equilibrium conditions $j^0(E,\mathbf{q}_{\parallel},T')$ is equal to the number of electrons departing from the surface in the reverse direction. The latter quantity consists of the emitted (from the metal) electrons $j(E,\mathbf{q}_{\parallel};T')$, and of electrons which have been reflected at the metal surface either elastically [represented by the second term in Eq. (10.18)] or inelastically [represented by the third term in Eq. (10.18)]. We have

$$j^0(E,\mathbf{q}_{\parallel};T') = j(E,\mathbf{q}_{\parallel};T') + \sum_{\mathbf{g}} r^e_{\mathbf{0g}}(E,\mathbf{q}_{\parallel})j^0(E,\mathbf{q}_{\parallel} + \mathbf{g};T')$$
$$+ \int r^i(E,E',\mathbf{q}_{\parallel},\mathbf{q}'_{\parallel})j^0(E',\mathbf{q}'_{\parallel},T')\, dE'\, dq'_{\parallel} \tag{10.18}$$

where r^e is the elastic current reflection-coefficient matrix and r^i the inelastic current reflection-coefficient matrix (some properties of these matrices are given in Appendix A). It is assumed that

$$r^i(E,E',\mathbf{q}_{\parallel},\mathbf{q}'_{\parallel}) = f(E,E',\mathbf{q}'_{\parallel})j(E,\mathbf{q}_{\parallel}) \tag{10.19}$$

where $f(E,E',\mathbf{q}'_{\parallel})$ varies slowly with energy but can be unknown otherwise. Using Eqs. (10.16)–(10.19) one can easily show that

$$j^{\text{SEE}}(E,\mathbf{q}_{\parallel}) = \left[1 - \sum_{\mathbf{g}} r^e_{\mathbf{0g}}(E,\mathbf{q}_{\parallel})\right] F(E) \tag{10.20}$$

$$F(E) \equiv C_0(E)/[1 + \alpha(E)] \tag{10.21a}$$

$$\alpha(E) \equiv \int f(E,E',\mathbf{q}'_{\parallel})C_0(E')\, dE'\, d^2q'_{\parallel} \tag{10.21b}$$

$F(E)$ is an (unknown) slowly varying function of the energy, and, therefore, the fine structure in the SEE spectra is contained in the factor

$$S(\mathbf{E},\mathbf{q}_{\parallel}) = 1 - \sum_{g} r^{e}_{\mathbf{O g}}(\mathbf{E},\mathbf{q}_{\parallel}) \qquad (10.22)$$

The calculation of $r^{e}_{\mathbf{O g}}$ is carried out as in LEED.

Feder and Pendry (1978) calculated the fine structure of the SEE spectrum from W(100) in the normal direction ($\mathbf{q}_{\parallel} = 0$) using a relativistic generalization of Eq. (10.20). A similar but improved calculation of the same spectrum has been performed by Schafer et al. (1981). The result of this calculation for $S(E) = S(\mathbf{E},\mathbf{q}_{\parallel} = 0)$ is shown in Fig. 10.9b. The experimental $S(E)$, shown in Fig. 10.9a, was obtained by dividing the experimental SEE spectrum (solid line in Fig. 10.5a) by the theoretical background curve (broken–solid line in the same figure). The calculation employed the ion-core (muffin-tin) potential for tungsten as evaluated by Matheiss (1965), and the phase shifts were corrected for room temperature (see Section 9.4) using the bulk Debye temperature $\Theta_D = 380$ K. The distance between the first (top) and the second atomic layers was assumed to be contracted by 7.5% with respect to the bulk interlayer distance. The surface barrier was approximated by an image potential barrier starting 1 Å above the center of the top atomic layer. The imaginary part of the potential was approximated as in Eq. (9.17). The values of the parameters α and γ, in this equation, were chosen so that the heights of two prominent peaks in the calculated reflection coefficient fitted the measured values of these quantities (Herlt, Feder, Meister, and Bauer, 1981). With V_{im} expressed in eV, the following values were obtained: $\alpha = 0.11$ and $\gamma = 0.83$.

It is evident that the theoretical SEE spectrum (Fig. 10.9b) is in very good agreement with the experimental spectrum (Fig. 10.9a). Schafer et al. note that the experimental peak at $E = 10.5$ eV is extremely sensitive to contamination and that the corresponding calculated peak is extremely sensitive to the choice of surface barrier, which suggests that this peak (denoted by M_3 and C in the experimental curves shown in Fig. 10.5) is due to a surface resonance and not to a peak in the (one-dimensional) bulk density of states as suggested by Christensen and Willis (1979). It is also worth noting that the theory does not show any peak at $E \simeq 18$ eV, in agreement with the experimental results of Schafer et al. and in disagreement with the Willis and Christensen results (broken line in Fig. 10.5). Figure 10.9c shows the fine structure in $(1 - r^{e})$ where r^{e} is the specular elastic reflection coefficient from W(100) at normal incidence measured by Herlt et al. (1981). For $E < 15$ eV only the specular beam emerges from the metal and, therefore $1 - r^{e} = 1 - r^{e}_{\mathbf{OO}}(E,0) = S(E)$. We see that the value of $S(E)$ measured in this way is also in very good agreement with the calculated value of $S(E)$. Finally, it is worth noting that the calculated SEE current from W(100) in the

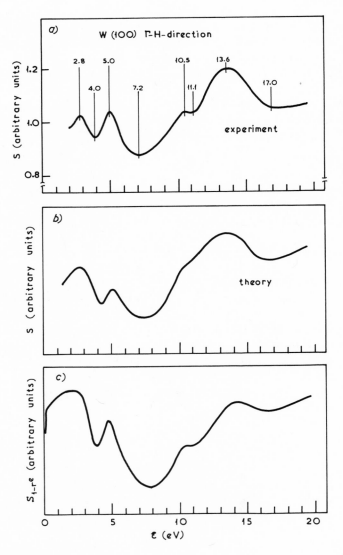

FIGURE 10.9

(a) Fine structure in the measured (solid line in Fig. 10.5a) SEE spectrum normal to W(100); (b) theoretical fine structure [Eq. (10.22)]; and (c) fine structure obtained from the reflection coefficient measured by Hertl *et al.* (1981). (From Schäfer *et al.*, 1981.)

normal direction is unpolarized. However, Feder and Pendry (1978) point out that for nonnormal emission spin polarization may be quite pronounced. No measurements of SEE spin polarization have been reported.

Angle and energy resolved SEE spectra from semiconductor surfaces exhibit the same kind of fine structure, originating from the bulk energy band structure or from surface resonances, as for metals. Best (1975, 1976, 1979) was able to identify both types of peaks in the measured spectra from a Si(111) 7 $\times$ 7 surface. A quantitative analysis of these spectra based on Eq. (10.20) is practically impossible because of the complicated geometry of this surface. Such a calculation, if it were possible, could give the spectral position structure, but *not* their relative intensity, which appears to depend, in contrast to the situation in tungsten, on the energy and direction of the incident beam even at relatively high primary energies. In this case the occupation of the "final" states is not determined by statistics alone, as assumed in the derivation of Eq. (10.20). A complete analysis of such spectra requires, therefore, a study in some detail of the "cascade" problem. This lies beyond the scope of the present book.

APPENDIX A

DETAILED BALANCE

A.1. INTRODUCTION

We consider a semi-infinite crystal, extending from $z = -\infty$ to $z \simeq 0$, in thermal equilibrium (at temperature T) with a noninteracting electron gas occupying the space from $z \simeq 0$ to $z = \infty$. We assume that the potential is constant (zero) sufficiently away ($z \gg 0$) from the surface. We have seen (see Section 9.1) that this assumption can be made without any loss in generality, even when an electric field is applied to the surface. The principle of detailed balance tells us that

$$j_0(E,\mathbf{q}_{\|}) = j(E,\mathbf{q}_{\|}) + \sum r^e_{\mathbf{q}_{\|} \, \mathbf{q}_{\|}+\mathbf{g}}(E)j_0(E,\mathbf{q}_{\|}+\mathbf{g})$$
$$+ \iint r^i(E,\mathbf{q}_{\|};E'\mathbf{q}'_{\|})j_0(E',\mathbf{q}'_{\|}) \, dE' \, d^2q'_{\|} \qquad \text{(A.1)}$$

$j_0(E,\mathbf{q}_{\|}) \, dE \, d^2q_{\|}$ is the number of electrons with energy between E and $E + dE$ and wave vector (parallel to the surface) between $\mathbf{q}_{\|}$ and $\mathbf{q}_{\|} + d^2q_{\|}$ arriving at the surface (in practice one should specify a plane: when there is no externally applied field this should lie just outside the range of the image forces; when there is an applied field and a well-defined potential barrier maximum this plane should lie a few angstroms to the right of the barrier maximum) per unit area per unit time. The expression on the right-hand side of Eq. (A.1) (when multiplied by dE $d^2q_{\|}$) gives the number of electrons in the same E and $\mathbf{q}_{\|}$ range departing from the surface per unit area per unit time in the reverse direction. $j(E,\mathbf{q}_{\|})$ represents the electrons emitted from the crystal. The second and third terms in the above expression represent (respectively) elastically and inelastically reflected electrons. We assume that the surface of the crystal corresponds to a crystallographic plane, so that for given E, only electrons incident along certain directions specified by $\mathbf{q}_{\|} + \mathbf{g}$, where $\mathbf{g}$ are the reciprocal vectors of the surface lattice, contribute to the elastically reflected current in the direction specified by $\mathbf{q}_{\|}$.

We have

$$j_0(E,\mathbf{q}_{\|}) = f(E)/4\pi^3\hbar \qquad \text{(A.2)}$$

which shows that this quantity is independent of $\mathbf{q}_{\|}$. $f(E)$ denotes, as usual, the Fermi–Dirac distribution function.

A.2. ELASTIC REFLECTION COEFFICIENT

Let an electron described by $|\mathbf{k}\rangle = L^{-3/2} \exp(i\mathbf{k} \cdot \mathbf{r})$ be incident on the crystal from the right (see Fig. A.1). The volume L^3 has been introduced for normalization purposes. The probability of transition from the state $|\mathbf{k}\rangle$ into a state $|\mathbf{q}\rangle$ representing a reflected electron is given by

$$W_{\mathbf{k}\to\mathbf{q}} = \frac{2\pi}{\hbar} \delta(E - E_{\mathbf{q}}) |T_{\mathbf{qk}}|^2 \tag{A.3}$$

$E = \hbar^2 k^2/2m$ is the initial and $E_{\mathbf{q}} = \hbar^2 q^2/2m$ the final energy of the electron and the δ function expresses the fact that energy is conserved in an elastic collision. When the collision between the electron and the crystal is described by a one-electron potential $V(\mathbf{r})$, the T-matrix element is given by

$$T_{\mathbf{qk}} = \frac{1}{L^3} \int e^{-i\mathbf{q}\cdot\mathbf{r}} T(\mathbf{r},\mathbf{r}') e^{i\mathbf{k}\cdot\mathbf{r}'} \, d^3r \, d^3r' \tag{A.4}$$

$$T(\mathbf{r},\mathbf{r}') = V(\mathbf{r}) \delta(\mathbf{r} - \mathbf{r}') + V(\mathbf{r})G_0(\mathbf{r},\mathbf{r}')V(\mathbf{r}') + \cdots \tag{A.5}$$

where G_0 is the free-electron Green's function (see Section 5.2.2).

The probability of transition from a given state $(E,\mathbf{k}_{\|})$ into a state with wave vector $\mathbf{q}_{\|}$ (we assume that $\hbar^2 q_{\|}^2/2m < E$) is given by

$$W(E,\mathbf{k}_{\|} \to E,\mathbf{q}_{\|}) = \frac{2\pi}{\hbar} \sum_{q_z} \delta\left(E - \frac{\hbar^2 q_{\|}^2}{2m} - \frac{\hbar^2 q_z^2}{2m} \right) |T_{\mathbf{qk}}|^2 \tag{A.6}$$

Replacing the sum $\Sigma_{q_z} \ldots$ by the integral $(L/2\pi)\int dq_z \ldots$, we obtain

$$W(E,\mathbf{k}_{\|} \to E,\mathbf{q}_{\|}) = (mL/\hbar^3) |T_{\mathbf{qk}}|^2/q_z \tag{A.7}$$

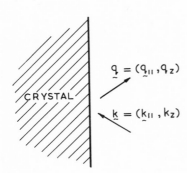

FIGURE A.1

Electron scattering at a crystal surface. $\mathbf{k}$ and $\mathbf{q}$ represent the incident and reflected electron waves, respectively.

where

$$q_z = [2mE/\hbar^2 - q_{\|}^2]^{1/2} \tag{A.8}$$

is real and positive. The contribution of the $\mathbf{q}_{\|}$th beam to the reflected probability current density is given by

$$j_{\text{ref}}(\mathbf{q}_{\|}) = W(E,\mathbf{k}_{\|} \to E,\mathbf{q}_{\|})/L^2 \tag{A.9}$$

and the corresponding elastic reflection coefficient matrix element by

$$r^e_{\mathbf{q}_{\|}\mathbf{k}_{\|}}(E) = j_{\text{ref}}(\mathbf{q}_{\|})/j_{\text{inc}} \tag{A.10}$$

where

$$j_{\text{inc}} = \hbar k_z/(mL^3) \tag{A.11}$$

is the incident current density [corresponding to the plane wave $|\mathbf{k}_{\|}\rangle = |\mathbf{k}_{\|},-k_z\rangle = |\mathbf{k}_{\|},-(2mE/\hbar^2 - k_{\|}^2)^{1/2}\rangle$]. Substituting Eqs. (A.7), (A.9), and (A.11) into Eq. (A.10) we obtain

$$r^e_{\mathbf{q}_{\|}\mathbf{k}_{\|}}(E) = (m^2L^2/\hbar^4)|T_{\mathbf{q}\mathbf{k}}|^2/(k_z q_z) \tag{A.12}$$

The elastic reflection coefficient, corresponding to an incident wave $|\mathbf{k}\rangle$, is given by

$$r^e(E,\mathbf{k}_{\|}) = \sum_{\mathbf{q}_{\|}} r^e_{\mathbf{q}_{\|}\mathbf{k}_{\|}}(E) \tag{A.13}$$

For a crystal surface the T-matrix elements are defined by Eq. (A.4), and therefore, $r^e_{\mathbf{q}_{\|}\mathbf{k}_{\|}}$ vanishes except when $\mathbf{q}_{\|} = \mathbf{k}_{\|} + \mathbf{g}$.

Time reversal symmetry tell us that when there are no magnetic field or spin–orbit interaction terms in the Hamiltonian

$$|T_{\mathbf{q}\mathbf{k}}|^2 = |T_{-\mathbf{k}-\mathbf{q}}|^2 \tag{A.14}$$

The validity of the above statement for a scattering potential $V(\mathbf{r})$ is obvious from Eqs. (A.4) and (A.5). For a discussion of time reversal symmetry and its consequences in the case of arbitrary interactions see, e.g., Landau and Lifshitz (1958). It follows from Eqs. (A.12) and (A.14) that

$$r^e_{\mathbf{q}_{\|}\mathbf{k}_{\|}}(E) = r^e_{-\mathbf{k}_{\|},-\mathbf{q}_{\|}}(E) \tag{A.15}$$

A.3. INELASTIC REFLECTION COEFFICIENT

Let us assume that, initially, the crystal is in a state, denoted by μ', with energy $E_{\mu'}$, and that the (incident) electron has a wave vector $\mathbf{q}'$ and energy $E' = \hbar^2 q'^2/2m$. We denote this state of the crystal plus electron system by $|i\rangle = |\mu',\mathbf{q}'\rangle$. The probability of transition to a state $|f\rangle = |\mu,\mathbf{q}\rangle$ is given by

$$w_{i \to f} = \frac{2\pi}{\hbar} \, \delta(E_{\mu'} + E' - E_\mu - E) |T_{fi}|^2 \qquad (A.16)$$

where E_μ and $E = \hbar^2 q^2/2m$ denote, respectively, the energy of the crystal and that of the electron after the collision. We do not need, for our purposes, to write an explicit expression for T_{fi}. When there are no magnetic field or spin–orbit interaction terms in the Hamiltonian we have, by time reversal symmetry,

$$|T_{fi}|^2 = |T_{i^*f^*}|^2 \qquad (A.17a)$$

where

$$|i^*\rangle = |\mu'^*,-\mathbf{q}'\rangle \qquad (A.17b)$$
$$|f^*\rangle = |\mu^*,-\mathbf{q}\rangle \qquad (A.17c)$$

We note that the wave function which describes the state μ^* is the complex conjugate of that which describes μ, and that

$$E_{\mu^*} = E_\mu \qquad (A.18)$$

The probability of transition of an incident electron with energy E' and wave vector (parallel to the surface) $\mathbf{q}'_{\parallel}$ into a state with energy between E and $E + dE$ and wave vector between $\mathbf{q}_{\parallel}$ and $\mathbf{q}_{\parallel} + d\mathbf{q}_{\parallel}$, when the crystal is initially in the state μ', is given by

$$\left(\frac{L}{2\pi}\right)^2 d^2 q_{\parallel} \, dE \sum_\mu \left(\frac{2\pi}{\hbar}\right) \delta(E_{\mu'} - E_\mu + E' - E)$$
$$\times \sum_{q_z} \delta(E - \hbar^2 q_{\parallel}^2/2m - \hbar^2 q_z^2/2m) |T_{fi}|^2$$
$$\qquad (A.19)$$
$$= \left(\frac{L}{2\pi}\right)^2 d^2 q_{\parallel} \, dE \sum_\mu \delta(E_{\mu'} - E_\mu + E' - E)(mL/\hbar^3)|T_{fi}|^2/q_z$$

with q_z defined by Eq. (A.8).

We denote the reflected current density of energy E and wave vector $\mathbf{q}_{\parallel}$ arising from an incident wave of energy E' and wave vector $\mathbf{q}'_{\parallel}$ by $j_{\text{ref}}(E,\mathbf{q}_{\parallel};E',\mathbf{q}'_{\parallel})$. To obtain $j_{\text{ref}}(E,\mathbf{q}_{\parallel};E',\mathbf{q}'_{\parallel}) \, dE \, d^2 q_{\parallel}$ we divide the expression (A.19) by L^2, multiply by $N \exp(-E_{\mu'}/k_B T)$ (this is the probability that the crystal is in the state μ'; N is a normalization constant), and sum over all μ'. The corresponding inelastic reflection coefficient matrix is defined by

$$r^i(E,\mathbf{q}_{\parallel};E',\mathbf{q}'_{\parallel}) \, dE \, d^2q_{\parallel} = j_{\text{ref}}(E,\mathbf{q}_{\parallel};E',\mathbf{q}'_{\parallel}) \, dE \, d^2q_{\parallel}/j_{\text{inc}} \qquad (\text{A.20})$$

where $j_{\text{inc}} = \hbar \, q'_z/(mL^3)$. We obtain

$$r^i(E,\mathbf{q}_{\parallel},E',\mathbf{q}'_{\parallel}) = \frac{m^2 L^4}{4\pi^2 \hbar^4} \sum_{\mu,\mu'} N e^{-E_{\mu'}/k_B T} \, \delta(E_{\mu'} - E_\mu + E' - E)$$

$$\times \, |T_{\mu,E,\mathbf{q}_{\parallel};\mu',E',\mathbf{q}'_{\parallel}}|^2/(q_z q'_z) \qquad (\text{A.21})$$

The inelastic reflection coefficient, corresponding to an incident wave $|\mathbf{q}'\rangle$, is given by

$$r^i(E',\mathbf{q}'_{\parallel}) = \iint r^i(E,\mathbf{q}_{\parallel};E',\mathbf{q}'_{\parallel}) \, dE \, d^2q_{\parallel} \qquad (\text{A.22})$$

Using Eqs. (A.17) and (A.18) and the fact that when the crystal is in thermal equilibrium with the electron gas on the vacuum side of the interface

$$e^{-E_{\mu'}/k_B T} f(E') = e^{-E_\mu/k_B T} f(E) \qquad \text{if } E_{\mu'} + E' = E_\mu + E \qquad (\text{A.23})$$

we obtain

$$r^i(E,\mathbf{q}_{\parallel};E',\mathbf{q}'_{\parallel}) f(E') = r^i(E',-\mathbf{q}'_{\parallel};E,-\mathbf{q}_{\parallel}) f(E) \qquad (\text{A.24})$$

Finally, using Eqs. (A.2), (A.13), (A.15), (A.22), and (A.24) in Eq. (A.1) we obtain

$$j(E,\mathbf{q}_{\parallel}) = \frac{f(E)}{4\pi^3 \hbar} [1 - r(E,-\mathbf{q}_{\parallel})] \qquad (\text{A.25})$$

where

$$r(E,\mathbf{q}_{\parallel}) = r^e(E,\mathbf{q}_{\parallel}) + r^i(E,\mathbf{q}_{\parallel}) \qquad (\text{A.26})$$

is the total reflection coefficient.

The energy distribution of the emitted electrons is obtained by integrating $j(E,\mathbf{q}_{\parallel})$ over all $\mathbf{q}_{\parallel}$ space. Replacing $r(E,-\mathbf{q}_{\parallel})$ by $r(E,\mathbf{q}_{\parallel})$ under the integral sign does not change the result and we obtain

$$j(E) = \frac{f(E)}{4\pi^3 \hbar} \iint [1 - r(E,\mathbf{q}_{\parallel})] \, d^2q_{\parallel} \qquad (\text{A.27})$$

APPENDIX B

PHYSICAL CONSTANTS, UNITS, AND CONVERSION FACTORS

Velocity of light: $c = 2.998 \times 10^{10}$ cm sec^{-1}

Electronic mass: $m = 9.11 \times 10^{-28}$ g

Electronic charge: $e = 4.80 \times 10^{-10}$ statcoulomb

Planck's constant: $\hbar = 1.054 \times 10^{-27}$ erg sec

Boltzmann's constant: $k_B = 1.38 \times 10^{-16}$ erg K^{-1}

Except where the contrary is explicitly stated, the formulas in this book are given in the absolute cgs system of units (for a definition of the electromagnetic units in this system see, e.g., Shockley, 1950). The equivalent expressions in atomic units (used in many articles and books dealing with the electronic properties of solids) are obtained by putting $m = e = \hbar = 1$.

Atomic unit of length: $\hbar^2/me^2 = 0.529 \times 10^{-8}$ cm

Atomic unit of energy (Hartree): $me^4/\hbar^2 = 4.36 \times 10^{-11}$ erg

One Hartree = 2 Rydbergs (Ry) = 27.21 eV

One electron volt (eV) = 1.6×10^{-12} erg

One angstrom (Å) = 10^{-8} cm

REFERENCES

Abrikosov, A. A., Gorkov, L. P., and Dzyaloshinski, T. E. *Methods of Quantum Field Theory in Statistical Physics*. Englewood Cliffs, New Jersey: Prentice-Hall, 1963.

Amelio, G. F. Theory for the energy distribution of secondary electrons. *J. Vac. Sci. Tech.*, 1970, 7, 593–604.

Anderson, J. R., Papaconstantopoulos, D. A., Boyer, L. L., and Schirber, J. E. Spin-polarized band-structure calculations for Ni. *Phys. Rev. B*, 1979, 20, 3172–3185.

Anderson, S., and Kasemo, B., Determination of energy gap edges in copper and nickel from specular electron reflectivities. *Solid State Commun.*, 1970, 8, 961–964.

Appelbaum, J. A., and Hamann, D. R., Variational calculation of the image potential near a metal surface. *Phys. Rev. B*, 1972a, 6, 1122–1130.

Appelbaum, J. A., and Hamann, D. R., Self-consistent electronic structure of solid surfaces. *Phys. Rev. B*, 1972b, 6, 2166–2177.

Appelbaum, J. A., and Hamann, D. R., Electronic structure of the Cu(111) surface. *Solid State Commun.*, 1978, 27, 881–883.

Appelbaum, J. A., Baraff, G. A., and Hamann, D. R., The Si(100) surface: A theoretical study of the unreconstructed surface. *Phys. Rev. B*, 1975, 11, 3822–3831.

Appelbaum, J. A., Baraff, G. A., and Hamann, D. R., The Si(100) surface. III. Surface reconstruction. *Phys. Rev. B*, 1976, 14, 588–601.

Arthur, J. R., Energy distribution of field emission from germanium. *Surf. Sci.*, 1964, 2, 389–395.

Arthur, J. R., Photosensitive field emission from *p*-type germanium. *J. Appl. Phys.*, 1965, 36, 3221–3227.

Azizov, U. V., and Shuppe, G. N., Emission and adsorption characteristics of faces of a tungsten single crystal. *Sov. Phys. Solid State*, 1966, 7, 1591–1594.

Bagchi, A., Theory of surface effect in photoassisted field emission. *Phys. Rev. B*, 1974, 10, 542–553.

Barbour, J. P., Dolan, W. W., Trolan, J. K., Martin, E. E., and Dyke, W. P., Space-charge effects in field emission. *Phys. Rev.*, 1953, 92, 45–51.

Baribeau, J. M., and Carette, J. D., Surface barrier of W(110) and surface resonances. *Surf. Sci.*, 1981, 112, 241–253.

Baskin, L. M., Lvov, O. I., and Fursey, G. N., General features of field emission from semiconductors. *Phys. Status Solidi B*, 1971, 47, 49–62.

Baskin, L. M., Lvov, O. I., and Fursey, G. N., On the theory of field emission from *p*-type semiconductors. *Phys. Status Solidi A*, 1977, 42, 757–767.

Beeby, J. L., The diffraction of low-energy electrons by crystals. *J. Phys. C*, 1968, 1, 82–87.

Belford, G. G., Kupperman, A., and Phipps, T. E., Application of numerical methods to the theory of the periodic deviations in the Schottky effect. *Phys. Rev.*, 1962, 128, 524–531.

Bell, A. E., and Swanson, L. W., Total energy distributions of field-emitted electrons at high current density. *Phys. Rev. B*, 1979, 19, 3353–3364.

Bennett, A. J., and Falicov, L. M., Theory of the electronic configuration of a metallic surface-adsorbate system. *Phys. Rev.*, 1966, 151, 512–518.

Bennett, M., and Inkson, J. C., Exchange and correlation potential in silicon. *J. Phys. C*, 1977, 10, 987–999.

Bermond, J. M., A measurement of the local electric field on field emitter crystals using T–F emission. *Surf. Sci.,* 1975, **50,** 311–328.

Bermond, J. M., Lenoir, M., Prulhiere, J. P., and Drechsler, M., Numerical data and experimental proof of the unified theory of electron emission (Christov). *Surf. Sci.,* 1974, **42,** 306–323.

Best, P. E., Angle-resolved secondary electron emission spectra from Si (111) 7 × 7 surface states. *Phys. Rev. Lett.,* 1975, **34,** 674–677.

Best, P. E., Energy and angular dependent secondary electron emission from a silicon (111) 7 × 7 surface. Emission from bulk states. *Phys. Rev. B,* 1976, **14,** 606–619.

Best, P. E., Energy and angular dependent secondary electron emission from a silicon (111) 7 × 7 surface II. Emission from surface-state resonances. *Phys. Rev. B,* 1979, **19,** 2246–2249.

Billington, R. L., and Rhodin, T. N., Surface resonances on W(100): Effects of Au and Cu adsorption. *Phys. Rev. Lett.,* 1978, **41,** 1602–1605.

Brown, A. A., Neelands, L. J., and Farnsworth, H. E., Thermionic work function of the (100) face of a tungsten single crystal. *J. Appl. Phys.,* 1950, 21, 1–4.

Buckingham, R. A., *Numerical Methods.* London: Pitman, 1962.

Bullett, D. W., and Cohen, M. L., Localized orbital approach to chemisorption I: H on W (100). *J. Phys. C,* 1977, **10,** 2083–2099.

Burgess, R. E., Kroemer, H., and Houston, J. M., Corrected values of Fowler–Nordheim field emission functions $v(y)$ and $s(y)$. *Phys. Rev.,* 1953, **90,** 515.

Campagna, M., Alvarado, S. F., and Kisker, E., Polarised electrons from metallic systems, in *Electrons in disordered metals and at metallic surfaces,* P. Phariseau, B. L. Györffy, and L. Scheire (Eds.). New York: Plenum Press, 1979.

Campuzano, J. C., Inglesfield, J. E., King, D. A., and Somerton, C., Surface states on W (001) (1 × 1) and ($\sqrt{2}$ × $\sqrt{2}$) R45° phases: Angle-resolved photoemission. *J. Phys. C,* 1981, **14,** 3099–3113.

Caroli, B., Field-emission in the presence of a magnetic field. *Surf. Sci.,* 1975, **51,** 237–248.

Caruthers, E., and Kleinman, L., Effects of different potentials on iron surface states. *Phys. Rev. Lett.,* 1975, **35,** 738–740.

Caruthers, E., Kleinman, L., and Aldredge, G. P., Effects of the potential on surface states. *Phys. Rev. B,* 1974 **10,** 1252–1254.

Chadi, D. J., Atomic and electronic structures of reconstructed Si (100) surfaces. *Phys. Rev. Lett.,* 1979, **43,** 43–47.

Chazalviel, J. N., and Yafet, Y., Theory of the spin polarization of field-emitted electrons from nickel. *Phys. Rev. B,* 1977, **15,** 1062–1071.

Chen, J. R., and Gomer, R., Mobility and two-dimensional compressibility of Xe on the (110) plane of tungsten. *Surf. Sci.,* 1980, **94,** 456–468.

Christensen, N. E., and Feuerbacher, B., Volume and surface photoemission from tungsten. I. Calculation of band structure and emission spectra. *Phys. Rev. B,* 1974, **10,** 2349–2372.

Christensen, N. E., and Willis, R. F., Secondary electron emission from tungsten. Observation of the electronic structure of the semi-infinite crystal. *J. Phys. C.,* 1979, **12,** 167–207.

Christov, S. G., General theory of electron emission from metals. *Phys. Status Solidi,* 1966, **17,** 11–26.

Christov, S. G., Recent tests and new applications of the unified theory of electron emission. *Surf. Sci.,* 1978, **70,** 32–51.

Chung, M. S., and Everhart, T. E., Simple calculation of energy distribution of low-energy secondary electrons emitted from metals under electron bombardment. *J. Appl. Phys.,* 1974, **45,** 707–709.

Clark, H. E., and Young, R. D., Field emission through single strontium atoms adsorbed on a tungsten surface. *Surf. Sci.,* 1968, **12,** 385–389.

Connolly, J. W. D., Energy bands in ferromagnetic nickel. *Phys. Rev.,* 1967, **159,** 415–426.

Cooper, B. R., Kreiger, E. L., and Segall, B., Determination of electron energy bands by phase-shift parametrization:application to silver. *Phys. Rev. B*, 1971, **4**, 1734–1748.

Cutler, P. H., and Gibbons, J. J., Model for the surface potential barrier and the periodic deviations in the Schottky effect. *Phys. Rev.*, 1958, **111**, 394–402.

Debe, M. K., and King, D.A., New evidence for a clean thermally induced c(2 × 2) surface structure on W(100). *J. Phys. C*, 1977, **10**, L303–L308.

Debe, M. K., and King, D. A., The clean thermally induced W(001) (1 × 1) → ($\sqrt{2}$ × $\sqrt{2}$) R45° surface structure transition and its crystallography. *Surf. Sci.*, 1979, **81**, 193–237.

Dempsey, D. G., and Kleinman, L., Surface properties and the photoelectron spin polarization of nickel. *Phys. Rev. Lett.*, 1977, **39**, 1297–1300.

Dempsey, D. G., Grise, W. R., and Kleinman, L., Energy bands of (100) and (110) ferromagnetic Ni films. *Phys. Rev. B*, 1978, **18**, 1270–1280.

Desjonqueres, M. C., and Cyrot-Lackmann, F., On the surface states of Mo and W. *J. Phys. F*, 1976, **6**, 567–572.

Dionne, N. J., and Rhodin, T. N., Field emission energy spectroscopy of the platinum-group metals. *Phys. Rev. B*, 1976, **14**, 322–340.

Domke, M., Jähnig, G., and Drechsler, M., Measurement of the adsorption energy on single cyrstal faces by field emission (hydrogen on tungsten). *Surf. Sci.*, 1974, **42**, 389–403.

Dousmanis, G. C., and Duncan, R. C., Jr., Calculations on the shape and extent of space charge regions in semiconductor surfaces. *J. Appl. Phys.*, 1958, **29**, 1627–1629.

Duke, C. B., and Alferieff, M. E., Field emission through atoms adsorbed on a metal surface. *J. Chem. Phys.*, 1967, **46**, 923–937.

Duke, C. B., and Laramore, G. E., Effect of lattice vibrations in a multiple-scattering description of low-energy electron diffraction. I. Formal perturbation theory. *Phys. Rev. B*, 1970, **2**, 4765–4782.

Durham, P., and Kar, N., Special features in photoemission from the s–p bands of copper. *Surf. Sci.*, 1981, **111**, L648–L656.

Dushman, S., Electron emission from metals as a function of temperature. *Phys. Rev.*, 1923, **21**, 623–636.

Dyke, W. P., and Dolan, W. W., Field emission, in *Advances in Electronics and Electron Physics*, Vol. 8, L. Marton (Ed.). New York: Academic Press, 1956.

Dyke, W. P., Trolan, J. K., Dolan, W. W., and Barnes, J., The field emitter: Fabrication, electron microscopy, and electric field calculations. *J. Appl. Phys.*, 1953, **24**, 570–576.

Eastman, D. E., and Himpsel, F. J., Energy band dispersion $E(\mathbf{k})$ of transition and noble metals using angle-resolved photoemission, in *Physics of Transition Metals, 1980*, P. Rhodes (Ed.). Bristol: Institute of Physics, Conference Series Number 55, 1981.

Echenique, P. M., and Pendry, J. B., The existence and detection of Rydberg states at surfaces. *J. Phys. C*, 1978, **11**, 2065–2075.

Edwards, D. M., and Hertz, J. A., Electron–magnon interactions in itinerant ferromagnetism. II. Strong ferromagnetism. *J. Phys. F*, 1973, **3**, 2191–2205.

Ehrlich, C. D., and Plummer, E. W., Measurement of the absolute tunnelling current density in field emission from tungsten (110). *Phys. Rev. B*, 1978, **18**, 3767–3771.

Eib, W., and Alvarado, S. F., Spin-polarized photoelectrons from nickel single crystals. *Phys. Rev. Lett.*, 1976, **37**, 444–446.

El-Kareh, A. B., Wolfe, J. C., and Wolfe, J. E., Contribution to the general analysis of field emission. *J. Appl. Phys.*, 1977, **48**, 4749–4753.

Engel, T., and Gomer, R. Adsorption of CO on tungsten: Field emission from single planes. *J. Chem. Phys.*, 1969, **50**, 2428–2437.

Engel, T., and Gomer, R., Adsorption of inert gases on tungsten: Measurements on single crystal planes. *J. Chem. Phys.*, 1970, **52**, 5572–5580.

Estrup, P. J., and Anderson, J., Chemisorption of hydrogen on tungsten (100). *J. Chem. Phys.*, 1966, **45**, 2254–2260.

Ewald, P. P., Die Berechnung Optischer und elektrostatischer Gitterpotentiale. *Ann. Phys. (Leipzig)*, 1921, **64**, 253–287.

Feder, R., Relativistic theory of low-energy electron diffraction: formalism and application to KKRZ-pseudopotentials. *Phys. Status Solidi B*, 1972, **49**, 699–710.

Feder, R., Relativistic theory of low energy electron diffraction: application to the (001) and (110) surfaces of tungsten. *Phys. Status Solidi B*, 1974, **62**, 135–146.

Feder, R., Spin polarization in low-energy electron diffraction from W(001). *Phys. Rev. Lett.*, 1976, **36**, 598–600.

Feder, R., and Pendry, J. B., Theory of secondary electron emission. *Solid State Commun.*, 1978, **26**, 519–521.

Feder, R., and Sturm, K., Spin–orbit effects in the electronic structure of the (001) surface of bcc 4*d* and 5*d* transition metals. *Phys. Rev. B*, 1975, **12**, 537–548.

Felter, T. E., Barker, R. A., and Estrup, P. J., Phase transition on Mo(100) and W(100) surfaces. *Phys. Rev. Lett.*, 1977, **38**, 1138–1141.

Feuchtwang, T. E., and Cutler, P. H., Effects of electronic energy-band structure on the energy distribution of field-emitted electrons. *Phys. Rev. B*, 1976, **14**, 5237–5253.

Feuchtwang, T. E., Cutler, P. H., and Nagy, D., A review of the theoretical and experimental analyses of electron spin polarization in ferromagnetic transition metals. II. New theoretical results for the analysis of ESP in field emission, photoemission, and tunnelling. *Surf. Sci.*, 1978, **75**, 490–528.

Feuchtwang, T. E., Cutler, P. H., and Schmit, J., A review of the theoretical and experimental analyses of electron spin polarization in ferromagnetic transition metals. I. Field emission, photoemission, magneto-optic Kerr effect and tunnelling. *Surf. Sci.*, 1978, **75**, 401–489.

Feuerbacher, B., and Willis, R. F., Photoemission and electron states at clean surfaces. *J. Phys. C*, 1976, **9**, 169–216.

Feuerbacher, B., Fitton, B., and Willis, R. F. (Eds.), *Photoemission and the Electronic Properties of Surfaces*. New York: Wiley, 1978.

Fine, J., Madey, T. E., and Scheer, M. D., The determination of work function from the ratio of positive to negative surface ionization of an alkali halide. *Surf. Sci.*, 1965, **3**, 227–233.

Fischer, R., Einfluss der effektiven Mass auf die Energieverteilung der bei der äusseren Feldemission aus Halb leitern emittierten Elektronen. *Phys. Status Solidi*, 1962, **2**, 1466–1470.

Flood, D. J., Molecular vibration spectra from field-emission energy distributions. *J. Chem. Phys.*, 1970, **52**, 1355–1360.

Fomenko, V. S., *Handbook of Thermionic Properties*. New York: Plenum, 1966.

Fowler, R. H., and Nordheim, L. W., Electron emission in intense electric fields. *Proc. R. Soc. London*, 1928, **A119**, 173–181.

Frankl, D. R., *Electrical Properties of Semiconductor Surfaces*. New York: Pergamon, 1967.

Froitzheim, H., Ibach, H., and Lehwald, S., Surface sites of H on W(100). *Phys. Rev. Lett.*, 1976, **36**, 1549–1551.

Fursey, G. N., and Egorov, N. V., Field emission from *p*-type Si. *Phys. Status Solidi*, 1969, **32**, 23–29.

Gadzuk, J. W., Resonance-tunnelling spectroscopy of atoms adsorbed on metal surfaces: theory. *Phys. Rev. B*, 1970, **1**, 2110–2129.

Gadzuk, J. W., Tunnelling from Cambridge surface states. *J. Vac. Sci. Tech.*, 1972, **9**, 591–596.

Gadzuk, J. W., and Plummer, E. W., Energy distributions for thermal-field emission. *Phys. Rev. B*, 1971a, **3**, 2125–2129.

Gadzuk, J. W., and Plummer, E. W., Hot-hole–electron cascades in field emission from metals. *Phys. Rev. Lett.*, 1971b, **26**, 92–95.

Gadzuk, J. W., and Plummer, E. W., Field emission energy distribution (FEED). *Rev. Modern Phys.,* 1973, **45,** 487–548.

Ganachaud, J. P., and Cailler, M., A Monte Carlo calculation of the secondary electron emission of normal metals. *Surf. Sci.,* 1979, **83,** 498–518.

Gartland, P. O., and Slagsvold, B. J., Transitions conserving parallel momentum in photoemission from the (111) face of copper. *Phys. Rev. B,* 1975, **12,** 4047–4058.

Gay, J. G., Smith, J. R., and Arlinghaus, F. J., Large surface-state/surface-resonance density on copper (100). *Phys. Rev. Lett.,* 1979, **42,** 322–335.

Glauber, R. J., Time-dependent displacement correlations and inelastic scattering by crystals. *Phys. Rev.,* 1955, **98,** 1692–1698.

Gomer, R., *Field emission and field ionization.* Cambridge, Massachusetts: Harvard University Press, 1961.

Gomer, R., Current fluctuations from small regions of adsorbate covered field emitters. *Surf. Sci.,* 1973, **38,** 373–393.

Gomer, R., Chemisorption on metals, in *Solid State Physics,* Vol. 30, H. Ehrenreich, F. Seitz, and D. Turnbull (Eds.). New York: Academic Press, 1975.

Gomer, R., Applications of field emission to chemisorption. *Surf. Sci.,* 1978, **70,** 19–31.

Good, R. H., and Müller, E. W., Field emission, in *Handbuch der Physik,* Vol. 21, S. Flügge (Ed.). Berlin: Springer-Verlag, 1956.

Griffiths, K., King, D. A., and Thomas, G., Hydrogen induced symmetry switching of the W(001) ($\sqrt{2} \times \sqrt{2}$)R45° low-temperature phase. *Vacuum,* 1981, **31,** 671–674.

Grise, W. R., Dempsey, D. G., Kleinman, L., and Mednick, K., Relativistic energy bands of (010) tungsten thin films. *Phys. Rev. B,* 1979, **20,** 3045–3050.

Gurman, S. J., Theory of surface states. II. The copper and tungsten (001) surfaces. *Surf. Sci.,* 1976a, **55,** 93–108.

Gurman, S. J., Surface states and the surface potential: The copper (111) face. *J. Phys. C,* 1976b, **9,** L609–L613.

Gurman, S. J., and Pendry, J. B., Existence of generalized surface states. *Phys. Rev. Lett.,* 1973, **31,** 637–639.

Guth, E., and Mullin, C. J., Electron emission of metals in electric fields. I. Explanation of the periodic deviations from the Schottky line. *Phys. Rev.,* 1941, **59,** 575–584.

Guth, E., and Mullin, C. J., Electron emission of metals in electric fields. III. The transition from thermionic to cold emission. *Phys. Rev.,* 1942, **61,** 339–348.

Haas, G. A., and Coomes, E. A., Schottky effect for SrO films on molybdenum. *Phys. Rev.,* 1955, **100,** 640–641.

Haas, G. A., and Thomas, R. E., Thermionic emission and work function, in *Techniques of Metal Research,* E. Passaglia (Ed.). Vol. 6. New York: Wiley, 1972.

Haas, G. A., and Thomas, R. E., Work function and secondary emission studies of various Cu crystal faces. *J. App. Phys.,* 1977, **48,** 86–93.

Hachenberg, O., and Brauer, W., Secondary electron emission from solids, in *Advances in Electronics and Electron Physics,* L. Marton (Ed.). Vol. 11. New York: Academic Press, 1959.

Handler, P., Energy level diagrams for germanium and silicon surfaces. *J. Phys. Chem. Solids,* 1960, **14,** 1–8.

Handler, P., and Portnoy, W. M., Electronic surface states and the cleaned germanium surface. *Phys. Rev.,* 1959, **116,** 516–526.

Harris, J., and Jones, R. O. Dynamical corrections to the image potential. *J. Phys. C,* 1973, **6,** 3585–3604.

Harris, J., and Jones, R. O., Image force for a moving charge. *J. Phys. C,* 1974, **7,** 3751–3754.

Harrison, W. A., *Solid State Theory.* New York: McGraw-Hill, 1970.

Harrison, W. A., Surface reconstruction on semiconductors. *Surf. Sci.,* 1976, **55,** 1–19.

Hedin, L., and Lundqvist, B. I., Explicit local exchange-correlation potentials. *J. Phys. C*, 1971, **4**, 2064-2083.

Hedin, L., and Lundqvist, S., Effects of electron–electron and electron–phonon interactions on the one-electron states of solids, in *Solid State Physics*, Vol. 23. F. Seitz, D. Turnbull, and H. Ehrenreich (Eds.). New York: Academic Press, 1969.

Heimann, P., Hermanson, J., Miosga, H., and Neddermeyer, H., *d*-like surface-state bands on Cu(100) and Cu(111) observed in angle-resolved photoemission spectroscopy. *Phys. Rev. B*, 1979, **20**, 3059-3066.

Heine, V., On the general theory of surface states and scattering of electrons in solids. *Proc. Phys. Soc. London*, 1963, **81**, 300-310.

Herlt, H. J., Feder, R., Meister, G., and Bauer, E. G., Experiment and theory of the elastic reflection coefficient from tungsten. *Solid State Commun.*, 1981, **38**, 973-976.

Herring, C., and Nichols, M. H., Thermionic emission. *Rev. Modern Phys.*, 1949, **21**, 185-270.

Himpsel, F. J., and Eastman, D. E., Observation of a Λ_1—symmetry surface state on Ni(111). *Phys. Rev. Lett.*, 1978, **41**, 507-511.

Himpsel, F. J., and Eastman, D. E., Photoemission studies of intrinsic surface states on Si(100). *J. Vacuum Sci. Technol.*, 1979, **16**, 1297-1299.

Ho, W., Willis, R. F., and Plummer, E. W., Observation of nondipole electron impact vibrational excitations: H on W(100). *Phys. Rev. Lett.*, 1978, **40**, 1463-1466.

Hohenberg, P., and Kohn, W., Inhomogeneous electron gas. *Phys. Rev.*, 1964, **136**, 864-871.

Holland, B.W., A simple model for temperature effects in LEED. *Surf. Sci.*, 1971, **28**, 258-266.

Holmes, M. W., and Gustafsson, T., Dispersion of surface states on W(100) and the surface reconstruction. *Phys. Rev. Lett.*, 1981, **47**, 443-446.

Holmes, M. W., King, D. A., and Inglesfield, J. E., Characterization and transferability of surface states on W(110). *Phys. Rev. Lett.*, 1979, **42**, 394-397.

Hölzl, J., and Schulte, F. K., Work function of metals, in *Springer Tracts in Modern Physics*. Vol. 85. G. Höhler (Ed.). Berlin: Springer-Verlag, 1979.

Houde, A. L., Ph.D. dissertation, University of Notre Dame, 1952 (unpublished).

Hughes, F. L., Levinstein, H., and Kaplan, R., Surface properties of etched tungsten single crystals. *Phys. Rev.*, 1959, **113**, 1023-1028.

Hughes, O. H., and White, P. M., The energy distribution of field-emitted electrons from GaAs. *Phys. Status Solidi*, 1969, **33**, 309-316.

Hutson, A.R., Velocity analysis of thermionic emission from single-crystal tungsten. *Phys. Rev.*, 1955, **98**, 889-901.

Inglesfield, J. E., The electronic structure of surfaces with the matching Green function method. II. fcc and bcc transition metal surfaces. *Surf. Sci.*, 1978, **76**, 379-396.

Inglesfield, J. E., and Holland, B. W., Electrons at surfaces, in *The Chemical Physics of Solid Surfaces and Heterogeneous Catalysis*. Vol. 1. D. A. King and D. P. Woodruff (Eds.). New York: Elsevier, 1981.

Inkson, J. C., The electron–electron interaction near an interface. *Surf. Sci.*, 1971, **28**, 69-76.

Itskovich, F. I., On the theory of field emission from metals. *Sov. Phys. JETP*, 1966, **23**, 945-953.

Itskovich, F. I., Contribution to the theory of field emission from metals. II. *Sov. Phys. JETP*, 1967, **25**, 1143-1153.

Jahnke, E., and Emde, F., *Tables of Functions*. New York: Dover, 1945.

Jeans, J., *The Mathematical Theory of Electricity and Magnetism*. Cambridge: University Press, 1941.

Jennings, P. J., An analysis of the surface barrier structure of W(001). *Surf. Sci.*, 1978, **75**, L773-L776.

Jennings, P. J., The surface barrier structure of Ni(001). *Surf. Sci.*, 1979, **88**, L25-L28.

Jones, J. P., Some developments in field emission techniques and their application, in *Chemical*

Physics of Solids and Their Surfaces. Vol. 8. M. W. Roberts and J. M. Thomas (Eds.). London: The Royal Society of Chemistry, 1980.

Jones, J. P., and Roberts, E. W., Field emission energy distributions from layer and crystal structures of copper on W(100). *Surf. Sci.*, 1977, **64**, 355-361.

Juenker, D. W., Surface barrier analysis for metals by means of Schottky deviations. *Phys. Rev.*, 1955, **99**, 1155-1160.

Juenker, D. W., Colladay, G. S., and Coomes, E. A., Surface barrier analysis for the highly refractory metals by means of the Schottky deviations. *Phys. Rev.*, 1953, **90**, 772-778.

Kambe, K., Theory of low-energy electron diffraction. I. Application of the cellular method to monatomic layers. *Z. Naturforsch.*, 1967a, **22a**, 322-330.

Kambe, K., Theory of electron diffraction by crystals. I. Green's function and integral equation. *Z. Naturforsch.*, 1967b, **22a**, 422-431.

Kambe, K., Theory of low-energy electron diffraction. II. Cellular method for complex monolayers and multilayers. *Z. Naturforsch.*, 1968, **23a**, 1280-1294.

Kar, N., Effect of ordered overlayer on field emission from a substrate. *Surf. Sci.*, 1978, **70**, 101-113.

Kar, N., and Soven, P., Field emission energy distribution from (111) copper. *Solid State Commun.*, 1976a, **19**, 1041-1043.

Kar, N., and Soven, P., Field emission energy distribution from adsorbate covered tungsten. *Solid State Commun.*, 1976b, **20**, 977-979.

Kasowski, R. V., New interpretation of surface states in W and Mo. *Solid State Commun.*, 1975, **17**, 179-183.

Kerker, G. P., Ho, K. M., and Cohen, M. L., Mo (001) surface: A self-consistent caluclation of the electronic structure. *Phys. Rev. Let.*, 1978, **40**, 1593-1596.

Kerker, G. P., Yin, M. T., and Cohen, M. L., Self-consistent electronic structure for a Mo (001) surface with saturated H adsorption. *Solid State Commun.*, 1979, **32**, 433-436.

Kerkides, P., Ph.D. Thesis, University of Salford, 1976 (unpublished).

Kingston, R. H., and Neustadter, S. F., Calculation of the space charge, electric field, and free carrier concentration at the surface of a semiconductor. *J. App. Phys.*, 1955, **26**, 718-720.

Kirtley, J., and Hall, J. T., Theory of intensities in inelastic-electron tunnelling spectroscopy orientation of adsorbed molecules. *Phys. Rev. B*, 1980, **22**, 848-856.

Kisker, E., Gudat, W., Campagna, M., Kuhlmann, E., Hopster, H., and Moore, I. D., Crossover from negative to positive spin polarization in the photoyield from Ni(111) near threshold. *Phys. Rev. Lett.*, 1979, **43**, 966-969.

Kleinman, L., Correlation effects on the energy band of Ni. *Phys. Rev. B*, 1979, **19**, 1295-1298.

Kleint, C., Surface diffusion model of adsorption induced field emission flicker noise. II. Experiments. *Surf. Sci.*, 1971, **25**, 411-434.

Kohn, W., and Sham, L. J., Self-consistent equations including exchange and correlation effects. *Phys. Rev.*, 1965, **140**, 1133-1138.

Kuyatt, C. E., and Plummer, E. W., Field emission deflection energy analyzer. *Rev. Sci. Instrum.*, 1972, **43**, 108-111.

Kuyatt, C. E., and Simpson, J. A., Electron monochromator design. *Rev. Sci. Instrum.*, 1967, **38**, 103-111.

Lambe, J., and Jaklevic, R. C., Molecular vibration spectra by inelastic electron tunnelling. *Phys. Rev.*, 1968, **165**, 821-832.

Landau, L. D., and Lifshitz, E. M., *Quantum Mechanics*. London: Pergamon Press, 1958.

Landolt, M., and Campagna, M., Spin polarization of field-emitted electrons and magnetism at the (100) surface of Ni. *Phys. Rev. Lett.*, 1977, **38**, 663-666.

Landolt, M., and Campagna, M., Electron spin polarization in field emission. *Surf. Sci.*, 1978, **70**, 197-210.

Landolt, M., and Yafet, Y., Spin polarization of electrons field emitted from single-crystal iron surfaces. *Phys. Rev. Lett.,* 1978, **40,** 1401-1403.

Landolt, M., Campagna, M., Chazalviel, J. N., Yafet, Y., and Wilkens, B., New tunnelling experiments from strong ferromagnets and predictions of the band theory: Resolution of a controversy? *J. Vacuum Sci. Technol.,* 1977, **14,** 468-470.

Lang, N. D., The density-functional formalism and the electronic structure of metal surfaces, in *Solid State Physics,* Vol 28. H. Ehrenreich, F. Seitz, and D. Turnbull (Eds.). New York: Academic Press, 1973.

Lang, N. D. Interation between closed-shell systems and metal surfaces. *Phys. Rev. Lett.,* 1981, **46,** 842-845.

Lang, N. D., and Kohn, W., Theory of metal surfaces: Charge density and surface energy. *Phys. Rev. B,* 1970, **1,** 4555-4568.

Lang, N. D., and Kohn, W., Theory of metal surfaces: Induced surface charge and image potential. *Phys. Rev. B,* 1973, **7,** 3541-3550.

Lang, N. D., and Williams, A. R., Theory of atomic chemisorption on simple metals. *Phys. Rev. B,* 1978, **18,** 616-636.

Laramore, G. E., and Duke, C. B., Effect of lattice vibrations in a multiple scattering description of low-energy electron diffraction. II. Double-diffraction analysis of the elastic scattering cross section. *Phys. Rev. B,* 1970, **2,** 4783-4795.

Laue, von M., Die Entropiekonstante der Glühelektronen. *Jahrb. Radioakt. Elektron.,* 1918a, **15,** 205-256.

Laue, von M., Unter welchen Bedingungen kann man von einem Elektronengas reden? *Jahrb. Radioakt. Elektron.,* 1918b, **15,** 257-270.

Lea, C., and Gomer, R., Evidence of electron-electron scattering from field emission. *Phys. Rev. Lett.,* 1970, **25,** 804-806.

Lea, C., and Gomer, R., Energy distribution in field emission from krypton covered tungsten. *J. Chem. Phys.,* 1971, **54,** 3349-3359.

Lee, M. J. G., and Reifenberger, R., Periodic field-dependent photocurrent from a tungsten field emitter. *Surf. Sci.,* 1978, **70,** 114-130.

Lee, M. J. K., Perz, J. M., and Fawcett, E. (Eds.), Panel discussion: Polarisation of electrons emitted from ferromagnets. *Physics of Transition Metals, 1977.* 276-286. Bristol: Institute of Physics, Conference Series Number 39, 1978.

Leighton, R. B., *Principles of Modern Physics.* New York: McGraw-Hill, 1959.

Lewis, B. F., and Fischer, T. E., Energy distributions of field-emitted electrons from silicon: Evidence for surface states. *Surf. Sci.,* 1974, **41,** 371-376.

Liebsch, A., Effect of self-energy corrections on the valence-band photoemission spectra of Ni. *Phys. Rev. Lett.,* 1979, **43,** 1431-1434.

Liebsch, A., Ni *d*-band self-energy beyond the low-density limit. *Phys. Rev. B,* 1981, **23,** 5203-5212.

Louie, S. G., Hydrogen on Pd(111): self-consistent electronic structure, chemical bonding, and photoemission spectra. *Phys. Rev. Lett.,* 1979, **42,** 476-479.

Louie, S. G., Electronic structure of *d*-band metal surfaces and adsorption systems, in *Physics of Transition Metals, 1980.* P. Rhodes (Ed.). Bristol: Institute of Physics Conference Series Number 55, 1981.

Madey, T. E., and Yates, J. T., Chemisorption on single crystals: H_2 on (100) tungsten, in *Structure et Proprietes des Solids.* Editions due centre national de la recherche scientifique. No. 187. Paris, 1970.

Margoninski, Y., Electrical measurements on clean and oxidized germanium surfaces. *Phys. Rev.,* 1963, **132,** 1910-1918.

Mattheiss, L. F., Energy bands for the iron transition series. *Phys. Rev.,* 1964, **134,** A970-A973.

Mattheiss, L. F., Fermi surface in tungsten. *Phys. Rev.*, 1965, **139**, A1893–A1904.

McRae, E. G., Electron diffraction at crystal surfaces. I. Generalization of Darwin's dynamical theory. *Surf. Sci.*, 1968, **11**, 479–491.

McRae, E. G., Calculation of absorptive potentials for low-energy electron scattering from optical data for solid Mg, Al, Cu, Ag, Au, Bi, C, and Al_2O_3. *Surf. Sci.*, 1976, **57**, 761–765.

McRae, E. G., Electronic surface resonances of crystals. *Rev. Modern Phys.*, 1979, **51**, 541–568.

McRae, E. G., and Caldwell, C. W., Very low energy electron reflection at Cu (100) surfaces. *Surf. Sci.*, 1976, **57**, 77–92.

Miller, S. C., Jr., and Good, R. H., Jr., A WKB-type approximation to the Schrödinger equation. *Phys. Rev.*, 1953a, **91**, 174–179.

Miller, S. C., Jr., and Good, R. H., Jr., Periodic deviations in the Schottky effect. *Phys. Rev.*, 1953b, **92**, 1367–1372.

Modinos, A., Field emission from surface states in semiconductors. *Surf. Sci.*, 1974, **42**, 205–227.

Modinos, A., A Green function technique for calculating total energy distributions of field-emitted electrons. *J. Phys. C*, 1976, **9**, 3867–3876.

Modinos, A., Field emission spectroscopy of transition metals. *Surf. Sci.*, 1978a, **70**, 52–91.

Modinos, A., The electronic work function of the different faces of tungsten. *Surf. Sci.*, 1978b, **75**, 327–341.

Modinos, A., Theory of thermionic emission. *Surf. Sci.*, 1982, **115**, 469–500.

Modinos, A., and Nicolaou, N., A method for the evaluation of tunnelling probabilitites through a slowly varying potential barrier containing potential holes. *J. Phys. C*, 1971, **4**, 2875–2893.

Modinos, A., and Nicolaou, N., Surface density of states and field emission. *Phys. Rev. B*, 1976, **13**, 1536–1547.

Modinos, A., and Oxinos, G., Electronic states of a semi-infinite crystal with a surface inpurity. *Can. J. Phys.*, 1978, **56**, 1531–1538.

Modinos, A., and Oxinos, G., Field emission from tungsten (110). *J. Phys. C*, 1981a, **14**, 1373–1380.

Modinos, A., and Oxinos, G., Spin-polarised field emission from nickel (100), in *Physics of Transition Metals, 1980*. P. Rhodes (Ed.). Bristol: Institute of Physics, Conference Series Number 55, 1981b.

Modinos, A., Aers, G. C., and Paranjape, B. V., Quantum-size effect in normal-metal tunnelling. *Phys. Rev. B*, 1979, **19**, 3996–4011.

Moore, I. D., and Pendry, J. B., Theory of spin polarised photoemission from nickel. *J. Phys. C*, 1978, **11**, 4615–4622.

Moruzzi, V. L., Janak, J. F., and Williams, A. R., *Calculated Electronic Properties of Metals*. New York: Pergamon, 1978.

Mott, N. F., and Jones, H., *The Theory of the Properties of Metals and Alloys*. New York: Dover, 1958.

Müller, E. W., Elektronenmikroskopische Beobachtungen von Feldkathoden. *Z. Phys.*, 1937, **106**, 541–550.

Müller, E. W., Work function of tungsten single crystal planes measured by the field emission microscope. *J. Appl. Phys.*, 1955, **26**, 732–737.

Munick, R. J., LaBerge, W. B., and Coomes, E. A., Periodic deviations in the Schottky effect for tantalum. *Phys. Rev.*, 1950, **80**, 887–891.

Murphy, E. L., and Good, R. H., Jr., Thermionic emission, field emission, and the transition region. *Phys. Rev.*, 1956, **102**, 1464–1473.

Muscat, J. P., and Newns, D. M., Valence electronic structure of alkalis adsorbed on free-electron like and transition metals. *Surf. Sci.*, 1978, **74**, 355–364.

Nagy, D., The polarization of electrons by tunnelling through a spin-dependent surface potential. *Surf. Sci.*, 1979, **90**, 102–108.

Nagy, D., Cutler, P. H., and Feuchtwang, T. E., Many-body and spin-dependent surface potential effects in spin-polarized field emission from metals. *Phys. Rev. B,* 1979, **19**, 2964–2974.

Nichols, M. H., The thermionic constants of tungsten as a function of crystallographic direction. *Phys. Rev.,* 1940, **57**, 297–306.

Nicolaou, N., and Modinos, A., Band-structure effects in field-emission energy distributions in tungsten. *Phys. Rev. B,* 1975, **11**, 3687–3696.

Nicolaou, N., and Modinos, A., Effect of surface potential on field emission energy distributions and on the surface density of states. *Surf. Sci.,* 1976, **60**, 527–539.

Noguera, C., Spanjaard, D., and Jepsen, D. W., Calculation of the density of states at the (100) and the (110) surfaces of molybdenum. *Phys. Rev. B,* 1978, **17**, 607–617.

Nordheim, L. W., The effect of the image force on the emission and reflection of electrons by metals. *Proc. R. Soc. London,* 1928, **A121**, 626–639.

Nottingham, W. B., Thermionic emission, in *Handbuch der Physik.* Vol. 21. S. Flügge (Ed.). Berlin: Springer-Verlag, 1956.

Nozawa, R., Bipolar expansion of screened Coulomb potentials, Helmholtz solid harmonics, and their addition theorems. *J. Math. Phys. N.Y.,* 1966, **7**, 1841–1860.

Oxinos, G., and Modinos, A., The polarization of physisorbed inert atoms due to the short range interaction with the metal substrate. *Surf. Sci.,* 1979, **89**, 292–303.

Oxinos, G., and Modinos, A., Green's function formalism for studying the electronic structure of atoms adsorbed on metals. *Can. J. Phys.,* 1980, **58**, 1126–1141.

Peierls, R. E., *Quantum Theory of Solids.* Oxford: Clarendon Press, 1955.

Pendry, J. B., Ion core scattering and low-energy electron diffraction—I. *J. Phys. C,* 1971, **4**, 2501–2513.

Pendry, J. B., *Low-Energy Electron Diffraction.* New York: Academic Press, 1974.

Pendry, J. B., and Gurman, S. J., Theory of surface states: General criteria for their existence. *Surf. Sci.,* 1975, **49**, 87–105.

Pendry, J. B., and Hopkinson, J. F. L., Photoemission from transition metal surfaces. *J. Phys. F,* 1978, **8**, 1009–1017.

Penn, D. R., Effect of bound hole pairs on the *d*-band photoemission spectrum of Ni. *Phys. Rev. Lett.,* 1979, **42**, 921–925.

Penn, D. R., and Plummer, E. W., Field emission as a probe of the surface density of states. *Phys. Rev. B,* 1974, **9**, 1216–1222.

Penn, D., Gomer, R., and Cohen, M. H., Energy distribution in field emission from adsorbate-covered surfaces. *Phys. Rev. B,* 1972, **5**, 768–778.

Perdew, J. P., and Zunger, A., Self-interaction correction to density-functional approximations for many-electron systems. *Phys. Rev. B,* 1981, **23**, 5048–5079.

Plummer, E. W., Photoemission and field emission spectroscopy, in *Interactions on Metal Surfaces: Topics in Applied Physics.* Vol. 4. R. Gomer (Ed.). Berlin: Springer-Verlag, 1975.

Plummer, E. W., and Bell, A. E., Field emission energy distributions of hydrogen and deuterium on the (100) and (110) planes of tungsten. *J. Vacuum Sci. Technol.,* 1972, **9**, 583–590.

Plummer, E. W., and Eberhardt, W., Magnetic surface states on Ni(100). *Phys. Rev. B,* 1979, **20**, 1444–1453.

Plummer, E. W. and Eberhardt, W., Angle-resolved photoemission as a tool for the study of surfaces, in *Advances in Chemical Physics.* Vol. 49. I. Prigogine, and S. A. Rice (Eds.), New York: Wiley, 1982.

Plummer, E. W., and Gadzuk, J. W., Surface states on tungsten. *Phys. Rev. Lett.,* 1970, **25**, 1493–1495.

Plummer, E. W., and Young, R. D., Field emission studies of electronic energy levels of adsorbed atoms. *Phys. Rev. B,* 1970, **1**, 2088–2109.

Politzer, B. A., and Cutler, P. H., Band-structure calculation of the electron spin polarization in field emission from ferromagnetic nickel. *Phys. Rev. Lett.*, 1972, **28**, 1330–1333.

Polizzotti, R. S., and Ehrlich, G., The work function of perfect W(110) planes: Fowler–Nordheim studies. *Surf. Sci.*, 1980, 24–36.

Posternak, M., Krakauer, H., Feeman, A. J., and Koelling, D. D., Self-consistent electronic structure of surfaces: Surface states and surface resonances on W(001). *Phys. Rev. B*, 1980, **21**, 5601–5612.

Protopopov, O. D., and Strigushchenko, I. V., Emission parameters of faces of a niobium single crystal. *Sov. Phys. Solid State*, 1968, **10**, 747–748.

Protopopov, O. D., Mikheeva, E. V., Sheinberg, B. N., and Shuppe, G. N., Emission parameters of tantalum and molybdenum single crystals. *Sov. Phys. Solid State*, 1966, **8**, 909–914.

Radon, T., Photo-field-emission spectroscopy of Γ-P and Γ-$\langle 013 \rangle$ bands of tungsten. *Surf. Sci.*, 1980, **100**, 353–367.

Radon, T., and Kleint, C., Photo-field-emission spectroscopy of optical transitions in the band structure of tungsten. *Surf. Sci.*, 1976, **60**, 540–560.

Radon, T., and Kleint, C., Periodische Stromabweichungen bei Photofeldemission aus Wolfram und ihre Wellenlängenabhängigkeit. *Ann. Phys. (Leipzig)*, 1977, **34**, 239–252.

Raether, H., Solid state excitations by electrons, in *Springer Tracts in Modern Physics*. Vol. 38. G. Höhler (Ed.), Berlin: Springer-Verlag, 1965.

Reifenberger, R., Haavig, D. L., and Egert, C. M., Numerical transmission probabilities and the oscillatory photo-induced field emission current: static and dynamic image charge effects. *Surf. Sci.*, 1981, **109**, 276–290.

Rhodin, T. N., and Ertl, G. (Eds.), *The Nature of the Surface Chemical Bond*. Amsterdam: North-Holland, 1979.

Richardson, O. W., Negative radiation from hot platinum. *Proc. Cambridge Philos. Soc.*, 1902, **11**, 286–295.

Richardson, O. W., The electron theory of contact electromotive force and thermoelectricity. *Philos. Mag.*, 1912, **23**, 263–278.

Richter, L., Ph.D. Thesis, University of Chicago, 1978 (unpublished).

Richter, L., and Gomer, R., Effect of Au adsorption on the tungsten (100) surface state. *Phys. Rev. Lett.*, 1976a, **37**, 763–765.

Richter, L., and Gomer, R., Field emission spectroscopy of gold on the (110) and (211) planes of tungsten. *Surf. Sci.*, 1976b, **59**, 575–580.

Richter, L., and Gomer, R., The effect of metallic adsorbates on the surface resonances of the tungsten and molybdenum (100) planes. *Surf. Sci.*, 1979, **83**, 93–116.

Rihon, N., Field emission energy distribution from zinc oxide: Observation of surface state emission. *Surf. Sci.*, 1978a, **70**, 92–100.

Rihon, N., Energy distribution study of field-emitted electrons from (100) and (111) nickel planes. *Phys. Status Solidi (a)*, 1978b, **49**, 697–703.

Rihon, N., Band bending in field emission spectroscopy from the zinc oxide (0001) polar face. *Phys. Status Solidi (a)*, 1981, **63**, 617–624.

Riviere, J. C., Work function: Measurements and results, in *Solid State Surface Science*. Vol. 1. M. Green (Ed.). New York: Marcel Dekker, 1969.

Roman, P., *Advanced Quantum Theory*. Reading, Massachusetts: Addison-Wesley, 1965.

Sacchetti, F., Single particle lifetime in copper. *J. Phys. F*, 1980, **10**, L231–L234.

Sakurai, T., and Müller, E. W., Field calibration using the energy distribution of field ionization. *Phys. Rev. Lett.*, 1973, **30**, 532–535.

Salehi, M., and Flinn, E. A., An experimental assessment of proposed universal yield curves for secondary electron emission. *J. Phys. D*, 1980, **13**, 281–289.

Salmon, L. T. J., and Braun, E., Energy distribution of electrons field-emitted from cadmium sulphide. *Phys. Status Solidi A,* 1973, **16,** 527–532.

Scalapino, D. J., and Marcus, S. M., Theory of inelastic electron–molecule interactions in tunnel junctions. *Phys. Rev. Lett.,* 1967, **18,** 459–461.

Schäfer, J., Schoppe, R., Hölzl, J., and Feder, R., Experimental and theoretical study of the angular resolved secondary electron spectroscopy (ARSES) for W(100) in the energy range $0 \leq E \leq 20$ eV. *Surf. Sci.,* 1981, **107,** 290–304.

Scheibner, E. J., and Tharp, L. N., Inelastic scattering of low-energy electrons from surfaces. *Surf. Sci.,* 1967, **8,** 247–265.

Schlier, R. E., and Farnsworth, H. E., Structure and adsorption characteristics of clean surfaces of germanium and silicon. *J. Chem. Phys.,* 1959, **30,** 917–926.

Schmidt, L., and Gomer, R., Adsorption of potassium on tungsten. *J. Chem. Phys.,* 1965, **42,** 3573–3598.

Schmit, J., and Good, R. H., Jr., Polarization effects in penetration of a barrier with a magnetic field applied. *Phys. Rev. B,* 1977, **16,** 1197–1200.

Schottky, W., Über den Einfluss von Structurwirkungen, besonders der Thomsonchen Bildkraft, auf die Elektronenemission der Metalle. *Physikalische Zeitschrift,* 1914, **15,** 872–878.

Schottky, W. Weitere Bemerkungen zum Elektronendampfproblem. *Phys. Z.,* 1919, **20,** 220–228.

Schottky, W., Über kalte und warme Elektronenentladungen. *Z. Phys.,* 1923, **14,** 63–106.

Schou, J., Transport theory for kinetic emission of secondary electrons from solids. *Phys. Rev. B,* 1980, **22,** 2141–2174.

Schwarz, L., Optimization of the statistical exchange parameter α for the free atoms H through Nb. *Phys. Rev. B,* 1972, **5,** 2466–2468.

Seah, M. P., Slow electron scattering from metals I. The emission of true secondary electrons. *Surf. Sci.,* 1969, **17,** 132–160.

Seifert, R. L. E., and Phipps, T. E., Evidence of a periodic deviation from the Schottky line. I. *Phys. Rev.,* 1939, **56,** 652–663.

Seiwatz, R., and Green, M., Space charge calculations for semiconductors. *J. Appl. Phys.,* 1958, **29,** 1034–1040.

Shcherbakov, G. P., and Sokolskaya, I. L., Experimental investigation of the energy distribution of field-emitted electrons from CdS single crystals. *Sov. Phys. Solid State,* 1963, **4,** 2581–2588.

Shelton, H., Thermionic emission from a planar tantalum crystal. *Phys. Rev.,* 1957, **107,** 1553–1557.

Shepherd, W. B., and Peria, W. T., Observation of surface-state emission in the energy distribution of electrons field-emitted from (100) orientated Ge. *Surf. Sci.,* 1973, **38,** 461–498.

Schockley, W., On the surface states associated with a periodic potential. *Phys. Rev.,* 1939, **56,** 317–323.

Schockley, W., *Electrons and Holes in Semiconductors, with Applications to Transistor Electronics.* New Jersey: Van Nostrand, 1950.

Slater, J. C., The ferromagnetism of nickel. *Phys. Rev.,* 1936, **49,** 537–545.

Slater, J. C., *Quantum Theory of Atomic Structure.* Vol. 1. New York: McGraw-Hill, 1960.

Slater, J. C., *Quantum Theory of Matter.* New York: McGraw-Hill, 1968.

Slater, J. C., *Symmetry and Energy Bands in Crystals.* New York: Dover, 1972.

Slater, J. C., *The Self-Consistent Field for Molecules and Solids: Quantum Theory of Molecules and Solids.* Vol. 4. New York: McGraw-Hill, 1974.

Smith, G. F., Thermionic and surface properties of tungsten crystals. *Phys. Rev.,* 1954, **94,** 295–308.

Smith, G. F., Velocity analysis of thermionic emission from single-crystal tungsten. *Phys. Rev.,* 1955, **100,** 1115–1116.

Smith, J. R., Beyond the local density approximation: Surface properties of (110)W. *Phys. Rev. Lett.,* 1970, **25,** 1023–1026.

Smith, J. R., Gay, J. G., and Arlinghaus, F. J., Self-consistent local-orbital method for calculating surface electronic structure: Application to Cu (100). *Phys. Rev. B,* 1980, **21,** 2201–2221.

Smith, N. V., and Mattheiss, L. F., Linear combination of atomic orbitals model for the electronic structure of H_2 on the W(001) surface. *Phys. Rev. Lett.,* 1976, **37,** 1494–1497.

Smith, R. J., Anderson, J., Hermanson, J., and Lapeyre, G. J., Study of W bulk bands with normal (001) photoemission using synchroton radiation. *Solid State Commun.,* 1976, **19,** 975–978.

Smoluchowski, R., Anisotropy of the electronic work function of metals. *Phys. Rev.,* 1941, **60,** 661–674.

Sommerfeld, A., Zur Elektronentheorie der Metalle auf grund der Fermischen Statistik. *Z. Phys.,* 1928, **47,** 1–60.

Sommerfeld, A., and Bethe, H., Zur Elektronentheorie der Metalle, in *Handbuch der Physik,* Vol. 24. S. Flügge (Ed.). Berlin: Springer-Verlag, 1933.

Soven, P., Plummer, E. W., and Kar, N., Field emission energy distribution (clean surfaces). *Crit. Rev. Solid State Sci.,* 1976, **6,** 111–131.

Spanjaard, D., Jepsen, D. W., and Marcus, P. M., Scattering matrix method for surface states applied to Al(001). *Phys. Rev. B,* 1979, **19,** 642–654.

Stafford, D. F., and Weber, A. H., Photoelectric and thermionic Schottky deviations for tungsten single crystals. *J. Appl. Phys.,* 1963, **34,** 2667–2670.

Stolz, H., Zur Theorie der Sekundärelektronenemission von Metallen. Der Transportprozess. *Ann. Phys. (Leipzig),* 1959, **3,** 197–210.

Stoner, E. C., Collective electron ferromagnetism. *Proc. R. Soc. London,* 1938, **A165,** 372–414.

Stoner, E. C., Collective electron ferromagnetism. II. Energy and specific heat. *Proc. R. Soc. London,* 1939, **A169,** 339–371.

Stratton, R., Theory of field emission from semiconductors. *Phys. Rev.,* 1962, **125,** 67–82.

Sratton, R., Energy distributions of field emitted electrons. *Phys. Rev.,* 1964, **135,** A794–A805.

Streitwolf, H. W., Zur Theorie der Sekundärelektronenemission von Metallen. Der Anregungsprozess. *Ann. Phys. (Leipzig),* 1959, **3,** 183–196.

Sultanov, V.M., Emission properties in different crystallographic directions of a monocrystalline tungsten sphere. *Radio Eng. Electron. Phys.,* 1964, **9,** 252–254.

Swanson, L. W., and Bell, A. E., Recent advances in field electron microscopy of metals, in *Advances in Electronics and Electron Physics.* Vol. 32. L. Marton (Ed.). New York: Academic Press, 1973.

Swanson, L. W., and Crouser, L. C., Total energy distribution of field-emitted electrons and single plane work functions for tungsten. *Phys. Rev.,* 1967a, **163,** 622–641.

Swanson, L. W., and Crouser, L. C., Anomalous total energy distribution for a molybdenum field emitter. *Phys. Rev. Lett.,* 1967b, **19,** 1179–1181.

Sykes, D. E., and Braun, E., Field emission from lead telluride. *Phys. Status Solidi B,* 1975, **69,** K137–K140.

Sytaya, E. P., Smorodinova, M. I., and Imangulova, N. I., Electron and ion emission from the (110) and (100) faces of a large tungsten monocrystal. *Sov. Phys. Solid State,* 1962, **4,** 750–753.

Tabor, D., and Wilson, J., The amplitude of surface atomic vibrations in the (100) plane of niobium. *Surf. Sci.,* 1970, **20,** 203–208.

Tamm, I., Über eine mögliche Art der Elektronenbindung an Kristalloberflächen. *Z. Phys.,* 1932, **76,** 849–850.

Tamm, P. W., and Schmidt, L. D., Interaction of H_2 with (100) W. I. Binding states. *J. Chem. Phys.,* 1969, **51,** 5352–5363.

Tamm, P. W., and Schmidt, L. D., Interaction of H_2 with (100) W. II. Condensation. *J. Chem. Phys.,* 1970, **52,** 1150–1160.

Tamm, P. W., and Schmidt, L. D., Condensation of hydrogen on tungsten. *J. Chem. Phys.,* 1971, **55**, 4253–4259.

Theophilou, A. K., An analytic estimate of the field penetration into metal surfaces. *J. Phys. F,* 1972, **2**, 1124–1136.

Theophilou, A. K., The energy density functional formalism for excited states. *J. Phys. C,* 1979, **12**, 5419–5430.

Theophilou, A. K., and Modinos, A., Metallic-field effect and its consequences in field emission, field ionization, and the capacitance of a capacitor. *Phys. Rev. B,* 1972, **6**, 801–812.

Thiry, P., Chandesris, D., Lecante, J., Guillot, C., Pinchaux, R., and Petroff, Y., *E* vs **k** and inverse lifetime of Cu(110). *Phys. Rev. Lett.,* 1979, **43**, 82–85.

Thurgate, S. M., and Jennings, P. J., Observations of LEED fine structure on the low-index planes of copper. *Surf. Sci.,* 1982, **114**, 395–404.

Todd, C. J., and Rhodin, T. N., Work function in field emission: The (110) plane of tungsten. *Surf. Sci.,* 1973, **36**, 353–369.

Tomasek, M., and Pick, S., Projected surface energy bands and Shockley surface states of transition metals. *Phys. Status Solidi B,* 1978, **89**, 11–26.

Topping, J., On the mutual potential energy of a plane network of doublets. *Proc. R. Soc. London,* 1927, **A114**, 67–72.

Treglia, G., Ducastelle, F., and Spanjaard, D., Perturbation treatment of correlations in transition metals. *J. Phys. (Paris),* 1980, **41**, 281–289.

Tsong, T. T., Field penetration and band bending near semiconductor surfaces in high electric fields. *Surf. Sci.,* 1979, **81**, 28–42.

Van der Ziel, A., A modified theory of production of secondary electrons in solids. *Phys. Rev.,* 1953, **92**, 35–39.

Van Hove, M. A., and Tong, S. Y., *Surface Crystallography by LEED.* Berlin: Springer-Verlag, 1979.

Van Oostrom, A., Temperature dependence of the work function of single crystal planes of tungsten in the range 78 K– 293 K. *Phys. Lett.,* 1963, **4**, 34–36.

Vorburger, T. V., Penn, D., and Plummer, E. W., Field emission work functions. *Surf. Sci.,* 1975, **48**, 417–431.

Wakoh, S., Band structure of metallic copper and nickel by a self-consistent procedure. *J. Phys. Soc. Jpn.,* 1965, **20**, 1894–1901.

Wakoh, S., and Yamashita, J., Band structure of ferromagnetic iron by a self-consistent procedure. *J. Phys. Soc. Jpn.,* 1966, **21**, 1712–1726.

Wakoh, S., and Yamashita, J., Band structure of cobalt by a self-consistent procedure. *J. Phys. Soc. Jpn.,* 1970, **28**, 1151–1156.

Wang, C.S., and Callaway, J., Energy bands in ferromagnetic nickel. *Phys. Rev. B,* 1977, **15**, 298–306.

Wang, C., and Gomer, R., Absolute coverage measurements: CO and oxygen on the (110) plane of tungsten. *Surf. Sci.,* 1978, **74**, 389–404.

Wang, C., and Gomer, R., Sticking coefficients of CO, O_2, and Xe on the (110) and (100) planes of tungsten. *Surf. Sci.,* 1979, **84**, 329–354.

Wang, C., and Gomer, R., Adsorption and coadsorption with oxygen of xenon on the (110) and (100) planes of tungsten. *Surf. Sci.,* 1980, **91**, 533–550.

Weling, F., The surface magnetisation of a tight-binding model ferromagnet. *J. Phys. F,* 1980, **10**, 1975–1993.

Weng, S. L., Surface resonances on the (100) plane of molybdenum. *Phys. Rev. Lett.,* 1977, **38**, 434–437.

Weng, S. L., Plummer, E. W., and Gustafsson, T., Experimental and theoretical study of the surface resonances on the (100) faces of W and Mo. *Phys. Rev. B,* 1978, **18**, 1718–1740.

Whitcutt, R. D. B., and Blott, B. H., Band edge at the (111) surface of copper measured by the total energy distribution of field-emitted electrons. *Phys. Rev. Lett.,* 1969, **23,** 639–640.

Whittaker, E. T., and Watson, G. N., *A Course of Modern Analysis.* Cambridge: University Press, 1952.

Willis, R. F., Angular-resolved secondary electron emission spectra from tungsten surfaces. *Phys. Rev. Lett.,* 1975, **34,** 670–674.

Willis, R. F., Structure determination of H chemisorbed on the reconstructed W(100) surface by vibrational modes analysis. *Surf. Sci.,* 1979, **89,** 457–466.

Willis, R. F., and Christensen, N. E., Secondary electron emission spectroscopy of tungsten: Angular dependence and phenomenology. *Phys. Rev. B,* 1978, **18,** 5140–5161.

Willis, R. F., Feuerbacher, B., and Fitton, B., Graphite conduction band states from secondary electron emission spectra. *Phys. Lett. A,* 1971a, **34,** 231–233.

Willis, R. F., Feuerbacher, B., and Fitton, B., Experimental investigation of the band structure of graphite. *Phys. Rev. B,* 1971b, **4,** 2441–2452.

Willis, R. F., Feuerbacher, B., and Fitton, B., Angular dependence of photoemission from intrinsic surface states on W(100). *Solid State Commun.,* 1976, **18,** 1315–1319.

Willis, R. F., Fitton, B., and Painter, G. S., Secondary electron emission spectroscopy and the observation of high-energy excited states in graphite: Theory and experiment. *Phys. Rev. B,* 1974, **9,** 1926–1937.

Wilson, R. G. Vacuum thermionic work functions of polycrystalline Be, Ti, Cr, Fe, Ni, Cu, Pt, and type 304 stainless steel. *J. Appl. Phys.,* 1966, **37,** 2261–2267.

Wolff, P. A., Theory of secondary electron cascade in metals. *Phys. Rev.,* 1954, **95,** 56–66.

Woodruff, D. P., Surface periodicity, crystallography and structure, in *The Chemical Physics of Solid Surfaces and Heterogeneous Catalysis.* Vol. 1. D. A. King and D.P. Woodruff (Eds.). New York: Elsevier, 1981.

Yang, J. J., and Yang, T. T., Investigation of thermionic emission in the region of periodic Schottky deviation. *Phys. Rev. B,* 1970, **1,** 3614–3623.

Yang, T. T., and Yang, J. J., Theoretical investigation of thermionic and photoelectric emission from semiconductors in the region of Schottky deviation. *Surf. Sci.,* 1971, **27,** 349–361.

Young, P. L., and Gomer, R., Field emission spectroscopy of gold on tungsten. *Surf. Sci.,* 1974, **44,** 268–274.

Young, R. D., Theoretical total-energy distribution of field-emitted electrons. *Phys. Rev.,* 1959, **113,** 110–114.

Young, R. D., and Clark, H. E., Effects of surface patch fields on field emission work function determinations. *Phys. Rev. Lett.,* 1966a, **17,** 351–353.

Young, R. D., and Clark, H. E., Anomalous work function of the tungsten (110) plane. *Appl. Phys. Lett.,* 1966b, **9,** 265–268.

Young, R. D., and Müller, E. W., Progress in field emission work function measurements of atomically perfect crystal planes. *J. Appl. Phys.,* 1962, **33,** 91–95.

Zavadil, J., and Modinos, A., Spin-polarised field emission from the (100) and (111) planes of nickel. *J. Phys. C,* 1982, **15,** 7255–7262.

INDEX

Adatom
  density of states, 212
  dipole moment associated with, 181
  field emission from site of, 224–225
  p resonance, 215
  s resonance, 214
Adsorption
  energy, measurement of, 203
  of gold on tungsten planes, 182–185
  of hydrogen on W(100), 188–189
  of inert atoms on tungsten planes, 170–182
  See also: Energy distribution, of field
    emitted electrons
Angular variables, 77
Atomic units, 353

Band bending, in semiconductors
  in the zero-emitted-current approximation,
    238–242
  in the zero-internal-current approximation,
    275–278
Band edge contraction, at the surface, 155
Band structure, 71, 94
  complex, 75
  of copper, 317, 319
  of nickel, 159, 160
  of tungsten, 143, 146, 338
Barrier: *see* Potential barrier
Bloch waves
  evanescent, 73, 75
  propagating, 71, 73, 75
Boltzmann's constant, 353

Cascade process, 327, 332
  energy distribution curve, 330, 333
Collisions
  electron–electron, 65, 327
  electron–plasmon, 66, 327
Complex band structure: *see* Band structure

Complex potential, 66, 88, 117, 141, 301,
  343
Correlation energy, 62

Debye temperature, 308
Debye–Waller factor, 308
Density of states
  deep inside the crystal, 110
  layer, 112
  local, 110
  one-dimensional, and its relation to the
    angle-resolved energy distribution
    of secondary emitted electrons,
    336–341
  surface, and its relation to the energy
    distribution of field-emitted
    electrons, 126–128, 179, 237
  See also: Adatom
Depletion region, in semiconductors, 286
  extension of, 292
Detailed balance, principle of, 347
Deviations from the Schottky line, 20–22,
  313–316, 322–323
Dipole layer, at the surface, 54, 173
Dipole potential, 200
Doubling layer method (scattering by a stack
  of atomic layers), 140
Effective mass
  of electrons, 240
  of holes, 240
Elastic reflection coefficient
  at a Cu(100) surface, 299, 320–321
  Herring–Nichols formula for, 303
  matrix, 349
  at the surface of a crystal, 349
  See also: Reflection coefficient
Electron affinity, in semiconductors, 254
Electron scattering: *see* Scattering of electrons
Electron states: *see* One-electron states

Energy analyzer
  electrostatic deflection, 57
  retarding, 56
Energy band structure: *see* Band structure
Energy distribution
  of emitted electrons in the thermal-field
      region, 30–31
  of field-emitted electrons, 24–29
    from Au covered W(100), 185–186,
      197–198
    from Cu (111) 154–155
    from Cu covered W(100), 187
    enhancement factor of, 26
    from ferromagnetic Ni(100), 161
    from ferromagnetic Ni(111), 162–163
    free-electron theory of, 24–25
    full width at half maximum (FWHM)
      of, 32, 259
    from Ge(100), 266–272, 275–280
    from hydrogen covered W(100), 189–
      192
    from Kr covered W(100), 183, 198
    many-body effects in, 164–167
    from Mo(100), 152–154
    from ordered overlayers, 193–199
    from the platinum group metals, 155–
      156
    from semiconductor conduction band,
      255, 256, 260
      surface band, 255, 264
      valence band, 255, 256, 262
    from singly adsorbed Ba on Mo(110),
      226, 232–235
      Ba on W(013), 228, 232–235
      Ba on W(110), 226, 232–233
      Ba on W(111), 227, 232–235
      Ca on W(013), 228, 232–233
      Ca on W(110), 226, 232–233
      Ca on W(111), 227, 232–233
      Sr on W(110), 226, 232
      Sr on W(111), 227
    and the surface density of states, 126–
      128, 179, 237
    from W(100), 27, 28, 141–148
    from W(110), 28, 148–152
    from W(111), 28
    from W(112), 27, 28
    Young's formula for, 25–26
    from ZnO(0001), 281
    See also: Green's function

Energy distribution (*cont.*)
  of secondary electrons
    angle-integrated, 329, 332
    angle-resolved, 331–332
      from semiconductor surfaces, 345
      from W(100), 333, 337–341,
        343–344
  of thermally-emitted electrons, 296–297
    from Cu(100), 321–322
    enhancement factor of, 321
    free-electron theory of, 33, 298
Energy gap
  absolute, 94, 99
  hybridizational (HG), 102
  symmetry, 99
Energy lines, 75
Exchange energy, 61, 62, 64

Fermi–Dirac distribution function, 2, 253,
    260
Fermi hole, 60
Fermi intergral, 240
Fermi level, 2
  effective, 253, 282–283, 286, 288
Fermi wavelength, 12
Ferromagnetism, band theory of, 157
Field emission, current density
  from adatom site, 224–225
  from semiconductor conduction band,
     257, 260, 262, 264
    surface band, 257, 264, 274
    valence band, 257, 263, 270
  diode, 44
  free-electron theory of, 13–18
  function $s(y)$, 12, 16
  function $t(y)$, 12, 14, 16
  function $v(y)$, 10, 12, 16
  micrograph, 46–47
  microscopy, 42–47
  region of temperature and applied field, 15
  work function measurements, 50–54
  See also: Energy distribution of field-
      emitted electrons
Field emitter
  geometry of, 43
  local field at the surface of, 48–50
  semiconductor, 237
Fowler–Nordheim
  equation, 16, 169
  plot, 16, 48, 52, 169, 270

Green's function, 131
    expansion of, 132
    method for calculating the energy
        distribution of field-emitted
        electrons, 129–136, 193
Ground-state energy, 63

Hartree–Fock method, 61, 64
Hohenberg and Kohn theorem, 62

Inelastic reflection coefficient
    at a Cu(100) surface, 299
    matrix, 351
    at the surface of a crystal, 351
Inelastic tunneling
    in M–I–M junctions, 199
    in field emission, 200–203
Internal voltage drop, in field emission from
        semiconductors, 282–287

Jellium model, 12, 207

Kronecker delta, 68

Lattice
    two-dimensional, 67, 194
    three-dimensional, 68
    reciprocal, two-dimensional, 68, 194
    reciprocal, three-dimensional, 70
Legendre functions of a complex argument, 82
Lifetime, of an electron state, 65, 117
Low energy electron diffraction (LEED),
        114–115, 118

Magnification factor (field emission
        microscope), 43
Many-body effects: *see* Energy distribution of
        field emitted electrons
Micrograph: *see* Field emission; Thermionic
        emission
Muffin-tin
    approximation, 66
    atom, 77

One-electron potential
    construction of
        by the $X\alpha$ method, 60–61
        by the Kohn and Sham method, 62–63
    for metal–adatom system, 207–208, 213,
        218

One-electron potential (*cont.*)
    self-consistent calculation of, 62, 67, 92,
        143, 185, 207, 245
    for semi-infinite metal, 91–93, 117, 121,
        300
    for tungsten, 109, 143, 146
One-electron states, 59–67, 117–119
    bound, of a semi-infinite crystal, 92–99
    density of: *see* Density of states
    of an infinite crystal, 70, 73, 94
    of a metal–adatom complex, 212
    scattering, of a semi-infinite crystal,
        113–114
    See also: Surface states
Overlayers, 192–193
    ordered, description of, 194–195

Periodic deviations from the Schottky line:
        see Deviations from the Schottky
        line
Photoemission spectroscopy, 106
Photofield emission, 167–168
Planck's constant, 353
Plasmon excitation, 66, 338
Poisson's equation, 239
Potential barrier
    at adsorbate covered metal surface, 175–
        176
    image, 7, 18, 175
    at metal surface, 7–8, 11, 17, 92–93, 151–
        152, 295, 300–301, 324
    at semiconductor surface, 238, 254
    triangular, 17
Probe hole, in field emission diode, 46
Projection of bulk band structure (PBS), 99

Q matrix elements for layer scattering, 74, 139
Quantum size effects in M–I–M tunneling,
        136

Reconstruction
    at metal surfaces, 91
    at semiconductor surfaces, 244–249
Reflection coefficient
    calculation of, in the parabolic
        approximation, 309–311
    in one-dimensional barriers, 5
    numerical calculation of, 5
    See also: Elastic reflection coefficient;
        Inelastic reflection coefficient

Resolution energy,
    in energy analysis, 57
    in field emission microscopy, 46
Richardson
    equation, 22, 314
    plot, 38
    work function ϕ^*, 41

Saturation current, in field emission from
        semiconductors, 285
Scattering of electrons
    by a layer of atoms, 80–89
        at finite temperatures, 309
    by a muffin-tin atom, 77–80
    by one-dimensional barriers, 3–7, 309–313
Scattering phase shifts, 79
    temperature-dependent, 308
Schottky formula, 20, 298
    plot, 20, 38, 313
Screen, fluorescent, 43
Screening charge, at a metal surface, 122
Secondary electrons, 327
Series expansion
    of plane into spherical waves, 78
    of spherical into plane waves, 84
Space charge effects, in field emission, 29
Spherical Bessel function, 78
Spherical Hankel function, 78
Spherical harmonic, 78
Spin–orbit interaction, 109, 147, 335
Spin polarization
    in secondary electron emission, 336, 345
    of field emission current, 156–157
        from (100) and (111) nickel, 158–164
    of photoemission current, 159
Step function, 334
Sticking coefficient, 171
Surface Brillouin zone (SBZ), 69
    central, 126
    of the (100) plane of a bcc lattice, 100
    of the (110) plane of a bcc lattice, 69
Surface diffusion coefficient, 204
Surface, macroscopic, 171
Surface resonances, 104–106, 116
Surface states, 99–102
    acceptor, 250
    barrier induced, 109
    donor, 250
    on Cu (100), 109

Surface states (*cont.*)
    on Cu (111), 155
    on Ge (100), 264, 267, 269–271, 276
    on Mo (100), 152
    on Ni (111), 162
    on Si (100), reconstructed, 248–249
    on Si (100), unreconstructed, 246–247
    replenishment of, 272–279
    Rydberg series of, 110
    Shockley, 102
    Tamm, 109
    on W (100), 106–109, 144, 147
    on W (110), 147, 150

Thermal-field emission, 23–24
Thermal vibrations, effect of
    on the elastic reflection coefficient, 306–
        309, 320–321
    on the energy distribution of thermally
        emitted electrons, 321–322
    on the periodic deviation from the Schottky
        line, 322–323
Thermionic emission
    current density, 297–298
    from Cu (100), 319–325
    diode, 36–37
    free-electron theory of, 18–22
    micrograph, 47
    region of temperature and applied field,
        19
    See also: Energy distribution of thermally
        emitted electrons
Time reversal symmetry, 349, 350
T matrix, 133
Topping's formula, 172
Total reflection coefficient, 296, 299
Trajectory of electron, in a field emission
        diode, 44–45
Transmission coefficient
    Miller and Good formula for, 6
    numerical calculation of, 5
    in one-dimensional barriers, 5
    WKB formula for, 7

Uncertainty principle, 65
Unit cell
    of three-dimensional reciprocal lattice,
        71
    of two-dimensional reciprocal lattice, 69

Vibrational-energy loss spectra, 202–203
Vibrational modes, of metal–admolecule
    complex, 200, 202

Weber's equation, 310
Width of energy of level, 65, 118
Width of surface resonance, 104–105
Work function, 7
   anisotropy, 54
   change due to adsorption, 170
   of different faces of tungsten, 51
   field emission measurements of, 50–54

Work function (*cont.*)
   temperature coefficient of, 39–40, 52
   thermionic data, 40–42

X matrix, 84
   properties of, 87
   Kambe's formula for, 89
X_α method, 60–61, 157

Zero-emitted-current approximation, 238
Zero-internal-current approximation, 252–
   253